SCIENCE
GOOD, BAD AND BOGUS

MARTIN GARDNER

SCIENCE
GOOD, BAD AND BOGUS

PROMETHEUS BOOKS
Buffalo, New York

SCIENCE: GOOD, BAD AND BOGUS. Copyright © 1981, 1989 by Martin Gardner. All rights reserved. Printed in the United States of America. No part of this book may be reproduced in any manner whatsoever without written permission, except in the case of brief quotations embodied in critical articles and reviews. Inquiries should be addressed to Prometheus Books, 700 East Amherst Street, Buffalo, New York 14215, 716-837-2475.

First paperback edition, 1989

To Persi, Randi, and Ray—friends,
magicians, and fellow warriors in
the never-ending battle against
dishonest and deluded science.

In science, "fact" can only mean "confirmed to such a degree that it would be perverse to withhold provisional assent." I suppose that apples might start to rise tomorrow, but the possibility does not merit equal time in physics classrooms.

Stephen Jay Gould

One horse-laugh is worth ten thousand syllogisms.

H. L. Mencken

Author's Note

The pieces in this collection were written at different times and for numerous different publications. Since many of the subjects appear and reappear throughout the book, there are some unintended repetitions of descriptive details. Because of their relevance in each case, I have allowed these to remain.

Part 1 consists of articles, Part 2 of book reviews, with the material in each part arranged chronologically.

Contents

Introduction

No one can define exactly what is meant by such words as *pseudoscience, crank,* and *crackpot.* The reason is simple. There is no exact way to define anything outside pure mathematics and logic, and even there some basic terms have extremely shaggy edges. It does not follow that colloquial terms assigned to portions of continua are not useful. As I have said many times, if we did not have words for extrema, such as black and white, night and day, hot and cold, we could not talk at all. I have just used the word *talk.* Give me a definition of *talk* and I will give you an example of something to which it is debatable whether the word applies. Indeed, Chapter 38 deals with the fuzziness of this common word. Fuzzy, yes, but also indispensable.

We all know there have been occasions when top scientists ridiculed ideas that later proved to be sound. We all know that great scientists have held opinions, both in and out of their specialized fields, that turned out to be hopelessly wrong. Let us not waste time belaboring the obvious. Nor must we forget that for every example of a crank who later became a hero there were thousands of cranks who forever remained cranks. We must not forget that for every outcast theory raised to respectability by a scientific revolution there were thousands of crazy theories that permanently bit the dust.

I take for granted that all scientific hypotheses are conjectures to which scientists and laymen alike assign degrees of belief that vary between one and zero. To make the point with extremes, the scientific community today—the "establishment," if you prefer—assigns a probability close to zero to the theory that the earth is hollow, open at the poles, and inhabited on the inside. No one would hesitate to call anyone a crank who believed such a theory. The science community today assigns a probability close to one to the belief that Venus was a planet long before the human race evolved. By the same token, it gives a probability close to zero to the theory that Venus originated as a comet that zoomed out of Jupiter and settled into its present orbit less than four thousand years ago. This proposed chain of events violates so many strongly confirmed facts and theories that the "establishment" has not hesitated to regard the late Immanuel Velikovsky as the very model of a crank.

Cranks by definition believe their theories, and charlatans do not, but this does not prevent a person from being both crank and charlatan. It is a familiar combination in the history of pseudoscience and occultism, and it is exemplified with different proportions of the two ingredients by many whose names appear in the pages to follow. Robert Browning's poem "Mr. Sludge, 'The Medium'" (Conan Doyle called it "doggerel") is a classic portrayal of such a mix, even though Browning based his Sludge on the British medium D. D. Home, whom I consider as total a charlatan as Arthur Ford (see Chapter 23). It was Elizabeth Browning's passionate faith in Spiritualism that almost wrecked an otherwise happy marriage.

I hope no one will imagine that I believe cranks should be silenced by any kind of legislation. In a free society every crank has a right to be heard, and no one can say that in our society they are not heard. Thanks to the freedom of our press and of the electronic media, the voices of cranks are often louder and clearer than the voices of genuine scientists. Crank books—on how to lose weight without cutting down on calories, on how to talk to plants, on how to cure your ailments by rubbing your feet, on how to apply horoscopes to your pets, on how to use ESP in making business decisions, on how to sharpen razor blades by putting them under little models of the Great Pyramid of Egypt—far outsell most books by reputable scientists.

I do not believe that books on worthless science, promoted into bestsellers by cynical publishers, do much damage to society except in areas like medicine, health, and anthropology. There are people who have died needlessly as a result of reading persuasive books recommending dangerous diets and fake medical cures. The idiocies of Hitler were strengthened in the minds of the German people by crackpot theories of anthropology. In recent years many children have become seriously disturbed by reading books and seeing movies about haunted houses and demon possession. Psychotic mothers have killed their children in attempts to exorcise devils. Although I am opposed to any kind of law that tells a publisher, or a movie or TV producer, what cannot be done, I reserve the right of moral indignation both as an individual and as a member of a pressure group.

I was among four representatives of the Committee for the Scientific Investigation of Claims of the Paranormal who met in 1977 with a group of

NBC officials to protest that network's outrageous pseudodocumentaries about the marvels of occultism. One official shouted in anger, "I'll produce anything that gets high ratings!" I thought to myself: this should be engraved on his tombstone. Of course he didn't mean it. A documentary on the adulteries of President John Kennedy, for instance, would achieve fantastic ratings. It would all be true, and one could even argue that it would perform a service for the American voter who is perpetually deceived by the carefully contrived images of political leaders. Why does NBC not produce such a show? Because it would be in bad taste; because in the long run it might damage NBC's public image. The sad fact was that not a single NBC official at our meeting knew enough about science to comprehend in the slightest the degree to which their moronic shows about the paranormal were in bad taste.

At about the time of this meeting a neighbor telephoned me for advice. An establishment doctor had told her daughter, then living in an Arkansas commune, that she needed an internal operation and needed it soon. But the young lady had decided that orthodox doctors could not be trusted. She wanted to fly to the Philippines, where she could be "operated" on painlessly and inexpensively by one of the "psychic surgeons" who had been prominently featured on an NBC show. Moreover, she had read books praising these "surgeons"—charlatans who perform miraculous operations without cutting open the flesh—published by otherwise reputable houses. (There is a stirring section on this subject, complete with color photographs, in *Roots of Consciousness,* by Jeffrey Mishlove, published by Random House. In 1980 the University of California, at Berkeley, gave Mishlove a doctorate in parapsychology!)

The mother was distraught when she called me. What could she give her daughter to read that might change her mind? The best I could think of was the debunking chapter on these Philippine quacks in Dr. William Nolen's excellent book *Healing.* But would the daughter believe Dr. Nolen? After all, was he not part of the hated medical establishment? These are the kinds of tragedies that are the direct result of the media's glamorizing of pseudomedicine. No government has a right to suppress such sleazy books and television shows, but those who understand and respect science have a right to be morally offended.

In discussing extremes of unorthodoxy in science I consider it a waste of time to give rational arguments. Those who are in agreement do not need to be educated about such trivial matters, and trying to enlighten those who disagree is like trying to write on water. People are not persuaded by arguments to give up childish beliefs; either they never give them up or they outgrow them. If a Protestant fundamentalist is convinced that the earth was created six thousand years ago and that all fossils are records of life that flourished until Noah's Flood, nothing you can say will have the slightest effect on his or her ignorant mind-set. As for those who have not yet made up their minds about evolution (and there are millions), the best advice you can give is to suggest that they go to a university and take some introductory courses in geology. Without such basic knowledge they will not even understand your arguments. Can you imagine a professional geologist sitting down for several days with Herbert Armstrong or Oral Roberts and convincing either preacher that the evidence for evolution is overwhelming?

For these reasons, when writing about extreme eccentricities of science, I have adopted H. L. Mencken's sage advice: one horse-laugh is worth ten thousand syllogisms. Concerning less extreme claims, such as those of parapsychology, I have occasionally called attention to poor experimental design and the prevalence of fraud; but even in this twilight area such arguments are unlikely to have any effect on the true believer.

Several pieces in this book are about topics I do not consider pseudoscience but which have what I regard as eccentric fringes. Black holes are certainly highly respected theoretical models, based as they are on classical relativity theory. I include a review of seven books about black holes only because two of the books seem to me irresponsible speculations by science journalists of the "gee-whiz" school. I do not regard the work on talking apes as pseudoscience, but I include a review of two books about this research because they suggest that much of it has been so crudely controlled that it borders on the crankish. Catastrophe theory is not pseudomathematics—it is elegant mathematics—but I include a review of four books on catastrophe theory because I think it has been carelessly applied to the behavioral sciences. Finally, although this book deals almost exclusively with modern pseudoscience, I include a chapter on the medieval theologian Ramon Lull because his "Great Art" has recently been revived as a technique for creative thinking. Lull's Great Art, in my opinion, is as worthless now as it was in the late Middle Ages and the Renaissance.

It goes without saying that some of my harsh judgments could be proved wrong by future science. I do not think many will be. Back in 1872 the British mathematician Augustus De Morgan wrote a two-volume work called *A Budget of Paradoxes*. It contains thousands of horse laughs at buffooneries of his day, most of them about mathematics and its applications. I know of no theory ridiculed in De Morgan's work that has since turned out to be viable. Since I swing my bat almost entirely at nonmathematical claims, I may not have as good a batting average, but perhaps I will come close.

If you are interested in keeping up with the latest trends in pseudoscience—and they seem to get wilder and funnier every year—let me recommend a subscription to a lively quarterly called the *Skeptical Inquirer*. It is published by the Committee for the Scientific Investigation of Claims of the Paranormal, which I mentioned earlier, and edited by Kendrick Frazier, former editor of *Science News*. For details, write to the magazine at Box 229, Central Park Station, Buffalo, New York 14215.

MARTIN GARDNER

Part One

1

Hermit Scientists

"The creation of dianetics is a milestone for man comparable to his discovery of fire and superior to his invention of the wheel and arch." This is the modest opening sentence of L. Ron Hubbard's book *Dianetics: The Modern Science of Mental Health.*

An engineer and writer of science fiction, with no status whatever in psychiatry, Hubbard has created all by himself what he and his followers believe to be a revolutionary science of mental therapy. Already, dianetics threatens to become a cult of wide proportions, especially in Los Angeles, and no less distinguished a scholar than Frederick L. Schuman, professor of political science at Williams College, has become an enthusiastic convert. In a letter to the *New Republic* (September 11, 1950) protesting an unfavorable review of *Dianetics,* Schuman wrote, "Not the book, but the review, is 'complete nonsense,' a 'paranoiac system' and a 'fantastic absurdity.' There are no authorities on dianetics save those who have tested it. All who have done so are in no doubt whatever as to who is here mistaken."

There is no need to go into the weird mixture of myths which form the core of Hubbard's book, except to point out in passing that it revives the ancient superstition that experiences of the mother can leave an impress on the mind of a foetus within a day after conception. "What's that chronic cough?" Hubbard asks in his first published article on dianetics (*Astounding Science Fiction,* May, 1950), and then answers, "That's

Reprinted with permission of the Editors from the Winter 1950–51 issue of the *Antioch Review.* Copyright 1951 by The Antioch Review, Inc.

mama's cough which compressed the baby into anaten [Hubbard's term for unconsciousness; derived from the words *analytical* and *attenuation*] when he was five days after conception. . . . What's arthritis? Foetal damage or embryo damage." And so on *ad nauseam*.

A few months before Hubbard's revelation, the Macmillan Company published Dr. Immanuel Velikovsky's *Worlds in Collision*. The book throws together a jumbled mass of data to support the preposterous theory that a giant comet once erupted from the planet Jupiter, passed close to the earth on two occasions, then settled down as Venus. The first visit to the earth of this erratic comet was precisely at the time Moses stretched out his hand and caused the Red Sea to divide. The manna which fell from the skies shortly thereafter was a precipitate, fortunately edible, of suspended elements in the celestial visitor's tail. Later the comet's return coincided with Joshua's successful attempt to make the sun and moon stand still. The miracles of both Moses and Joshua were the result, Velikovsky informs us, of a temporary cessation of the earth's spin.

Although Velikovsky's work is a tissue of absurdities, and has been recognized as such by every geologist and astronomer in the country, it is astonishing how many who reviewed the book were caught off guard by the author's persuasive rhetoric. John J. O'Neill, science editor of the *New York Herald Tribune*, described the book as "a magnificent piece of scholarly historical research." Horace Kallen, a distinguished educator and author, wrote, "The vigor of the scientific imagination, the boldness of construction and the range of inquiry and information fill me with admiration." Ted Thackrey, editor of the *New York Compass*, suggested that Velikovsky's discoveries "may well rank him in contemporary and future history with Galileo, Newton, Kepler, Darwin, Einstein. . . ." And the book received enthusiastic endorsements by Clifton Fadiman and Fulton Oursler.

In view of the astonishing sales of the Velikovsky and Hubbard books, both totally without scientific merit, we may well ask ourselves if we are slipping back into an era of lurid and irresponsible science reporting. Perhaps the most alarming indication of this trend is the current widespread acceptance of the theory that flying saucers are spaceships from another planet. *True* magazine broke the news that the discs were piloted by Martians, but Frank Scully's recent best-seller, *Behind the Flying Saucers*, argues elaborately that they were flown here with the speed of light by inhabitants from Venus, who are exact duplicates of earthlings except that they are midgets three feet tall.

Although one may censure publishers and magazine editors for printing such incredible nonsense without first seeking evaluation by competent scientists, the primary cause of the new flowering of pseudoscience seems to be a hunger on the part of a gullible public for sensational science news. The sudden success of atomic research, hitherto the subject matter of science fiction, is certainly a major factor in this trend. After splitting the atom, nothing seems surprising any more. In addition, widespread anxiety caused by fear of atomic war, together with other factors, seems to be turning the minds of countless frightened people toward religion and/or mental therapy. It is not hard to understand the mass appeal of dianetics, which offers a quick, relatively inexpensive, and painless shortcut to psychoanalysis; or the widespread interest in Velikovsky's theories which reestablish the historical accuracy of the Old Testament for orthodox Catholics, Protestants, and Jews.

What about the authors of these two masterpieces of pseudoscience? Are they deliberate hoaxers, out to make a dishonest dollar, or are they sincere in believing their own theories? In Velikovsky's case, unquestionably the latter is the truth. Occasionally a carefully planned hoax has fooled the public for a time, such as the famous Moon Hoax of the *New York Sun* in 1835, but such pranks are short-lived and soon exposed. Of a different character altogether is the work of the self-styled scientist, incompetent in his field, but living under a delusion of greatness and driven by unconscious compulsions to create off-trail theories of incredible complexity and ingenuity.

When Renaissance science first began to free itself from metaphysical biases, it was the rule rather than exception for courageous pioneers to find their work greeted with derision by their colleagues. Galileo had to battle not only church authorities but fellow scientists who were more preoccupied with Aristotle than with an experimental determination of how the world did, in fact, behave. As Aristotle's scientific authority declined, however, opposition to new ideas in science became more and more confined to areas where science clashed with Christian doctrine. Since the turn of the century, even this area of conflict has become remarkably small, and widespread opposition by scientists to a legitimate theory, based on verifiable evidence and cogent reasoning, is an increasing rarity. For a contemporary scientist, often the quickest way to fame is to overturn a widely held theory. Einstein's work on relativity is an excellent illustration of how easily a revolutionary hypothesis can meet with almost immediate serious response, careful testing, and ultimate acceptance. Of course there are exceptions, and there are always borderline areas where confirming evidence remains so debatable as to leave eccentric theories in legitimate dispute (for example, Sheldon's work on body types and large sections of psychiatry). But, if anything, science today leans backward in the friendly consideration of bizarre hypotheses.

Outside and quite apart from the cooperative process of communication and testing that goes on constantly within every branch of science, there are the lonely, isolated, hermit scientists. If their knowledge is meager and their I.Q. low—as in the case of the late Wilbur Glenn Voliva of Zion City, Illinois, who believed the earth shaped like a pancake—they seldom achieve a following among the general public and are widely recognized as crackpots. If they are victims of sufficiently intense paranoid drives, they may be confined to mental institutions where they putter away their days perfecting perpetual motion machines and methods of trisecting angles; or writing unreadable, neologistic treatises on the inner secrets of the universe.

Occasionally, however, a milder paranoia combines with a brilliant, creative intellect. In such cases, the self-styled scientist's belief in his own greatness, together with his tendency to interpret lack of recognition as a form of persecution by stubborn and prejudiced authorities, effectively bars him from the social give and take of the scientific process. He retires like a hermit within his laboratory or study, to emerge later with tomes of vast erudition, usually written in a complex jargon of invented terms and phrases. Around the Master will cluster a group of ardent admirers—either disciples whose own psychological demands find identification with those of the Master, or simply naïve cultists who lack the knowledge to penetrate the Master's self-deceptions.

Classic works in the genre of pseudoscience fall broadly into two classes. There are those which have as a major purpose the rationalization of a religious dogma (such as Velikovsky's defense of the orthodox Jewish interpretation of Old Testament history) and the nonreligious theories (such as Hubbard's) which are a pure product of the author's delusions of scientific competence. Because the fantastic views of Velikovsky and Hubbard have been, and will continue to be, dissected elsewhere, it may be of interest to take a look at the works of two other hermit scientists, one religious and one nonreligious, whose contemporary theories in many ways resemble those of Hubbard and Velikovsky but which are even more ingenious examples of scientific self-delusion. In doing so, we may catch something of the pretentious atmosphere and the paranoid flavor which pervade such works.

As an illustration of the hermit scientist's rationalization of religious dogma, no better example can be found than the impressive geological speculations of George McCready Price. According to *Who's Who*, Price is at present a retired professor of geology at Walla Walla College, a Seventh Day Adventist school in Washington. He enjoys the distinction of being the last, perhaps the greatest, of Protestant opponents of evolution.

Price's views are set forth at length in *The New Geology*, a weighty college textbook published in 1923. So carefully reasoned is his approach that thousands of Protestant fundamentalists accept it today as the final word on the subject, and even the skeptical reader will find it difficult to refute without considerable background in geology.

The heart of Price's objection to traditional paleontology can be stated in a few words. The chief evidence for evolution, he points out, is the fact that fossils proceed from simple to more complex forms as you move from older to younger geological strata. Unfortunately, there is no adequate method of dating the ages of strata except by means of the fossils they contain. Thus a vicious circularity is involved. The theory of evolution is assumed in order to classify the fossils in evolutionary order. The fossils are used to date the beds. Then the succession of fossils from "old" to "young" strata is cited as "proof" of evolutionary development.

Price's own opinion is that all the beds were deposited simultaneously by the Great Flood described in *Genesis,* in turn caused by an astronomical disturbance which sent a huge tidal wave around the earth. Fossils are the records of antediluvian flora and fauna. (The flood theory of fossils, by the way, has a long, distinguished history, having been defended by such authorities as Philo, Chrysostom, Tertullian, St. Augustine, St. Jerome, Martin Luther, and innumerable eighteenth and nineteenth century scientists. Addison once penned a Latin ode to it.) If this is true, then in outcrops where several or more fossil-bearing beds are found in one place, one would expect the fossils to be in reverse of the evolutionary order as often as conforming to it. This, Price declares, is precisely the case, and much space in his books is devoted to descriptions of "upside down" areas. To explain away these embarrassing beds, Price asserts, traditional geologists invent imaginary faults and folds. The following quotation on this point is a sample of Price's pleasant style:

> . . . there is scarcely an artificial geological section made within recent years that does not contain one or more of these "thrust faults," or "thrusts." But the really important thing to remember in this connection is that it is solely because the fossils are found occurring in the wrong order of sequence that any such devices are thought to be necessary—devices which, as has already been suggested of similar expedients to explain away evidence, deserve to rank with the famous "epicycles" of Ptolemy, and will do so some day.

It would be a mistake to think of Price's scientific knowledge as on the same level with, say, William Jennings Bryan's. It was, in fact, Price who was cited as the leading geological authority by Bryan during the famous Scopes trial. His books are well written, packed with impressive erudition and indisputable evidence of sound geologic information. They are, of course, rationalizations of the Protestant fundamentalist's interpretation of the Old Testament, just as Velikovsky's book is a rationalization of traditional Judaism; but the religious motive is hardly sufficient to force a man of Price's intelligence into the curious role he has played. Other compulsions creep out when he refers to his sad task of "reforming the science of geology almost single-handed," and in such passages as:

> Twenty-five years ago, when I first made some of my revolutionary discoveries in geology, I was confronted with this very problem of how these

new ideas were to be presented to the public. And it was only after I found that the regular channels of publication were denied me, that I decided to use the many other doors which stood wide open. Perhaps I made a mistake. Perhaps I should have had more regard to the etiquette of scientific pedantry, and should have stood humbly hat in hand before the editorial doors which had been banged in my face more than once. But I decided otherwise, with a full realization of the consequences; and I have not yet seen any reason for thinking that I really made a mistake. Some day it may appear that the reigning clique of "reputable" scientists have never had a monopoly of the facts of nature.

But enough of Price. Let us turn to a more colorful scientist whose work has recently become a lively cult among the more Bohemian intellectuals of New York and elsewhere—the psychiatrist Wilhelm Reich. Like Hubbard's dianetics, Reich's "orgone therapy" has no connection with religious dogma but is presented simply as a revolutionary discovery in biology and psychology.

Reich began his curious career in Austria as an orthodox Freudian but later broke with the psychoanalysts, founding his own publishing house in Germany in 1931. He also severed his ties with the Austrian Communist Party, having served in the same cell with the writer Arthur Koestler.[1] Five years later, Reich opened an institute at Oslo, where he met with furious attack by Scandinavian biologists who insisted his knowledge was less than that of an undergraduate. Expelled from Norway, he came to New York in 1939 at the invitation of Dr. Theodore P. Wolfe, an associate professor of psychiatry at Columbia University, and lectured for a brief term at New York's New School for Social Research. He now maintains a press in Greenwich Village, and research laboratories in Forest Hills, New York, and Organon, Maine.

In Reich's best-known work, *The Function of the Orgasm,* he compares himself to Peer Gynt, i.e., the unconventional genius, out of step with society, misunderstood, ridiculed. Society has the last laugh, he writes, until the Peer Gynts are proved right. In his latest publication, *Listen, Little Man,* 1949, Reich likens himself to such persecuted figures as Jesus and Karl Marx. "Whatever you have done to me or will do to me in the future," he declares, "whether you glorify me as a genius or put me in a mental institution, whether you adore me as your savior or hang me as a spy, sooner or later necessity will force you to comprehend that *I have discovered the laws of the living. . . .*"

A pamphlet by Dr. Wolfe, published by Reich's Orgone Institute in 1948, is called *Emotional Plague Versus Orgone Biophysics.* The purpose of the booklet is stated on the cover:

A vicious campaign of slander and distortion against Wilhelm Reich and his work was begun early in 1947. There is no telling where it will lead. This campaign has not been confined to magazine and newspaper articles, but an agency of the United States Government has been dragged into it.

Chief signs of this "emotional plague" (Reich's term for the slander campaign) are two articles by Mildred Brady, one in *Harper's* (April, 1947), the other in *The New Republic* (May 26, 1947). The government agency is the Food and Drug Administration, at that time investigating Reich's "orgone accumulators." These are large boxes of wood on the outside and metal inside. Patients rent them from the Institute, then sit inside them to build up their orgone potential by absorbing the box's abnormally high concentration of orgone energy (a nonelectromagnetic radiant energy coming from outer space which Reich discovered in Norway in 1939). "The Orgone Accumulator is the most important single discovery in the history of medicine, bar none," Wolfe writes.

The following paragraph from a letter of Reich's, published in the pamphlet, is revealing:

It is an old story. It is older than the ancient Greeks whom we consider the bearers of a flourishing culture. . . . It was no different two thousand years later. Giordano Bruno, who fought for scientific knowledge and against astrological superstition, was condemned to death by the Inquisition. It is the same psychic pestilence which delivered Galileo to the Inquisition, let Copernicus die in misery, made Leeuwenhoek a recluse, drove Nietzsche into insanity, Pasteur and Freud into exile. It is the indecent, vile attitude of contemporaries of all times. This has to be said clearly once and for all. One cannot give in to such manifestations of the pestilence.

A word about orgone energy. Reich regards his discovery of it as comparable to the Copernican Revolution. A failure to accept it on the part of other psychiatrists is, of course, "resistance to a new concept."[2] In *Character Analysis* he interprets Freud's "Id" as the action of orgone energy in the body. The energy provides a biological and physical base for psychiatry, and to operate with the old Freudian drives is, Reich asserts, like trying to drink from a mirror image of a glass of water. In *The Function of the Orgasm* he describes orgone energy as blue in color (it has been photographed on Kodachrome film, Wolfe tells us), and adds that it is responsible for the Northern Lights, St. Elmo's Fire, lightning, the blue of the sky, electric disturbances during sun-spot activity, and the blue coloration of sexually excited frogs. "Cloud formations and thunder storms," he writes—"phenomena which to date have remained unexplained—depend on changes in the concentration of atmospheric orgone." In 1947 Reich measured the energy with a Geiger counter.

It is interesting to note in passing that Reich also attributes the flickering of stars to orgone energy. Another hermit scientist, Dr. William H.

Bates, in his medical opus, *Cure of Imperfect Eyesight by Treatment Without Glasses,* had this to say about the same topic:

> The idea that the stars twinkle has been embodied in song and story, and it is generally accepted as a part of the natural order of things, but it can be demonstrated that the supposed twinkling is simply an illusion of the mind. . . .
>
> While persons with imperfect eyesight usually see the stars twinkle, they do not necessarily do so. Therefore it is evident that the strain which causes the twinkling is different from that which causes the error of refraction. If one can look at a star without trying to see it, it does not twinkle. . . . On the other hand, one can make the planets or even the moon twinkle, if one strains sufficiently to see them.

Reich's most astounding discovery is reported in the article "The Natural Organization of Protozoa from Orgone Energy Vesicles," in the November, 1942, issue of his *International Journal of Sex Economy and Orgone Research.* In this paper, accompanied by microphotographs, Reich describes his observations of protozoa being formed spontaneously from aggregates of bions. The bion is another Reich discovery. It is the unit of living matter, consisting of a membrane surrounding a liquid and pulsating with orgone energy. Bions are constantly being formed in nature by the disintegration of both organic and inorganic matter. Under his microscope Reich observed bions grouping together to form various types of protozoa, and he has the photographs to prove it. Cancer cells, incidentally, are protozoa which develop from tissue bions.[3] To charges of critics that protozoa get into his cultures from the air, or were already on the disintegrating material in the form of dormant cysts, Reich simply answers that it isn't so, though he gives no evidence of taking adequate precautions against either possibility.

Disciples of Reich frequently defend him by saying, "Granted that his biological work is highly suspect, you'll have to admit he's made great contributions to the field of mental therapy." This may be true. But it has somewhat the same plausibility as a statement like the following: "Granted that Professor Ludwig von Hoofenmeister errs in his theory that stars are holes in an opaque sphere surrounding the earth, you'll have to admit he has made magnificent discoveries in his study of cosmic rays."

The reader may wonder why a competent scientist does not publish a detailed refutation of Reich's absurd biological speculations. The answer is that the informed scientist doesn't care, and would, in fact, damage his or her reputation by taking the time to undertake such a thankless task.[4] For the same reasons, scarcely a single classic in the field of modern scientific curiosa has prompted an adequate reply. The one exception is the work of the Russian geneticist Lysenko, unimportant in itself but with a wider significance because it strengthens a cultural paranoia and dramatically highlights the conflict between a relatively free and a rigidly controlled science.

The hermit scientist is usually ignored. No eminent authorities bothered to "refute" Ignatius Donnelly's *Ragnarok*[3] or his even more painstaking work on Atlantis. No one has refuted Piazzi Smyth's brilliant volume on the Great Pyramid of Egypt, Captain John Symmes' hollow earth theory, or Philip Gosse's *Omphalos*. This latter work, by Edmund Gosse's father, argued that just as Adam and Eve were created with navels, which indicated a past event which had not occurred, so the world was created with a fossil record of a past geologic history that never took place. The theory is, in fact, irrefutable and consequently much sounder than the views of Velikovsky or Price. It has afforded Bertrand Russell with several happy illustrations of epistemological principles.

Occasionally a philosopher or writer is taken in by the work of a brilliant hermit and produces a book or essay in defense of it (e.g., Aldous Huxley's *The Art of Seeing,* 1942, a defense of Dr. Bates), but the professional scientist prefers to ignore it, or perhaps study it with tolerant amusement. Such neglect, of course, only strengthens the convictions of the self-declared genius. "My previous larger treatise on this subject," writes Price in a later work, *The Phantom of Organic Evolution,* "has not been answered. It will not be answered. But it has been ignored, and probably will still be ignored, because very few even among men of science, have the patience to follow carefully a completely new line of argument based on unfamiliar facts." And Velikovsky has patronizingly remarked (*New York Times Book Review,* April 2, 1950, p. 12), "If I had not been psychoanalytically trained I would have had some harsh words to say to my critics."

Thus it is that probably no scientist of importance will present the bewildered public with detailed proofs that the earth did not twice stop whirling in Old Testament times, or that neuroses bear no relation to the experiences of an embryo in the mother's womb. The current flurry of discussion about Velikovsky and Hubbard will soon subside, and their books will begin to gather dust on library shelves. Perhaps Tiffany Thayer will appoint them honorary members of the Fortean Society, that remarkable institution devoted to the writings of Charles Fort, dedicated to the frustration of science, and the haven of lost causes.

NOTES

1. See Koestler's reference to Reich in his contribution to *The God That Failed,* edited by Richard Crossman, 1949, p. 43.

2. Hubbard likewise regards opposition to his views as unconscious resistance. "Anyone

attempting to stop an individual from entering dianetic therapy," he writes, "either has a use for the aberrations of that individual or has something to hide."

3. Both Reich and Hubbard are concerned with cancer therapy. Hubbard writes: "At the present time dianetic research is scheduled to include cancer and diabetes. There are a number of reasons to suppose that these may be engramatic ["engram" is Hubbard's term for a mental record of a consciously forgotten experience] in cause, particularly malignant cancer."

4. It should be emphasized that the isolation of a scientist, the novelty of his theories, or the psychological motivations behind his research provide no grounds whatever for the rejection of his work by other scientists. The rejection must be solely on the basis of the failure of his work to meet standards of scientific adequacy. It is not within the scope of this paper, however, to discuss technical criteria by which hypotheses are given high, low, or negative degrees of confirmation. Our purpose is simply to glance at several examples of a type of scientific activity which fails completely to conform to scientific standards but at the same time is the result of such intricate mental activity that it wins temporary acceptance by many laymen insufficiently informed to recognize the scientist's incompetence. Although there obviously is no sharp line separating competent from incompetent research, and there are occasions when a scientific "orthodoxy" may delay the acceptance of novel views, the fact remains that the distance between the work of competent scientists and the speculations of a Voliva or Velikovsky is so great that a qualitative difference emerges which justifies the label of "pseudoscience." Since the time of Galileo the history of pseudoscience has been so completely outside the history of science that the two streams touch only in the rarest of instances.

5. Velikovsky's brief footnote reference to this book does not indicate how remarkably similar it is to his own work. Donnelly's thesis is that earth's Glacial Age, and earlier epochs of diastrophism, were due to brushes with a comet. Over 200 pages are devoted to legends which Donnelly, like Velikovsky, believed to be memories of the last brush. Concerning Joshua's miracle, Donnelly writes: "And even that marvelous event, so much mocked at by modern thought, the standing-still of the sun at the command of Joshua, may be, after all, a reminiscence of the catastrophe. . . . In the American legends, we read that the sun stood still, and Ovid tells us that 'a day was lost.' Who shall say what circumstances accompanied an event great enough to crack the globe itself into immense fissures? It is at least, a curious fact that in Joshua (Chapter X) the standing still of the sun was accompanied by a fall of stones from heaven by which multitudes were slain."

Postscript

This was my first piece of writing about pseudoscience. It prompted a literary agent to call me and persuade me to expand the theme into a book, *In the Name of Science,* published by Putnam in 1952. The book was quickly remaindered, then soon reprinted by Dover as a paperback with the title *Fads and Fallacies in the Name of Science.* The Dover edition became something of a best-seller, mainly as a result of repeated attacks on the book by guests on the Long John Nebel radio talk-show. I recall turning on the show at 3 A.M. one morning, when I was giving a bottle of milk to my newborn son, and being startled to hear a voice say, "Mr. Gardner is a liar." It was John Campbell, Jr., editor of *Astounding Science Fiction,* expressing his anger over the book's chapter on dianetics.

In my *Antioch Review* paper I was clearly wrong in predicting that interest in Velikovsky and Hubbard would "soon subside." Today, thirty years later, the late Immanuel Velikovsky still has a loyal band of addle-pated acolytes, and dianetics, which became part of Hubbard's new "religion," Scientology, is the backbone of one of the nation's biggest cockamamie cults. Many books have attacked Scientology, and the church has done its best to discredit them and defame the authors. For the terrible, incredible story of how the church framed Paulette Cooper to punish her for her paperback, *The Scandal of Scientology,* see the *New York Times,* January 22, 1979, and "Scientology: Anatomy of a Frightening Cult," by Eugene H. Methvin, in *Reader's Digest,* May 1980. The cult continues to attract show-biz types, like John Travolta, who know even less about science than Ronald Reagan.

Because Scientologists believe in reincarnation and paranormal powers, the cult appeals strongly to self-styled psychics and to investigators of psi forces. Ingo Swann and the late Pat Price, two of the top psychics "authenticated" by Harold Puthoff and Russell Targ at Stanford Research International, were ardent Scientologists (Swan still is). Puthoff himself was once active in Scientology (see chapter 30) and was married in the church. Now that this cult is in serious trouble with the federal government, and other governments around the world (France has a warrant out for Hubbard's arrest), Puthoff has been trying to minimize his early enthusiasm for it.

Although George M. Price's name is unfamiliar to most people, his infantile geological views continue to underpin the new books by Protestant creationists. *The Genesis Flood,* by John C. Whitcomb, Jr., and Henry M. Morris (1961), is an outstanding example. This awesome 518-page monograph is pure Price throughout, although Price is almost never credited.

Velikovsky was enormously impressed by Price, and the two cranks corresponded. In *Earth in Upheaval* you'll find a number of quotations from, and references to, Price's foolish books. Most readers would assume that Velikovsky was quoting a reputable geologist. Price had no formal education in geology. He began his career as janitor and handyman for an Adventist college in Loma Linda, California, where he helped lay bricks for the buildings. The college eventually gave him a B.A. degree. The best biographical reference to date on Price is *Crusader for Freedom,* by Harold W. Clark, published by the church's Pacific Press in 1966.

It would be interesting to know just what Velikovsky believed about evolution. His books give only vague hints. Perhaps someday there will be a posthumous publication of his weighty opinions in this area. Admirers of Velikovsky raise howls of protest whenever I suggest that his orthodox Judaism played a role in the shaping of his theories, but I still find his motivations strikingly similar to those of Price. (For more on this see Chapter 37). Marcello Truzzi, in the April 1979 issue (Nos. 3–4) of his *Zetetic Scholar,* presented a "dialogue" on Velikovsky by ten authors, some defending Velikovsky, some attacking him. One would suppose from this scholarly debate that Velikovsky continued to present the scientific community with a significant challenge!

Reich's orgonomy cult *seems* to be waning (I could be wrong!), though most of his books are back in print, and his followers are still to be found among writers, artists, and show people like Orson Bean. Numerous books about him, favorable and otherwise, have been written in recent years. His daughter, Eva Reich, a pediatrician in Hancock, Maine, is active as a lecturer on orgonomy. Her father's rain-making device—huge tubes that squirt orgone energy into the clouds—is in her front yard. For a while she was using orgone energy accumulators to treat infants at a hospital for premature babies in Harlem; but, after the director asked her to cease or resign, she chose the latter.

Eva is firmly persuaded that human auras are orgone energy. See Lynn Franklin's long, sad interview, "Like Father, Like Daughter," in the *Maine Sunday Telegram,* June 22, 1980. According to *Newsweek* (December 13, 1976): "For twenty years, Eva Reich has been hiding microfilms of portions of Reich's papers in a mushroom cave in the Catskill Mountains. Unless the courts intervene, she says, she may make these secrets available to the world."

A silly book has just crossed my desk: *The Quest for Wilhelm Reich,* by Colin Wilson (Doubleday, 1981). Poor Colin. He had great promise as a young writer in Britain before he went crackers over the paranormal. Wilson sees Reich as crazy, but nevertheless a genius whose discovery of orgone energy puts him in the company of Semmelweis, Mendel, and all those other great scientists who were unappreciated in their day. No book on Reich is less worth reading.

2

"Bourgeois Idealism" in Soviet Nuclear Physics

The controversy in Russian genetics, which in 1948 resulted in the political victory of Lysenko, has been widely publicized. Not nearly so well known is a remarkably similar controversy in theoretical nuclear physics. The conflict came into the open about 1949. Since then a clear Party line on the subject has emerged, and physicists holding opposing views have made the usual confessions of error, with promises to revise their published writing at the earliest opportunity.

To understand the controversy it will be necessary to review briefly its historical background. Shortly after the Bolshevik revolution, the atom was widely regarded as a kind of miniature solar system. The nucleus corresponded to the sun around which electrons, like planets, were whirling in obedience to fixed, determinable laws. But as scientists probed more deeply into the atom, an astonishing state of affairs came to light. It was found impossible to predict an individual electron's behavior. If you were successful in determining its position, there would be a large error in the measurement of its velocity. And if you measured the electron's velocity correctly, you had no way of knowing where it was. Werner Heisenberg, Germany's leading nuclear physicist, formulated his well-known "Principle of Uncertainty." Without going into technical details, the principle says that the relationship between an electron and the observer is such that there is an irreducible element of randomness about the electron's actions. At first, this randomness was thought to be

Reprinted with permission from the *Yale Review*, March 1954. Copyright 1954 by Yale University.

due solely to the influence of observational techniques, but gradually physicists came to the conclusion that it was an intrinsic characteristic of the electron's behavior, even when no one was observing it.

Other lines of research underlined this basic ambiguity. If the electron were regarded as a particle, it was impossible to explain a wide variety of experiments which indicated it to be fundamentally a wave. On the other hand, no wave theory could explain the experiments which proved the electron to be corpuscular. Niels Bohr, the famous Danish atomic physicist, adopted an attitude toward this exasperating dilemma which rapidly became the accepted one. Instead of seeking a metaphysically coherent description of the electron's "real" structure, he proposed that physicists accept both the wave and corpuscular theories, even though they contradict each other. This became known as Bohr's "Principle of Complementarity." The ultimate nature of the electron is, from this point of view, a mystery. The best we can do is describe it by two incommensurable approaches, neither of which can be reduced to the other. (Students of philosophy will recognize the curious similarity between Bohr's complementarity principle and the "double truth" doctrine of Averroës, and other theologians both Christian and Moslem. The difference is that the theologians' double truth arose from the attempt to harmonize reason with logically paradoxical doctrines derived from revelation. Bohr's double truth is merely a description, in the simplest terms, of an unavoidable *experimental* predicament. In both cases, however, there is acceptance of seemingly contradictory views as equally valid.)

Both Heisenberg's and Bohr's principles were announced as nothing more than convenient descriptions of experimental data. But philosophers of idealistic bent (using the word "idealism" to mean an emphasis on mind or spirit as more "real" than the material universe), and a few physicists with similar views, were quick to seize upon both principles as ammunition for the defense of their metaphysical preferences. In America, for example, Arthur H. Compton wrote a book called *The Freedom of Man,* in which his arguments ran in a chain from Heisenberg's uncertainty principle to indeterminism in nature, to free will, and thence to God, immortality, and Protestantism. In England, Sir Arthur Eddington voiced similar arguments.

The leading nuclear physicists of Russia, older men of undisputed competence, had no hesitancy in accepting the views of Heisenberg and Bohr. But the younger scientists, saturated with the philosophical dogmas of the Marx-Engels-Lenin tradition, were seized with alarm. The randomness of the electron did not jibe with the rigid determinism of dialectical materialism, just as in biology the randomness of mutations seemed to Lysenko and his followers to destroy the operation of natural laws in evolution. Similarly, the principle of complementarity seemed to assert a kind of mystical, transcendental electron which could not be captured by materialist procedures.

In opposition to Heisenberg, the younger Soviet scientists maintained that the so-called "caprice" of the electron was due simply to the imperfection of measuring instruments. In reply to Bohr they argued that, since the wave and corpuscular theories contradicted each other, it was impossible to hold both theories simultaneously. One or the other must be abandoned, they insisted, or new facts discovered for harmonizing what now appears to be contradictory evidence.

In his pamphlet *Dialectical Materialism and Science,* 1949, Maurice Cornforth, the British Communist writer, expresses the Soviet attitude toward Bohr's principle as follows:

> This is a simple logical contradiction between contradictory propositions, of the sort that was analyzed more than two thousand years ago by Aristotle. Aristotle taught that if a theory contains logical contradictions then that theory cannot be accepted; and dialectical materialists agree with him. The contradictions in bourgeois physical theory are symptoms of the profound crisis of that theory, not signs that it is becoming "dialectical."
>
> The task of dialectics is not to accept the contradictory proposition that an electron is both a wave and a particle. Its task is to disclose the real dialectical contradiction in physical processes—the objective contradiction in the physical world, not a formal contradiction between propositions—and to show how the wave-like and particle-like properties manifested by electrons come into being on the basis of that real contradiction. This has not been done, but remains to be done. It is a question of physical research.
>
> So far as bourgeois physical theory is concerned, some of its main difficulties center around the theory of the atomic nucleus. The atomic nucleus constitutes, as it were, the central knot of contradictions of the physical world, just as the simple commodity constituted the central knot of contradictions in the sphere of economics. Bourgeois theory in physics is no more capable of understanding the nature of the atomic nucleus than bourgeois theory in economics was capable of understanding the nature of commodities.

A better and more detailed defense in English of the Communist approach to modern atomic theory is a book called *The Crisis in Physics.* It was written by Christopher Caudwell, pseudonym of Christopher St. John Sprigg, a young British poet and writer who joined the International Brigade and was killed in Spain in 1937. The book was posthumously published in 1939 and has recently been reprinted.

Caudwell reasoned as follows. Theoretical physics, like everything else, reflects the ideology of a culture, which is in turn a product of the society's economic structure. Modern capitalism is breaking down; consequently physics in bourgeois nations is in a similar state of anarchy. Only dialectical materialism, the ideology of a classless society, can provide the proper guide for physics. It avoids the errors of idealism and positivism by asserting the reality of matter and denying all mystical, unknowable entities. At the same time it avoids the errors of old-fashioned mechanism and determinism by emphasizing human "freedom"—not in the sense that causal laws are violated by "free will," but in the sense that the proletariat can develop an awareness of its ability to reshape history. Although Einstein is praised for his opposition to idealistic tendencies, he is regarded as the last of the old-fashioned determinists. "There seems no doubt that Einstein's world represents the final productive development of the bourgeois world-view—Nature as the object in pure contemplation. It is the climax of mechanism."

Caudwell's book is an accurate presentation of the views of the younger, politically oriented physicists in Russia. These views clashed sharply, of course, with the views of the older scientists. Until the end of the Second World War, however, nuclear physics was a relatively unimportant field of research in the USSR, and so the conflict was correspondingly unimportant. It merely simmered.

In the United States, a voluntary censorship on nuclear research was begun in 1941. Russian periodicals in the field continued to be exported, however, until the bomb fell on Hiroshima. The effect of the bomb on Russian physics was hardly less explosive. While America issued the Smyth report, and shifted to a more liberal policy in the classification of documents relating to nuclear physics, the Soviets clamped down a censorship even more rigid than America's during the war. At present all our technical journals in physics are sent to the USSR. No similar publications are exchanged; and the English abstracts that were formerly printed in Russian technical periodicals have been discontinued. *The Journal of Physics,* a Soviet publication in English, was suspended shortly after the close of the war. (This journal was founded and for a time edited by an Austrian physicist, Alexander Weissberg. He was arrested by the NKVD in 1937. After three years in Soviet prisons he managed to escape; later he wrote a highly informative account of his experiences, *The Accused,* published in 1951.)

Inside Russia, of course, enormous grants were made to the nuclear physicists. Young and obscure scientists, many of them members of the Party, were given high administrative posts. And this, as can be imagined, suddenly brought to a crisis the controversy which had previously been of no great moment.

The story of what has since taken place parallels in many respects the story of what happened in Russian genetics. Motivated by inflexible devotion to dogmas contained in discussions of theoretical physics by Marx, Engels, and Lenin (written, of course, long before the research which led to the Heisenberg and Bohr principles), by a desire to exercise newly acquired power, and also, perhaps, by professional jealousy, the Party physicists declared open war on what they called the "reactionary bourgeois idealistic tendencies" of their older colleagues. And the younger scientists are winning without a struggle.

An early hint of the approaching storm was an article by M. Mitin, a member of the Academy of Sciences (the highest scientific body of the USSR), in the November 20, 1948, issue of *Literaturnaya Gazeta*. After discussing the triumph of Lysenko's views over reactionary bourgeois biology, Mitin turns his attention toward the similar struggle in physics. The principle of uncertainty is attacked as an obscurantist doctrine promulgated by "reactionary flunkies of Anglo-American imperialism in the field of science." The struggle has become irreconcilable, he declares, and "only consistent materialism can purge physics of idealistic tendencies."

In the same issue, another article quotes, as a basis for a "militant program of action," the words of Andrei Zhdanov: "All the forces of obscurantism and reaction are now established in the service of the struggle against Marxism. . . . The subterfuges of contemporary bourgeois atomic physicists lead them to conclusions about the 'freedom of the will' of electrons. Who, then, if not we—the land of victorious Marxism and her philosophers—are to stand at the head of the struggle against depraved and infamous bourgeois ideology! Who then, if not we, are to deliver the shattering blows!" (This translation is from the excellent chapter on Soviet science in *The Country of the Blind,* by George S. Counts and Nucia Lodge.)

Soviet atomic scientists were quick to take the hint. In 1949 a revised edition appeared of D. I. Blokhintsev's *Introduction to Quantum Mechanics,* an outstanding Soviet text in the field. A new chapter had been added to deal with the relationship between physics and dialectical materialism. It contained a violent attack on the "obscurantism" and "subjective idealism" of what it termed the reactionary Copenhagen school of Bohr.

Other nuclear physicists lost no time in falling in step with the rapidly emerging Party line. Before we judge these men too harshly let us recall, as we can be sure that they recalled, what happened to the Soviet geneticists who opposed Lysenkoism. In 1936, when the drive against the gene theory of heredity began, geneticist Agol vanished from the scene. He had been convicted of "Menshevik idealism" in biology. The following year the Medico-genetical Institute, finest of its kind in the world, was attacked by *Pravda,* and soon dissolved. Solomon Levit, the institute's head, confessed his guilt and has not been heard from since.

In 1939 Trofim Lysenko (whose views, according to Professor H. J. Muller, Nobel Prize geneticist, are the "merest drivel") replaced Nikolai I. Vavilov in all the latter's administrative posts. Vavilov, Russia's leading geneticist, internationally respected, persisted in clinging to the views which outside the Soviet Union were universally accepted. He was arrested in 1940 as a British spy and sent to a Siberian labor camp, where he died in disgrace. During the war, death from unknown causes came to four distinguished geneticists—Karpechenko, Koltsov, Serebrovsky, and Levitsky.

"Certain it is," writes Professor Muller, "that from 1936 on Soviet geneticists of all ranks lived a life of terror. Most of them who were not imprisoned, banished, or executed were forced to enter other lines of work. The great majority of those who were allowed to remain in their laboratories were obliged to redirect their researches in such a way as to make it appear that they were trying to prove the correctness of the officially approved antiscientific views. During the chaotic period toward the close of the war, some escaped to the West. Through it all, however, a few have remained at work, retained as showpieces to prove that the USSR still has some working geneticists."

After the historic conference in 1948, when Lysenko announced his formal endorsement by the Communist government, the last remaining geneticist of recognized competence, Dubinin, was dismissed from his post and his excellent laboratory liquidated.

In view of these events, it is not difficult to understand the letter by Professor S. E. Khaikin, one of the outstanding older physicists of the USSR, which was published in March, 1950, in *Uspekhi Fizicheskikh Nauk* ("Progress of the Physical Sciences"), a nontechnical journal still obtainable in this country. This communication is typical of many recent "confessions" by leading Soviet physicists who have been accused of bourgeois idealistic tendencies.

Professor Khaikin begins by saying that his textbook on mechanics, especially the second edition, has recently been criticized by members of the physics faculty at Moscow University and by the Party organ of the Physical Institute of the Academy of Sciences. Unfortunately, he says, the criticisms are true and "have helped me to see distinctly the methodological deficiencies of my textbook." He considers it his duty, he continues, to make this as clear as possible, in the hope that it will help students to escape the errors caused by his "incomplete, inexact, and false" exposition.

The philosophy of dialectical materialism, he writes, and the science of physics mutually strengthen each other. Physics deepens the understanding of the philosophy, and conversely "dialectical materialism is the only philosophy which allows one correctly to reflect upon and understand physics." It is for this reason, he says, that Lenin devoted so much of his writing to physics. It is the duty of a physics textbook to strengthen the students' belief in dialectical materialism.

His mistake, he goes on, led him to a point of view which reflected the "idealistic aggression" of capitalist nations against the philosophy of Marx, Engels, Lenin, and Stalin. In his book he failed to make clear the materialistic foundations of physical laws. This may lead easily to idealism, and he hopes his letter will check any damaging influence the book may have had.

Why did he make these errors? It was "because when I wrote the textbook I did not guide myself by the Lenin principle of Party loyalty in science." In addition, he also failed to pay sufficient attention to the great role of Soviet scientists in the field of mechanics. A long list follows of men whom he failed to mention — names completely unknown to scientists outside the Soviet Union.

In reply to criticism by a physicist named Korolev, Professor Khaikin says that his critic was right in all cases but failed to make his criticisms as cogent as he might have. Two statements of Lenin's are quoted, statements that Professor Khaikin believes put the argument more convincingly. He hastens to add that he does not mean to minimize the errors of his book by this criticism of his critic. "I feel very clearly the defects," he concludes. "I consider it my duty to eliminate them at the first opportunity."

There is no space here for detailing the specific criticisms (all by obscure, lesser men) of Professor Khaikin's textbook. They appear in the same issue of the journal, and in previous issues, and are along the lines explained earlier in this article. It is worth noting, however, that the editor of the periodical apologizes for his mistake in permitting the magazine, in 1948, to publish a review that praised Khaikin's book. He fears the professor has many ideological friends in the USSR who have slavishly accepted the viewpoints of foreign scientists.

On January 2, 1950, *Pravda* printed a speech by an almost unknown physicist, Professor D. N. Nasledov, which is even more revealing. Delivered at a Party conference in Leningrad, it is a bitter attack on one of Russia's most famous theoretical physicists, Professor J. I. Frenkel. The quotations below are from an English translation which appeared in the American periodical *Physics Today*.

"In our Physio-Technical Institute of the Academy of Sciences of the USSR in Leningrad," Nasledov declared, "for a long time idealistic thoughts were openly expressed and feasibility of a fruitful application of the Marxist method to the natural sciences was denied. These idealistic errors permitted by some scientists were not subjected to a deep and earnest criticism and they were not opposed by vigorous materialistic thought, exposing the various attacks of our enemies abroad against the advanced Soviet science of physics."

The leading offender, he continues, was Professor Frenkel, who had a "negative attitude" toward dialectical materialism, and whose writings reflected the opinions of bourgeois physicists. It was necessary, therefore, for Professor Frenkel's opinions to be subjected to sharp criticism.

"The result of this criticism was that Professor Frenkel admitted his ideological errors and in his declaration stated that he had come to the conclusion that the Marxist-Leninist theory in the natural sciences, and particularly in the science of physics, is of foremost importance. Professor Frenkel has promised to correct the admitted errors in all his subsequent work and to rewrite some of his textbooks in the materialistic spirit. We regard this declaration as a great achievement of the work of our

Party organization, which found the way to set such a prominent scientist as Professor Frenkel right. Our duty in the future is to help Professor Frenkel not to stray any more."

The orthodox line in Soviet nuclear physics, unlike the Lysenko line in genetics, is not confined solely to the USSR. The vast majority of physicists throughout the world, it is true, accept the new "statistical" view of quantum theory. But a few Western scientists, notably Einstein, favor the older deterministic approach. A religious man, in the sense in which Spinoza was religious, it is impossible for Einstein to imagine a cosmos in which ultimate units are not dancing to rigid, predictable laws. "Many of us regard this as a tragedy," physicist Max Born has recently written, "—for him, as he gropes his way in loneliness, and for us who miss our leader and standard-bearer." It is interesting to note that Bertrand Russell, whom no one can accuse of Soviet sympathies, believes with Einstein and the Russian scientists that randomness in quantum theory is due to temporary ignorance, and will be removed when more is learned about the electron. Max Planck, Louis de Broglie, and David Bohm also share this hope.

The fact that Einstein is almost alone among Western scientists in his opposition to prevailing opinions in quantum theory, but finds himself strongly supported by the orthodox Soviet physicists, may be a psychological factor in blinding him to the monstrous tyranny of the Soviet regime. Another factor may be the influence of his good friend Leopold Infeld, the Polish mathematician who collaborated with him in 1942 on a book called *The Evolution of Physics*. In 1950, Professor Infeld resigned from the University of Toronto to remain in Poland (he had been lecturing at the University of Warsaw) to, as he put it, "work for peace."

Yet Einstein is no hero in the USSR. Although Soviet physicists have accepted many aspects of his relativity theories, there have always been important reservations. In recent years, attacks on relativity theory in general, and Einstein in particular, have increased markedly. Professor M. S. Eigenson, writing on "The Crisis of Bourgeois Cosmology" in the July, 1950, issue of *Priroda* ("Nature"), attacks Einstein's concept of a "closed cosmos"—that is, a finite cosmos, closing back on itself through the fourth dimension. Professor Eigenson considers the finite universe a product of capitalistic idealism intimately related to the decline of bourgeois culture and a return to the outmoded cosmic models of Ptolemy and Aristotle. Even worse, he writes, is the "expanding universe" of Eddington, which limits the cosmos in time as well as space, suggesting a moment of "creation" and thereby violating the principle of the conservation of matter discovered by the Russian physicist M. V. Lomonosov (outside the USSR scientists credit this principle to the French chemist Lavoisier).

In 1952, the Soviet Navy newspaper, *Krasny Flot,* printed an article by A. Maximov which called Einstein's views "reactionary, antiscientific, antimaterialistic, and idealistic" and added that there was no excuse for Einstein because, unlike other "bourgeois physicists," he was familiar with the writings of Marx, Engels, and Lenin. Maximov likened Einstein's views to those of Ernst Mach, of whom Lenin had written, "Mach's philosophy is to science what Judas' kiss was to Christ." The new edition of the *Soviet Philosophical Dictionary*, published in 1952, calls Einstein's theory of relativity "a reactionary, antiscientific distortion of the truth" and heaps scorn on the "mystics and obscurantists who babble about the fourth dimension, the finity of the universe, and similar absurdities."

One important question remains to be considered. Will Russian research on atomic weapons be retarded by the pressure of a Party line on matters relating to theoretical physics? It is hard to say. The conflict is on such an abstract level that it may have little effect on the lower, more technical areas where work is in progress. On the other hand, political meddling by scientifically illiterate commissars may lead indirectly to the same type of bungling which characterized Germany's attempt to make an atom bomb, and, more recently, Perón's fantastic temporary sponsorship of the Austrian pseudoscientist Ronald Richter.

Samuel Goudsmit, in his remarkable book *Alsos*, paints an amusing picture of what he calls the "misorganization" of German science under Hitler. *Alsos* was the code name of the mission that entered Germany after the war to determine how far the Nazis had advanced in atomic research. To the surprise of most American physicists, the answer proved to be "practically nowhere." In Professor Goudsmit's opinion (he was the scientific head of the mission), it was the elevation of second-rate, but ardently pro-Nazi, scientists to posts of authority which played the major role in the deterioration of German physics.

Research in the German Army, for example, was headed by a mediocre physicist whose only published works were a few papers on the vibrations of piano strings. Although the Nazis had no tradition of materialism to which they felt they must adhere in their science, their distrust of relativity (Einstein was non-Aryan) forced the better physicists to waste endless hours explaining to their inferiors that relativity had been amply verified and could be abandoned only with serious damage to their work. The ignorance of administrators diverted funds and energies into worthless projects. Only in the Air Force was research efficiently organized, due largely, Goudsmit believes, to the relatively greater freedom which Goering permitted the scientists.

An indication of Soviet confusions in the field of atomic research was revealed in 1951 by an article in *Bolshevik*, written by the son of the late Andrei Zhdanov. (See the *New York Times*, December 24, 1951, page 1.) The article was the first official repudiation of a previous Soviet "discovery" that cosmic rays contained basic particles called "varitrons." Two Russian physicists had received the Stalin Prize First Class in 1948 for this discovery, which was widely publicized as proof of the superiority of Soviet science. Zhdanov attacks Russian physicists who defended the varitron theory, and also repudiates another atomic "discovery" for which Professor Georgi Latyshev had received the Stalin Prize in 1949. Both repudiations accord with the views of Western physicists. One may conclude that, if a scientific theory is sufficiently preposterous, even the Soviet commissars may ultimately abandon it. There is even the possibility that Lysenkoism may be undergoing some sort of modification. Lysenko has been denounced several times in Soviet publications since Stalin's death, though chiefly for personal incompetence and minor deviations rather than for basic theoretical errors.

A frontispiece in *Alsos* pictures Germany's "Oak Ridge" laboratory. It was about the size of a small cottage. We know, of course, that Russian atomic laboratories are much larger. The explosion at Hiroshima convinced them that the bombs were no bourgeois idealistic dreams, and the success of their espionage agents has saved them many years of difficult research. It would be foolhardy to suppose the Russian physicists incapable of duplicating the achievements of Western physicists. Nevertheless, there are grounds for hope that, as the Russian scientists grope for more efficient weapons, the iron hand of Party control may have the same deadening effect it had in Germany on new and novel lines of research.

Postscript

The great Soviet campaign against Niels Bohr began in 1947 when A. A. Zhdanov made his famous speech about the "corrupt and vile" influence on science of bourgeois idealism, especially the "Kantian distortions" of the quantum physicists. Bohr's view was singled out as "rubbish" to be "thrown down the drain" in a battle that would deliver "fatal blows" against idealism.

It was not until several years after Stalin's death in 1953 that Soviet science began to recover. Relativity theory was officially accepted in the mid-fifties, then the expanding universe, and finally Bohr's complementarity. Einstein and Bohr became great heroes in the USSR and still are today. In 1961 Leopold Infeld strongly denounced those Marxist theoreticians who had earlier condemned Einstein, Bohr, Linus Pauling, and others without knowing anything about their work. In 1962 Peter Kapitza made similar remarks. If Soviet scientists had listened to the Marxist philosophers, Kapitza said, Soviet space exploration would have been impossible.

Trofim Lysenko, a pseudobotanist to whom I devoted a chapter in *Fads and Fallacies,* was eventually stripped of his power and died in disgrace in 1976. Although philosophers in the USSR, as elsewhere, continue to argue about how to interpret quantum mechanics, there are no longer vitriolic denunciations of the Copenhagen view. Bureaucratic controls may still hamper research, as they always do, but this is balanced by massive government support of research, especially in work related to war technology. Contemporary philosophy of science in the USSR is complex, and I know of no better reference in English than Loren R. Graham's *Science and Philosophy in the Soviet Union* (Knopf, 1972). See also S. Muller-Markus's excellent paper "Niels Bohr in the Darkness and Light of Soviet Philosophy," in *Inquiry,* vol. 1, Spring 1966, pp. 73-93.

I have been criticized for calling Einstein "blind" to the realities of Stalinism. Much as I revere Einstein, it must be said that he, like so many other intellectuals of the thirties, made only the feeblest effort to learn the truth about Stalin. Here is a quotation from a letter he wrote to Max Born, and which you will find on page 130 in *The Born-Einstein Letters* (1971), edited by Born:

> By the way, there are increasing signs that the Russian trials are not faked, but that there is a plot among those who look upon Stalin as a stupid reactionary who has betrayed the ideas of the revolution. Though we find it difficult to imagine this kind of internal thing, those who know Russia best are all more or less of the same opinion. I was firmly convinced to begin with that it was a case of a dictator's despotic acts, based on lies and deception, but this was a delusion.

Born's comments are in my opinion just:

> The Russian trials were Stalin's purges, with which he attempted to consolidate his power. Like most people in the West, I believed these show trials to be the arbitrary acts of a cruel dictator. Einstein was apparently of a different opinion: he believed that when threatened by Hitler the Russians had no choice but to destroy as many of their enemies within their own camp as possible. I find it hard to reconcile this point of view with Einstein's gentle, humanitarian disposition.

3

The *Ars Magna* of Ramon Lull

Near the city of Palma, on the island of Majorca, largest of the Balearic isles off the eastern coast of Spain, a huge saddle-shaped mountain called Mount Randa rises abruptly from a monotonously level ridge of low hills. It was this desolate mountain that Ramon Lull, Spanish theologian and visionary, climbed in 1274 in search of spiritual refreshment. After many days of fasting and contemplation, so tradition has it, he experienced a divine illumination in which God revealed to him the Great Art by which he might confound infidels and establish with certainty the dogmas of his faith. According to one of many early legends describing this event, the leaves of a small lentiscus bush (a plant still flourishing in the area) became miraculously engraven with letters from the alphabets of many languages. They were the languages in which Lull's Great Art was destined to be taught.

After his illumination, Lull retired to a monastery where he completed his famous *Ars magna,* the first of about forty treatises on the working and application of his eccentric method. It was the earliest attempt in the history of formal logic to employ geometrical diagrams for the purpose of discovering nonmathematical truths, and the first attempt to use a mechanical device—a kind of primitive logic machine—to facilitate the operation of a logic system.

Throughout the remainder of Lull's colorful, quixotic life, and for centuries after his death, his Art was the center of stormy controversy.

Reprinted with permission from *Logic Machines and Diagrams,* by Martin Gardner (McGraw-Hill, New York, 1958). Copyright 1958 by McGraw-Hill Book Company.

Franciscan leaders (Lull belonged to a lay order of the movement) looked kindly upon his method, but Dominicans tended to regard it as the work of a madman. Gargantua, in a letter to his son Pantagruel (Rabelais, *Gargantua and Pantagruel,* Book II, Chapter 8), advises him to master astronomy "but dismiss astrology and the divinitory art of Lullius as but vanity and imposture." Francis Bacon similarly ridiculed the Art in two passages of almost identical wording, one in *The Advancement of Learning* (Book II), the other in *De augmentis scientiarum,* a revised and expanded version of the former book. The passage in *De augmentis* (Book VI, Chapter 2) reads as follows:

> And yet I must not omit to mention that some persons, more ostentatious than learned, have laboured about a kind of method not worthy to be called a legitimate method, being rather a method of imposture, which nevertheless would no doubt be very acceptable to certain meddling wits. The object of it is to sprinkle little drops of science about, in such a manner that any sciolist may make some show and ostentation of learning. Such was the Art of Lullius: such the Typocosmy traced out by some; being nothing but a mass and heap of the terms of all arts, to the end that they who are ready with the terms may be thought to understand the arts themselves. Such collections are like a fripper's or broker's shop, that has ends of everything, but nothing of worth.

Swift is thought to have had Lull's Art in mind when he described a machine invented by a professor of Laputa (*Gulliver's Travels,* Part III, Chapter 5). This contrivance was a 20-foot square frame containing hundreds of small cubes linked together by wires. On each face of every cube was written a Laputan word. By turning a crank, the cubes were rotated to produce random combinations of faces. Whenever a few words happened to come together and make sense, they were copied down; then from these broken phrases erudite treatises were composed. In this manner, Swift explained, "the most ignorant person at a reasonable charge, and with a little bodily labour, may write books in philosophy, poetry, politics, law, mathematics, and theology, without the least assistance from genius or study."

On the other hand we find Giordano Bruno, the great Renaissance martyr, speaking of Lull as "omniscient and almost divine," writing fantastically elaborate treatises on the Lullian Art, and teaching it to wealthy noblemen in Venice, where it had become a fashionable craze. Later we find young Leibniz fascinated by Lull's method. At the age of nineteen he wrote his *Dissertio de arte combinatoria* (Leipzig, 1666), in which he discovers in Lull's work the germ of a universal algebra by which all knowledge, including moral and metaphysical truths, can some day be brought within a single deductive system.[1] "If controversies were to arise," Leibniz later declared in an oft-quoted passage, "there would be no more need of disputation between two philosophers than between two accountants. For it would suffice to take their pencils in their hands, to sit down to their slates, and to say to each other (with a friend to witness, if they liked): Let us calculate."

These speculations of Leibniz's have led many historians to credit Lull with having foreshadowed the development of modern symbolic logic and the empiricist's dream of the "unity of science." Is such credit deserved? Or was Lull's method little more than the fantastic work of a gifted crank, as valueless as the geometric designs of medieval witchcraft? Before explaining and attempting to evaluate Lull's bizarre, now forgotten Art, it will perhaps be of interest to sketch briefly the extraordinary, almost unbelievable career of its inventor.[2]

Ramon Lull was born at Palma, probably in 1232. In his early teens he became a page in the service of King James I of Aragon and soon rose to a position of influence in the court. Although he married young and had two children, his life as a courtier was notoriously dissolute. "The beauty of women, O Lord," he recalled at the age of forty, "has been a plague and tribulation to my eyes, for because of the beauty of women have I been forgetful of Thy great goodness and the beauty of Thy works."

The story of Lull's conversion is the most dramatic of the many picturesque legends about him, and second only to Saint Augustine's as a celebrated example of a conversion following a life of indulgence. It begins with Lull's adulterous passion for a beautiful and pious married woman who failed to respond to his overtures. One day as he was riding a horse down the street he saw the lady enter church for High Mass. Lull galloped into the cathedral after her, only to be tossed out by irate worshippers. Distressed by this scene the lady resolved to put an end to Lull's campaign. She invited him to her chamber, uncovered the bosom that he had been praising in poems written for her, and revealed a breast partially consumed by cancer. "See, Ramon," she exclaimed, "the foulness of this body that has won thy affection! How much better hadst thou done to have set thy love on Jesus Christ, of Whom thou mayest have a prize that is eternal!"

Lull retired in great shame and agitation. Shortly after this incident, while he was in his bedroom composing some amorous lyrics, he was startled by a vision of Christ hanging on the Cross. On four later occasions, so the story goes, he tried to complete the verses, and each time was interrupted by the same vision. After a night of remorse and soul searching, he hurried to morning confession as a penitent, dedicated Christian.

Lull's conversion was followed by a burning desire to win nothing less than the entire Moslem world for Christianity. It was an obsession that dominated the remainder of his life and eventually brought about his violent death. As the first necessary step in this ambitious missionary project, Lull began an intensive study of the Arabic language and theology. He purchased a Moorish slave who lived in his home for nine years, giving him instruction in the language. It is said that one day Lull struck the slave in the face after hearing him blaspheme the name of Christ. Soon thereafter the Moor retaliated by attacking Lull with a knife. Lull succeeded in disarming him and the slave was jailed while Lull pondered the type of punishment he should receive. Expecting to be put to death, the Moor hanged himself with the rope that bound him.

Before this unfortunate incident, Lull had managed to finish writing, probably in Arabic, his first book, the *Book of Contemplation*. It is a massive, dull work of several thousand pages that seeks to prove by "necessary reasons" all the major truths of Christianity. Thomas Aquinas had previously drawn a careful distinction between truths of natural theology that he believed could be established by reason, and truths of revelation that could be known only by faith. Lull found this distinction unnecessary. He believed that all the leading dogmas of Christianity, including the trinity and incarnation, could be demonstrated by irrefutable arguments, although there is evidence that he regarded "faith" as a valuable aid in understanding such proofs.

Lull had not yet discovered his Great Art, but the *Book of Contemplation* reveals his early preoccupation with a number symbolism that was characteristic of many scholars of his time. The work is divided into five books in honor of the five wounds of Christ. Forty subdivisions signify the forty days Christ spent in the wilderness. The 366 chapters are designed to be read one a day, the last chapter to be consulted only in leap years. Each chapter has ten paragraphs (the ten commandments); each paragraph has three parts (the trinity), making a total of thirty parts per chapter (the thirty pieces of silver). Angles, triangles, and circles are occasionally introduced as metaphors. Of special interest to modern logicians is Lull's practice of using letters to stand for certain words and phrases so that arguments can be condensed to almost algebraic form. For example, in Chapter 335 he employs a notation of 22 symbols and one encounters passages such as this:

> If in Thy three properties there were no difference . . . the demonstration would give the D to the H of the A with the F and the G as it does with the E, and yet the K would not give significance to the H of any defect in the F or the G; but since diversity is shown in the demonstration that the D makes of the E and the F and the G with the I and the K, therefore the H has certain scientific knowledge of Thy holy and glorious Trinity.[3]

There are unmistakable hints of paranoid self-esteem in the value Lull places on his own work in the book's final chapter. It will not only prove to infidels that Christianity is the one true faith, he asserts, but it will also give the reader who follows its teaching a stronger body and mind as well as all the moral virtues. Lull expresses the wish that his book be "disseminated throughout the world," and he assures the reader that he has "neither place nor time sufficient to recount all the ways wherein this book is good and great."

These immodest sentiments are characteristic of most eccentrics who become the founders of cults, and it is not surprising to hear similar sentiments echoed by disciples of the Lullian Art in later centuries. The Old Testament was regarded by many Lullists as the work of God the Father, the New Testament, of God the Son, and the writings of Lull, of God the Holy Spirit. An oft-repeated jingle proclaimed that there had been three wise men in the world—Adam, Solomon, and Ramon:

> *Tres sabios hubo en el mundo,*
> *Adán, Solomón y Raymundo.*

Lull's subsequent writings are extraordinarily numerous, although many of them are short and there is much repetition of material and rehashing of old arguments. Some early authorities estimated that he wrote several thousand books. Contemporary scholars consider this an exaggeration, but there is good reason to think that more than two hundred of the works attributed to him are his (the alchemical writings that bear his name are known to be spurious). Most of his books are polemical, seeking to establish Christian doctrines by means of "necessary reasons," or to combat Averroism, Judaism, and other infidel doctrines. Some are encyclopedic surveys of knowledge, such as his 1,300-page *Tree of Science,* in which he finds himself forced to speak "of things in an abbreviated fashion." Many of his books are in the form of Socratic dialogues. Others are collections of terse aphorisms, such as his *Book of Proverbs,* a collection of some 6,000 of them. Smaller treatises, most of which concern the application of his Great Art, are devoted to almost every subject matter with which his contemporaries were concerned— astronomy, chemistry, physics, medicine, law, psychology, mnemonics, military tactics, grammar, rhetoric, mathematics, zoology, chivalry, ethics, politics.

Very few of these polemical and pseudoscientific works have been translated from the original Catalan or Latin versions, and even in Spain they are now almost forgotten. It is as a poet and writer of allegorical romances that Lull is chiefly admired today by his countrymen. His Catalan verse, especially a collection of poems on *The Hundred Names of God,* is reported to be of high quality, and his fictional works contain such startling and imaginative conceptions that they have become an imperishable part of early Spanish literature. Chief of these allegorical books is *Blanquerna,* a kind of Catholic *Pilgrim's Progress.*[4] The protagonist, who closely resembles the author, rises through various levels of church organization until he becomes Pope, only to abandon the office, amid much weeping of cardinals, to become a contemplative hermit.

The Book of the Lover and the Beloved, Lull's best-known work, is contained within *Blanquerna* as the supposed product of the hermit's pen.[5] More than any other of Lull's works, this book makes use of the phrases of human love as symbols for divine love—a practice as common in the Moslem literature prior to Lull's time as it was later to become in the writings of Saint Theresa and other Spanish mystics. Amateur analysts who enjoy looking for erotic symbols will find *The Book of the Lover and the Beloved* a fertile field. All of Lull's passionate temperament finds an outlet here in his descriptions of the intimate relationship of the lover (himself) to his Beloved (Christ).

In Lull's other great work of fantasy, *Felix, or the Book of Marvels,* we find him describing profane love in scenes of such repulsive realism that they would shock even an admirer of Henry Miller's fiction. It is difficult not to believe that Lull's postconversion attitude toward sex had much to do with his vigorous defense of the doctrine of the immaculate conception at a time when it was opposed by the Thomists and of course long before it became church dogma.

After Lull's illumination on Mount Randa, his conviction grew steadily that in his Art he had found a powerful weapon for the conversion of the heathen. The failure of the Crusades had cast doubt on the efficacy of the sword. Lull was convinced that rational argument, aided by his method, might well become Gòd's new means of spreading the faith. The remainder of his life was spent in the restless wandering and feverish activity of a missionary and evangelical character. He gave up the large estate he had inherited from his father, distributing his possessions to the poor. His wife and children were abandoned, though he set aside funds for their welfare. He made endless pilgrimages, seeking the aid of popes and princes in the founding of schools and monasteries where his Great Art could be taught along with instruction in heathen languages. The teaching of Oriental languages to missionaries was one of Lull's dominant projects, and he is justly regarded as the founder of Oriental studies in European education.

The esoteric character of his Art seems to have exerted a strong magic appeal. Schools and disciples grew so rapidly that in Spain the Lullists became as numerous as the Thomists. Lull even taught on several occasions at the great University of Paris—a signal honor for a man holding no academic degree of any kind. There is an amusing story about his attendance, when at the Sorbonne, of a class taught by Duns Scotus, then a young man fresh from triumphs at Oxford. It seems that Scotus became annoyed by the old man in his audience who persisted in making signs of disagreement with what was being said. As a rebuke, Scotus asked him the exceedingly elementary question, "What part of speech is 'Lord'?" Lull immediately replied, "The Lord is no part, but the whole," and then proceeded to stand and deliver a loud and lengthy oration on the perfections of God. The story is believable because Lull always behaved as a man possessed by inspired, irrefutable truth.

On three separate occasions Lull made voyages to Africa to clash verbal swords with Saracen theologians and to preach his views in the streets of Moslem cities. On the first two visits he barely escaped with his life. Then at the age of eighty-three, his long beard snow white and his eyes burning with desire for the crown of martyrdom, he set sail once more for the northern shore of Africa. In 1315, on the streets of Bugia, he began expounding in a loud voice the errors of Moslem faith. The Arabs were understandably vexed, having twice ousted this stubborn old man from their country. He was stoned by the angry mob and apparently died on board a Genoese merchant ship to which his bruised body had been carried.[6] A legend relates that before he died he had a vision of the American continent and prophesied that a descendant (i.e., Columbus) of one of the merchants would some day discover the new world.

". . . no Spaniard since," writes Havelock Ellis (in a chapter on Lull in his *The Soul of Spain,* 1908), "has ever summed up in his own person so brilliantly all the qualities that go to the making of Spain. A lover, a soldier, something of a heretic, much of a saint, such has ever been the typical Spaniard. Lull's relics now rest in the chapel of the Church of San Francisco, at Palma, where they are venerated as those of a saint, in spite of the fact that Lull has never been canonized.

In turning now to an examination of the Great Art itself,[7] it is impossible, perhaps, to avoid a strong sense of anticlimax. One wishes it were otherwise. It would be pleasant indeed to discover that Lull's method had for centuries been unjustly maligned and that by going directly to the master's own expositions one might come upon something of value that deserves rescue from the oblivion into which it has settled. Medieval scholars themselves sometimes voice such hopes. "We have also excluded the work of Raymond Lull," writes Philotheus Boehner in the introduction to his *Medieval Logic,* 1952, "since we have to confess we are not sufficiently familiar with his peculiar logic to deal with it adequately, though we suspect that it is much better than the usual evaluation by historians would lead us to believe." Is this suspicion justified? Or shall we conclude with Etienne Gilson (*History of Christian Philosophy in the Middle Ages,* 1955) that when we today try to use Lull's tables "we come up against the worst difficulties, and one cannot help wondering whether Lull himself was ever able to use them"?

Essentially, Lull's method was as follows. In every branch of knowledge, he believed, there are a small number of simple basic principles or categories that must be assumed without question. By exhausting all possible combinations of these categories we are able to explore all the knowledge that can be understood by our finite minds. To construct tables of possible combinations we call upon the aid of both diagrams and rotating circles. For example, we can list two sets of categories in two

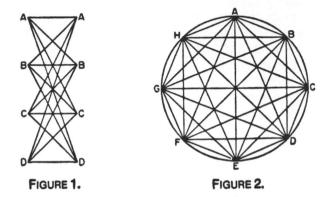

FIGURE 1.   FIGURE 2.

vertical columns (Figure 1), then exhaust all combinations simply by drawing connecting lines as shown. Or we can arrange a set of terms in a circle (Figure 2), draw connecting lines as indicated, then by reading around the circle we quickly obtain a table of two-term permutations.

A third method, and the one in which Lull took the greatest pride, is to place two or more sets of terms on concentric circles as shown in Figure 3. By rotating the inner circle we easily obtain a table of combinations. If there are many sets of terms that we wish to combine, this mechanical method is much more efficient than the others. In Lull's time these circles were made of parchment or metal and painted vivid colors to distinguish different subdivisions of terms. There is no doubt that the use of such strange, multicolored devices threw an impressive aura of mystery around Lull's teachings that greatly intrigued men of little learning, anxious to find a short-cut method of mastering the intricacies of scholasticism. We find a similar appeal today in the "structural differential" invented by Count Alfred Korzybski to illustrate principles of general semantics. Perhaps there is even a touch of the same awe in the reverence with which some philosophers view symbolic logic as a tool of philosophical analysis.

Before going into the more complicated aspects of Lull's method, let us give one or two concrete illustrations of how Lull used his circles. The first of his seven basic "figures" is called *A*. The letter *"A,"* representing God, is placed in the center of a circle. Around the circumference, inside sixteen compartments (or "camerae" as Lull called them), we now place the sixteen letters from *B* through *R* (omitting *J*, which had no existence in the Latin of the time). These letters stand for sixteen divine attributes—*B*

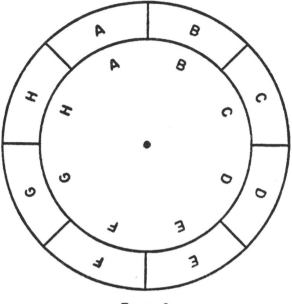

FIGURE 3.

for goodness (*bonitas*), *C* for greatness (*magnitudo*), *D* for eternity (*eternitas*), and so on. By drawing connecting lines (Figure 4) we obtain 240 two-term permutations of the letters, or 120 different combinations that can be arranged in a neat triangular table as shown below.

BC	BD	BE	BF	BG	BH	BI	BK	BL	BM	BN	BO	BP	BQ	BR
	CD	CE	CF	CG	CH	CI	CK	CL	CM	CN	CO	CP	CQ	CR
		DE	DF	DG	DH	DI	DK	DL	DM	DN	DO	DP	DQ	DR
			EF	EG	EH	EI	EK	EL	EM	EN	EO	EP	EQ	ER
				FG	FH	FI	FK	FL	FM	FN	FO	FP	FQ	FR
					GH	GI	GK	GL	GM	GN	GO	GP	GQ	GR
						HI	HK	HL	HM	HN	HO	HP	HQ	HR
							IK	IL	IM	IN	IO	IP	IQ	IR
								KL	KM	KN	KO	KP	KQ	KR
									LM	LN	LO	LP	LQ	LR
										MN	MO	MP	MQ	MR
											NO	NP	NQ	NR
												OP	OQ	OR
													PQ	PR
														QR

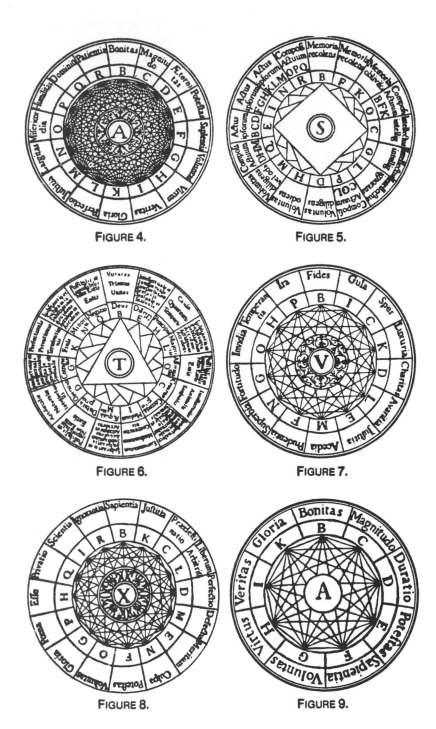

FIGURE 4.

FIGURE 5.

FIGURE 6.

FIGURE 7.

FIGURE 8.

FIGURE 9.

(From the *Enciclopedia universal ilustrada*, Barcelona, 1923.)

Each of the combinations above tells us an additional truth about God. Thus we learn that His goodness is great (*BC*) and also eternal (*BD*), or to take reverse forms of the same pairs of letters, His greatness is good (*CB*) and likewise His eternity (*DB*). Reflecting on these combinations will lead us toward the solution of many theological difficulties. For example, we realize that predestination and free will must be combined in some mysterious way beyond our ken; for God is both infinitely wise and infinitely just; therefore He must know every detail of the future, yet at the same time be incapable of withholding from any sinner the privilege of choosing the way of salvation. Lull considered this a demonstration *"per aequiparantium,"* or by means of equivalent relations. Instead of connecting ideas in a cause-and-effect chain, we trace them back to a common origin. Free will and predestination sprout from equally necessary attributes of God, like two twigs growing on branches attached to the trunk of a single tree.

Lull's second figure concerns the soul and is designated by the letter *S*. Four differently colored squares are used to represent four different states of the soul. The blue square, with corners *B, C, D, E*, is a normal, healthy soul. The letters signify memory that remembers (*B*), intellect that knows (*C*), will that loves (*D*), and the union of these three faculties (*E*). The black square (*FGHI*) is the condition that results when the will hates in a normal fashion as, for example, when it hates sin. This faculty is symbolized by the letter *H*. *F* and *G* stand for the same faculties as *B* and *C*, and *I* for the union of *F*, *G*, and *H*. The red square (*KLMN*) denotes a condition of soul in which the memory forgets (*K*), the mind is ignorant (*L*), and the will hates in an abnormal fashion (*M*). These three degenerate faculties are united in *N*. The green square (*OPQR*) is the square of ambivalence or doubt. *R* is the union of a memory that retains and forgets (*O*), a mind that both knows and is ignorant (*P*), and a will that loves and hates (*Q*). Lull considered this last state the unhealthiest of the four. We now superimpose the four squares (Figure 5) in such a way that their colored corners form a circle of sixteen letters. This arrangement is more ingenious than one might at first suppose. For in addition to the four corner letters *E, I, N, R*, which are unions of the other three corners of their respective squares, we also find that the faculties *O, P*, and *Q* are unions of the three faculties that precede them as we move clockwise around the figure. The circle of sixteen letters can now be rotated within a ring of compartments containing the same faculties to obtain 136 combinations of faculties.

It would be impossible and profitless to describe all of Lull's scores of other figures, but perhaps we can convey some notion of their complexity. His third figure, *T*, concerns relations between things. Five equilateral triangles of five different colors are superimposed to form a circle of fifteen letters, one letter at each vertex of a triangle (Figure 6). As in the previous figure, the letters are in compartments that bear the same color as the polygon for which they mark the vertices. The meanings of the letters are: God, creature, and operation (blue triangle); difference, similarity, contrariety (green); beginning, middle, end (red); majority, equality, minority (yellow); affirmation, negation, and doubt (black). Rotating this circle within a ring bearing the same fifteen basic ideas (broken down into additional elements) gives us 120 combinations, excluding pairs of the same term (*BB, CC,* etc). We are thus able to explore such topics as the beginning and end of God, differences and similarities of animals, and so on. Lull later found it necessary to add a second figure *T,* formed of five tinted triangles whose vertices stand for such concepts as before, after, superior, inferior, universal, particular, etc. This likewise rotated within a ring to produce 120 combinations. Finally, Lull combined the two sets of concepts to make thirty in all. By placing them on two circles he obtained 465 different combinations.

Lull's fourth figure, which he called *V,* deals with the seven virtues and the seven deadly sins. The fourteen categories are arranged alternately around a circle in red (sinful) and blue (virtuous) compartments (Figure 7). Drawing connecting lines, or rotating the circle within a similarly labeled ring, calls our attention to such questions as when it might be prudent to become angry, when lust is the result of slothfulness, and similar matters. Lull's figure *X* employs eight pairs of traditionally opposed terms, such as being (*esse*) and privation (*privatio*), arranged in alternate blue and green compartments (Figure 8). Figures *Y* and *Z* are undivided circles signifying, respectively, truth and falsehood. Lull used these letters occasionally in connection with other figures to denote the truth or falsehood of certain combinations of terms.

This by no means exhausts Lull's use of rotating wheels. Hardly a science or subject matter escapes his analysis by this method. He even produced a book on how preachers could use his Art to discover new topics for sermons, supplying the reader with 100 sample sermons produced by his spinning wheels! In every case the technique is the same: find the basic elements, then combine them mechanically with themselves or with the elements of other figures. Dozens of his books deal with applications of the Art, introducing endless small variations of terminology and symbols. Some of these works are introductions to more comprehensive treatises. Some are brief, popular versions for less intellectual readers who find it hard to comprehend the more involved figures. For example, the categories of certain basic figures are reduced from sixteen to nine (see Figure 9). These simpler ninefold circles are the ones encountered in the writings of Bruno, Kircher, and other Renaissance Lullists, in Hegel's description of the Art (*Lectures on the History of Philosophy,* vol. 3), and in most modern histories of thought that find space for Lull's method. Two of Lull's treatises on his Art are written entirely in Catalan verse.

One of Lull's ninefold circles is concerned with objects of knowledge—God, angel, heaven, man, the imagination, the sensitive, the negative, the elementary, and the instrumental. Another asks the nine questions—whether? what? whence? why? how great? of what kind? when? where? and how? Many of Lull's books devote considerable space to questions suggested by these and similar circles. *The Book of the Ascent and Descent of the Intellect,* using a twelvefold and fivefold circle in application to eight categories (stone, flame, plant, animal, man, heaven, angel, God) considers such scientific posers as: Where does the flame go when a candle is put out? Why does rue strengthen the eyes and onions weaken them? Where does the cold go when a stone is warmed?

In another interesting work Lull uses his Art to explain to a hermit the meaning of some of the *Sentences* of Peter Lombard. The book takes up such typical medieval problems as: Could Adam and Eve have cohabited before they ate their first food? If a child is slain in the womb of a martyred mother, will it be saved by a baptism of blood? How do angels speak to each other? How do angels pass from one place to another in an instant of time? Can God make matter without form? Can He damn Peter and save Judas? Can a fallen angel repent? In one book, the *Tree of Science,* over four thousand such questions are raised! Sometimes Lull gives the combination of terms in which the answer may be found, together with a fully reasoned commentary. Sometimes he merely indicates the figures to be used, letting the reader find the right combinations for himself. At other times he leaves the question entirely unanswered.

The number of concentric circles to be used in the same figure varies from time to time—two or three being the most common. The method reaches its climax in a varicolored metal device called the *figura universalis,* which has no less than fourteen concentric circles! The mind reels at the number and complexity of topics that can be explored by this fantastic instrument.

Before passing on to an evaluation of Lull's method, it should be mentioned that he also frequently employed the diagrammatic device of the tree to indicate subdivisions of genera and species. For Lull it was both an illustrative and a mnemonic device. His *Principles of Medicine,* for example, pictures his subject matter as a tree with four roots (the four humors) and two trunks (ancient and modern medicine). The trunks branch off into various boughs on which flowers bloom, each flower having a symbolic meaning (air, exercise, food, sleep, etc.). Colored triangles, squares, and other Lullian figures also are attached to the branches.

None of Lull's scientific writings, least of all his medical works, added to the scientific knowledge of his time. In such respects he was neither ahead nor behind his contemporaries. Alchemy and geomancy he rejected as worthless. Necromancy, or the art of communicating with the dead, he accepted in a sense common in his day and still surviving in the attitude of many orthodox churchmen; miraculous results are not denied, but they are regarded as demonic in origin. Lull even used the success of necromancers as a kind of proof of the existence of God. The fallen angels could not exist, he argued, if God had not created them.

There is no doubt about Lull's complete acceptance of astrology. His so-called astronomical writings actually are astrological, showing how his circles can be used to reveal various favorable and unfavorable combinations of planets within the signs of the zodiac. In one of his books he applies astrology to medicine. By means of the Art he obtains sixteen combinations of the four elements (earth, air, fire, water) and the four properties (hot, cold, moist, dry). These are then combined in various ways with the signs of the zodiac to answer medical questions concerning diet, evacuation, preparation of medicines, fevers, color of urine, and so on.

There is no indication that Ramon Lull, the Doctor Illuminatus, as he was later called, ever seriously doubted that his Art was the product of divine illumination. But one remarkable poem, the *Desconort* ("Disconsolateness"), suggests that at times he may have been tormented by the thought that possibly his Art was worthless. The poem is ingeniously constructed of sixty-nine stanzas, each consisting of twelve lines that end in the same rhyme. It opens with Lull's bitter reflections on his failure for the past thirty years to achieve any of his missionary projects. Seeking consolation in the woods, he comes upon the inevitable hermit and pours out to him the nature of his sorrows. He is a lonely, neglected man. People laugh at him and call him a fool. His great Art is ridiculed and ignored. Instead of sympathizing, the hermit tries to prove to Ramon that he deserves this ridicule. If his books on the Art are read by men "as fast as a cat that runs through burning coals," perhaps this is because the dogmas of the church cannot be demonstrated by reason. If they could be, then what merit would there be in believing them? In addition, the hermit argues, if Lull's method is so valuable, how is it that the ancient philosophers never thought of it? And if it truly comes from God, what reason has he to fear it will ever be lost?

Lull replies so eloquently to these objections that we soon find the hermit begging forgiveness for all he has said, offering to join Ramon in his labors, and even weeping because he had not learned the Art earlier in life!

Perhaps the most striking illustration of how greatly Lull valued his method is the legend of how he happened to join the third order of Franciscans. He had made all necessary arrangements for his first missionary trip to North Africa, but at the last moment, tormented by doubts and fears of imprisonment and death, he allowed the boat to sail without him. This precipitated a mental breakdown that threw him into a state of profound depression. He was carried into a Dominican church and while praying there he saw a light like a star and heard a voice speaking from above: "Within this order thou shalt be saved." Lull hesitated to join the order because he knew the Dominicans had little interest in his Art, whereas the Franciscans had found it of value. A second time the voice spoke from the light, this time threateningly. "And did I not tell thee that only in the order of the Preachers thou wouldst find salvation?" Lull finally decided it would be better to undergo personal damnation than risk the loss of his Art whereby others might be saved. Ignoring the vision, he joined the Franciscans.

It is clear from Lull's writings that he thought of his method as possessing many values. The diagrams and circles aid the understanding by making it easy to visualize the elements of a given argument. They have considerable mnemonic value, an aspect of his Art that appealed strongly to Lull's Renaissance admirers. They have rhetorical value, not only arousing interest by their picturesque, cabalistic character, but also aiding in the demonstration of proofs and the teaching of doctrines. It is an investigative and inventive art. When ideas are combined in all possible ways, the new combinations start the mind thinking along novel channels and one is led to discover fresh truths and arguments, or to make new inventions. Finally, the Art possesses a kind of deductive power.

Lull did not, however, regard his method as a substitute for the formal logic of Aristotle and the schoolmen. He was thoroughly familiar with traditional logic and his writings even include the popular medieval diagrams of immediate inference and the various syllogistic figures and moods. He certainly did not think that the mere juxtaposition of terms provided in themselves a proof by "necessary reasons." He did think, however, that by the mechanical combination of terms one could discover the necessary building blocks out of which valid arguments could then be constructed. Like his colleagues among the schoolmen, he was convinced that each branch of knowledge rested on a relatively few self-evident principles which formed the structure of all knowledge in the same way that geometrical theorems were formed out of basic axioms. It was natural for him to suppose that by exhausting the combinations of such principles one might explore all possible structures of truth and so obtain universal knowledge.

There is a sense, of course, in which Lull's method of exploration does possess a formal deductive character. If we wish to exhaust the possible combinations of given sets of terms, then Lull's method obviously will do this for us in an irrefutable way. Considered mathematically, the technique is sound, though even in its day it was essentially trivial. Tabulating combinations of terms was certainly a familiar process to mathematicians as far back as the Greeks, and it would be surprising indeed if no one before Lull had thought of using movable circles as a device for obtaining such tables. Lull's mistake, in large part a product of the philosophic temper of his age, was to suppose that his combinatorial method had useful application to subject matters where today we see clearly that it does not apply. Not only is there a distressing lack of "analytic" structure in areas of knowledge outside of logic and mathematics, there is not even agreement on what to regard as the most primitive, "self-evident" principles in any given subject matter. Lull naturally chose for his categories those that were implicit in the dogmas and opinions he wished to establish. The result, as Chesterton might have said, was that Lull's circles led him in most cases into proofs that were circular. Other schoolmen were of course often guilty of question begging, but it was Lull's peculiar distinction to base this type of reasoning on such an artificial, mechanical technique that it amounted virtually to a satire of scholasticism, a sort of hilarious caricature of medieval argumentation.

We have mentioned earlier that it was Leibniz who first saw in Lull's method the possibility of applying it to formal logic.[8] For example, in his *Dissertio de arte combinatoria* Leibniz constructs an exhaustive table of all possible combinations of premises and conclusions in the traditional syllogism. The false syllogisms are then eliminated, leaving no doubt as to the number of valid ones, though of course revealing nothing that was not perfectly familiar to Aristotle. A somewhat similar technique of elimination was used by Jevons in his "logical alphabet" and his logic machine and is used today in the construction of matrix tables for problems in symbolic logic. Like Lull, however, Leibniz failed to see how restricted was the application of such a technique, and his vision of reducing all knowledge to composite terms built up out of simple elements and capable of being manipulated like mathematical symbols is certainly as wildly visionary as Lull's similar dream. It is only in the dimmest sense that Leibniz can be said to anticipate modern symbolic logic. In Lull's case the anticipation is so remote that it scarcely deserves mention.

Still, there is something to be said for certain limited applications of Lull's circles, though it must be confessed that the applications are to subject matters which Lull would have considered frivolous. For example, parents seeking a first and middle name for a newborn baby might find it useful to write all acceptable first names in one circle and acceptable middle names in a larger circle, then rotate the inner circle to explore the possible combinations. Ancient coding and decoding devices for secret ciphers make use of Lullian-type wheels. Artists and decorators sometimes employ color wheels for exploring color combinations. Anagram puzzles often can be solved quickly by using Lullian circles to permute the required letters. A cardboard toy for children consists of a rotating circle with animal pictures around the circumference, half of each animal on the circle and half on the sheet to which the wheel is fastened. Turning the circle produces amusing combinations—a giraffe's head on the body of a hippopotamus, and so on. One thinks also of Sam Loyd's famous

"Get off the earth" paradox. Renan once described Lull's circles as "magic," but in turning Loyd's wheel the picture of an entire Chinese warrior is made to vanish before your very eyes.[9] It is amusing to imagine how Lull would have analyzed Loyd's paradox, for his aptitude for mathematical thinking was not very high.

Even closer to the spirit of Lull's method is a device that was sold to fiction writers many years ago and titled, if I remember correctly, the "Plot Genii." By turning concentric circles one could obtain different combinations of plot elements. (One suspects that Aldous Huxley constructed his early novels with the aid of wheels bearing different neurotic types. He simply spun the circles until he found an amusing and explosive combination of house guests.) Mention also should be made of the book called *Plotto,* privately published in Battle Creek, Michigan, in 1928, by William Wallace Cook, a prolific writer of potboilers. Although *Plotto* did not use spinning wheels, it was essentially Lullian in its technique of combining plot elements, and apparently there were many writers willing to pay the seventy-five dollar price originally asked by the author.

In current philosophy one occasionally comes upon notions for which a Lullian device might be appropriate. For instance, Charles Morris tells us that a given sign (e.g., a word) can be analyzed in terms of three kinds of meaning: syntactic, semantic, and pragmatic. Each meaning in turn has a syntactic, semantic, and pragmatic meaning, and this threefold analysis can be carried on indefinitely. To dramatize this dialectical process one might use a series of rotating circles, each bearing the words "syntactic," "semantic," and "pragmatic," with the letter S in the center of the inner wheel to signify the sign being analyzed.

In science there also are rare occasions when a Lullian technique might prove useful. The tree diagram is certainly a convenient way to picture evolution. The periodic table can be considered a kind of Lullian chart that exhausts all permissible combinations of certain primitive principles and by means of which chemists have been able to predict the properties of elements before they were discovered. Lull's crude anticipation was a circle bearing the four traditional elements and rotated within a ring similarly labeled.

There may even be times when an inventor or researcher might find movable circles an aid. Experimental situations often call for a testing of all possible combinations of a limited number of substances or techniques. What is invention, after all, except the knack of finding new and useful combinations of old principles? When Thomas Edison systematically tested almost every available substance as a filament for his light bulb, he was following a process that Lull would probably have considered an extension of his method. One American scientist, an acoustical engineer and semi-professional magician, Dariel Fitzkee, actually published in 1944 a book called *The Trick Brain,* in which he explains a technique for combining ideas in Lullian fashion for the purpose of inventing new magic tricks.

If the reader will take the trouble to construct some Lullian circles related to a subject matter of special interest to himself, and play with them for a while, he will find it an effective way of getting close to Lull's mind. There is an undeniable fascination in twisting the wheels and letting the mind dwell on the strange combinations that turn up. Something of the mood of medieval Lullism begins to pervade the room, and one comprehends for the first time why the Lullian cult persisted for so many centuries.

For persist it did. [10] Fifty years after Lull's death it was strong enough to provoke a vigorous campaign against Lullism, led by Dominican inquisitors. They succeeded in having Lull condemned as a heretic by a papal bull, though later church officials decided that the bull had been a forgery. Lullist schools, supported chiefly by Franciscans, flourished throughout the late Middle Ages and Renaissance, mostly in Spain but also in other parts of Europe. We have already cited Bruno's intense interest in the Art. This great ex-Dominican considered Lull's method divinely inspired though badly applied. For example, he thought Lull mad to suppose that such truths of faith as the incarnation and trinity could be established by necessary reasons. Bruno's first and last published works, as well as many in between, were devoted to correcting and improving the method, notably *The Compendious Building and Completion of the Lullian Art.*

In 1923 the British Museum acquired a portable sundial and compass made in Rome in 1593 in the form of a book (Figure 10). On the front

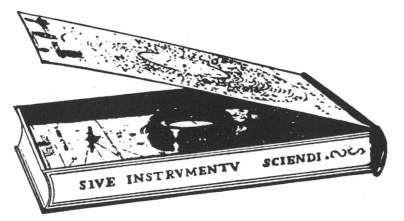

FIGURE 10. Sixteenth-century portable sundial engraved with Lullian figures. (From *Archaeologia,* Oxford, 1925.)

and back of the two gilt copper "covers" are engraved the Lullian circles shown in Figures 11 to 14. For an explanation of these circles the reader

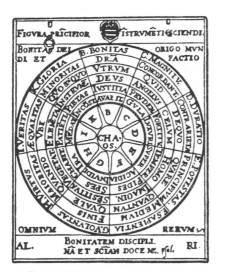

FIGURE 11. Upper cover, outer side

FIGURE 12. Upper cover, inner side

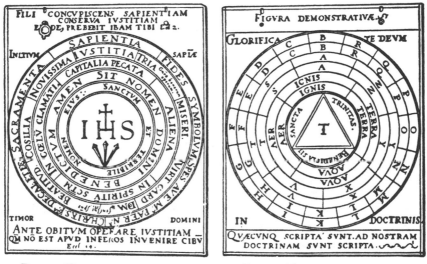

FIGURE 13. Lower cover, outer side

FIGURE 14. Lower cover, inner side

Figures 11 to 14. Circles used by Renaissance Lullists. (From *Archaeologia*, Oxford, 1925.)

is referred to O. M. Dalton's article, "A Portable Dial in the Form of a Book, with Figures Derived from Ramon Lul," *Archaeologia*, vol. 74, second series, Oxford, 1925, pp. 89–102.

The seven smaller diagrams in Figure 12 are all from Lull's writings[11] and perhaps worth a few comments. The square in the upper left corner is designed to show how the mind can conceive of geometrical truths not apparent to the senses. A diagonal divides the square into two large triangles, one of which is subdivided to make the smaller triangles *B* and *C*. Each triangle contains three angles; so that our senses immediately perceive nine angles in all. However, we can easily imagine the large triangle to be subdivided also, making four small triangles or twelve angles in all. The three additional angles exist "potentially" in triangle *A*. We do not see them with our eyes, but we can see them with our imagination. In this way our intellect, aided by imagination, arrives at new geometrical truths.

The top right square in Figure 12 is designed to prove that there is only one universe rather than a plurality of worlds. The two circles represent two universes. We see at once that certain parts of *A* and *B* are nearer to each other than other parts of *A* and *B*. But, Lull argues, "far" and "near" are meaningless concepts if nothing whatever exists in the space between *A* and *B*. We are forced to conclude that two universes are impossible.

I think what Lull means here, put in modern terms, is that we cannot conceive of two universes without supposing some sort of space-time relation between them, but once we relate them, we bring them into a common manifold; so we can no longer regard them as separate universes. Lull qualifies this by saying that his argument applies only to actual physical existence, not to higher realms of being which God could create at will, since His power is infinite.

The four intersecting circles are interesting because they anticipate in a vague way the use of circles to represent classes in the diagrammatic methods of Euler and Venn. The four letters which label the circles stand for *Esse* (being), *Unum* (the one), *Verum* (the true), and *Bonum* (the good). *Unum, verum,* and *bonum* are the traditional three "transcendentales" of scholastic philosophy. The overlapping of the circles indicates that the four qualities are inseparable. Nothing can exist without possessing unity, truth, and goodness.

The circle divided into three sectors represents the created universe, but I am not sure of the meaning of the letters which apparently signify the parts. The lower left square illustrates a practical problem in navigation. It involves a ship sailing east, but forced to travel in a strong north wind. The lower right square is clearly a Lullian table displaying the twelve permutations of *ABCD* taken two letters at a time.

The remaining diagram, at the middle of the bottom, is a primitive method of squaring the circle and one fairly common in medieval pseudo-mathematical works. We first inscribe a square and circumscribe a square; then we draw a third square midway between the other two. This third square, Lull mistakenly asserts, has a perimeter equal to the circumference of the circle as well as an area equal to the circle's area. Lull's discussion of this figure (in his *Ars magna et ultima*) reveals how far behind he was of the geometry of his time.[12] His method does not provide even a close approximation of the perimeter or area of the desired square.[13]

Books on the Lullian art proliferated throughout the seventeenth century, many of them carrying inserted sheets of circles to be cut out, or actual rotating circles with centers attached permanently to the page. Wildly exaggerated claims were made for the method. The German Jesuit Athanasius Kircher (1601–1680), scientist, mathematician, cryptographer, and student of Egyptian hieroglyphics, was also a confirmed Lullist. He published in Amsterdam in 1669 a huge tome of nearly 500 pages titled *Ars magna sciendi sive combinatoria*. It abounds with Lullian figures and circles bearing ingenious pictographic symbols of his own devising.[14]

The eighteenth century witnessed renewed opposition to Lull's teachings in Majorca and the publication of many Spanish books and pamphlets either attacking or defending him. Benito Feyjóo, in the second volume of his *Cartas eruditas y curiosas* ("Letters erudite and curious"), ridiculed Lull's art so effectively that he provoked a two-volume reply in 1749–1750 by the Cistercian monk Antonio Raymundo Pasqual, a professor of philosophy at the Lullian University of Majorca. This was followed in 1778 by Pasqual's *Vinciciae Lullianae,* an important early biography and defense of Lull. The nineteenth and twentieth centuries saw a gradual decline of interest in the Art and a corresponding increase of attention toward Lull as a poet and mystic. A periodical devoted to Lullian studies, the *Revista luliana,* flourished from 1901 to 1905. Today there are many enthusiastic admirers of Lull in Majorca and other parts of Spain, though the practice of his Art has all but completely vanished.

The Church has approved Lull's beatification, but there seems little likelihood he will ever be canonized. There are three principal reasons. His books contain much that may be considered heretical. His martyrdom seems to have been provoked by such rash behavior that it takes on the coloration of a suicide. And finally, his insistence on the divine origin of his Art and his constant emphasis on its indispensability as a tool for the conversion of infidels lends a touch of madness, certainly of the fantastic, to Lull's personality.

Lull himself was fully aware that his life was a fantastic one. He even wrote a book called *The Dispute of a Cleric and Ramon the Fantastic,* in which he and a priest each try to prove that the other has had the more preposterous life. At other times he speaks of himself as "Ramon the Fool." He was indeed a Spanish *joglar* of the faith, a troubadour who sang his passionate love songs to his Beloved and twirled his colored circles as a juggler twirls his colored plates, more to the amusement or annoyance of his countrymen than to their edification. No one need regret that the controversy over his Great Art has at last been laid to rest and that the world is free to admire Lull as the first great writer in the Catalan tongue and a religious eccentric unique in medieval Spanish history.

NOTES

1. In later years Leibniz was often critical of Lull, but he always regarded as sound the basic project sketched in his *Dissertio de arte combinatoria*. In a letter written in 1714 he makes the following comments:

> When I was young, I found pleasure in the Lullian art, yet I thought also that I found some defects in it, and I said something about these in a little schoolboyish essay called *On the Art of Combinations*, published in 1666, and later reprinted without my permission. But I do not readily disdain anything—except the arts of divination, which are nothing but pure cheating—and I have found something valuable, too, in the art of Lully and in the *Digestum sapientiae* of the Capuchin, Father Ives, which pleased me greatly because he found a way to apply Lully's generalities to useful particular problems. But it seems to me that Descartes had a profundity of an entirely different level. [*Gottfried Wilhelm von Leibniz: Philosophical Papers and Letters*, edited and translated by Leroy E. Loemker, University of Chicago Press, 1956, vol. 2, p. 1067.]

2. In sketching Lull's life I have relied almost entirely on E. Allison Peers's magnificent biography, *Ramon Lull*, London, 1929, the only adequate study of Lull in English. An earlier and briefer biography, *Raymond Lull, the Illuminated Doctor*, was published in London, 1904, by W. T. A. Barber, who also contributed an informative article on Lull to the *Encyclopedia of Religion and Ethics*. Other English references worth noting are: Otto Zöckler's article in the *Religious Encyclopedia*; William Turner's article in the *Catholic Encyclopedia*; George Sarton, *Introduction to the History of Science*, 1931, vol. 2; and Lynn Thorndike, *A History of Magic and Experimental Science*, 1923, vol. 2.

A voluminous bibliography of Lull's works, with short summaries of each, may be found in the *Histoire littéraire de la France*, Paris, 1885, vol. 29, pp. 1-386, an indispensable reference for students of Lull. There also is an excellent article on Lull by P. Ephrem Langpré in vol. 9 of the *Dictionnaire de théologie Catholique*, Paris, 1927. It is interesting to note that a 420-page novel based on the life of Lull, *Le Docteur illumine*, by Lucien Graux, appeared in Paris in 1927.

The most accessible Spanish references are the articles on Lull in the *Enciclopedia universal ilustrada*, Barcelona, 1923, and vol. 1 of *Historia de la filosofía española*, by Tomás Carreras y Artau, Madrid, 1939.

3. Quoted by Peers, *op. cit.*, p. 64.

4. An English translation by Peers was published in 1926.

5. Separately issued in English translation by Peers in 1923.

6. Lull's death is the basis of a short story by Aldous Huxley. "The Death of Lully," in his book *Limbo*, 1921.

7. The only satisfactory description in English of Lull's method is in vol. 1 of Johann Erdmann's *History of Philosophy*, English translation, London, 1910. There are no English editions of any of Lull's books dealing with his Art. Peers's biography may be consulted for a list of Latin and Spanish editions of Lull's writings.

8. See *La logique de Leibniz,* by Louis Couturat, Paris, 1901, chap. 4, and *Leibniz,* by Ruth Lydia Shaw, London, 1954, chap. 8.

9. Chapter 7 of my *Mathematics, Magic, and Mystery,* 1956, contains a reproduction and analysis of Loyd's "Get off the earth" puzzle and several related paradoxes.

10. *Historia del Lulisme,* by Joan Avinyó, a history of Lullism to the eighteenth century, was published in Barcelona in 1925. My quick survey of Lullism draws largely on Peers's account.

11. With the exception of the table of permutations, all these diagrams are reproduced and discussed in Zetzner's one-volume Latin edition of several of Lull's works, first printed in Strasbourg, 1598.

12. Bryson of Heraclea, a pupil of Socrates, had recognized that, if you keep increasing the number of sides of the inscribed and circumscribed polygons, you get increasingly closer approximations of the circle. It was through applying this method of limits that Archimedes was able to conclude that pi was somewhere between 3.141 and 3.142.

13. It has been called to my attention that, if a diagonal line *AB* is drawn on Lull's figure as shown in Figure 15, it gives an extremely good approximation to the side of a square with an area equal to the area of the circle.

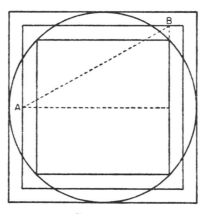

FIGURE 15.

14. Kircher's enormous books are fascinating mixtures of science and nonsense. He seems to have anticipated motion pictures by constructing a magic lantern that threw images on a screen in fairly rapid succession to illustrate such events as the ascension of Christ. He invented (as did Leibniz) an early calculating machine. On the other hand, he devoted a 250-page treatise to details in the construction of Noah's Ark!

Kircher's work on the Lullian art appeared three years after Leibniz's youthful treatise of similar title (see Note 1). Leibniz later wrote that he had hoped to find important matters discussed in Kircher's book but was disappointed to discover that it "had merely revived the Lullian art or something similar to it, but that the author had not even dreamed of the true analysis of human thoughts" (vol. 1, p. 352, of the edition of Leibniz's papers and letters cited in Note 1).

Postscript

Interest in Lull, though less in his *Ars magna* than in other aspects of his life and writings, continues to be greater in Europe than here. The Second

International Lullist Congress took place at El Encinar and Miramar, Mallorca, in 1976, bringing together eminent Lullist scholars from all over the world. In 1972, Oxford University Press brought out J. N. Hillgarth's 504-page monograph, *Ramon Lull and Lullism in Fourteenth-Century France*, though it has little to say about Lull's art. Another recent reference is the article on Lull, by Robert D. F. Pring-Mill, in Scribner's *Dictionary of Scientific Biography*, vol. 7, 1973.

David Kahn, writing in *Isis* (vol. 71, no. 256, 1980), "On the Origin of Polyalphabetic Substitution," credits Lull's circles with suggesting to Alberti, in 1289, the idea of using rotating disks to generate substitution ciphers.

Another method of exploring all combinations of sets of basic concepts is to label the sides of a rectangular, brick-shaped, or higher-dimensional matrix with the concepts. Each cell on the matrix then represents an *n*-tuplet combination of the concepts. In his book *The Fourth Dimension* (1904), Charles Howard Hinton uses a $4 \times 4 \times 4 \times 4$ hypercube in this way to model syllogisms. Leibniz would, I think, have been intrigued by Hinton's model, though I suspect Lull would have found it difficult to understand.

In recent decades the use of boxlike charts of this sort for displaying what is called the "Cartesian product" of *n*-dimensional coordinate systems has become popular among some writers on creative thinking. The idea is that such a display allows you to inspect combinations more rapidly and efficiently than with rotating wheels. Meditating on the bizarre combinations is then supposed to lead your mind into offbeat paths. The leading exponent of this technique is Fritz Zwicky, an American astrophysicist, who calls it the "morphological method." He began singing its praises in the forties, and eventually devoted an entire book to it, *Discovery, Invention, Research: Through the Morphological Approach* (Macmillan, 1969). In Zwicky's words, the book attempts to show "how the morphological approach inspires the imagination to ever new visions and advances, and how almost automatically the surest ways to discovery, invention, and new avenues of research reveal themselves."

A more whimsical Lullian device, called the Think Tank, was widely advertised in 1975 by the World Future Society, whatever that is, in Washington, D.C. For $45 you obtained a rotating plastic sphere, eight inches across, containing 13,000 different words printed on tiny plastic slivers. Suppose you have a problem you want to solve in some novel creative way. First you rotate the tank to tumble the words about; then you look through its circular window and see a set of seemingly irrelevant words.

"But the irrelevant becomes relevant," one advertisement read, "as soon as you start making associations with the word and applying them to the problem situation. . . . You do not seek out suitable words. You take what you get and you use what you get; otherwise you would destroy the effect of randomness and look for an obvious solution. And this is not the way new ideas are generated. The object is to produce numerous and diverse ideas until a EUREKA idea comes. Later you can evaluate and select ideas which are feasible or commercially interesting."

With the tank comes a 90-page instruction booklet by Edward de Bono, a British author of many books on creativity. He explains how the tank serves as an invaluable tool for "lateral" or "zig-zag" thinking that twists your mind into new channels. The tank itself is manufactured by its inventor, Savo Bojicic, a Canadian businessman who emigrated from Yugoslavia. A profile of him entitled "Yes, You Too Can Think Your Way to Greater Riches, Happiness and Success" appeared in the Canadian magazine *Quest*, November, 1975.

As I said earlier, there is a trivial sense in which all creative thinking, in the arts as well as in the sciences and in everyday life, can be thought of as combinatorial. Music is a combination of tones, a novel is a combination of words, and so on. Einstein once put it this way: "The psychical entities which seem to serve as elements in thought are certain signs and more or less clear images which can be voluntarily reproduced and combined. This combinatory play seems to be the essential feature in productive thought."

The big question—and it was raised by Lull—is whether the mechanical production of all combinations, or a selection of random combinations, is a genuine aid to creative thought. In most cases the number of possible combinations of primitive concepts is enormous. Except for a minute percentage, the combinations are nonsensical and useless. The difficult task, and it is here that the dark mystery of creativity lies, is to sift out the extremely small set of useful combinations.

There are, of course, some instances in which the mechanical generation of combinations may be an aid to creativity. A poet, for example, may be unable to find in his or her memory a word that has a suitable combination of meaning, meter, and rhyme. A rhyming dictionary enables the poet to go quickly over all possibilities. And it is surely true, as the mathematician Stanislaw Ulam says in his autobiography, that the necessity of finding a word that rhymes "forces novel associations and almost guarantees deviations from routine chains or trains of thought. It becomes paradoxically a sort of automatic mechanism of originality." But aside from such special cases, and a few others mentioned in the preceding pages, the Lullian combinatorial technique seems to me to be of extremely limited value.

4

Some Trends in Pseudoscience

In 1952, in a book called *In the Name of Science,* I surveyed various aspects of contemporary pseudoscience. The book was published by Putnam, remaindered, and then reissued by Dover as a paperback entitled *Fads and Fallacies.* A new chapter dealt with the Bridey Murphy case. Remember Bridey? Millions of readers took Morey Bernstein's book seriously until a Chicago newspaper discovered that Mrs. Virginia Tighe, who under hypnosis seemed to be a reincarnation of an Irish girl named Bridey Murphy, was simply dredging up memories of a real Bridey Murphy who lived in Chicago across the street from where Virginia had lived as a little girl. That was the end of Bridey.

But not, of course, the end of pseudoscience. Enough has happened since the fading away of Bridey to provide material for another book. I have no desire to write such a book, but your editor's request for an article gives me a chance to comment briefly on some of the more interesting recent developments in this curious twilight zone.

John Campbell, editor of *Analog Science Fiction,* likes to boost the circulation of his magazine by introducing new scientific nonsense from time to time. His readers (if we trust Campbell's own surveys) lap it up. First it was dianetics, that great new approach to psychiatry dreamed up by L. Ron Hubbard, in which neuroses are traced back to what an embryo hears in the mother's womb. Then it was psionics: electronic machines that perform feats of extrasensory perception. After exhausting the psionic

Reprinted with permission from *The Quid,* Winter 1963–64 (Jefferson High School, Elizabeth, N.J.).

gimmick, Campbell turned to the Dean space drive. This whimsical device rotates weights in such a way that, it is claimed, a space thrust results. When the device is put on a bathroom scale and turned on, it appears to lose weight: a true "bootstrap lifter," Campbell called it. The thing was invented by Norman L. Dean, a man who appraised mortgages for the Federal Housing Administration in Washington, and first publicized by Campbell in an article in the June, 1960, issue of *Analog*. "I do not insist that I am incontrovertibly right," Campbell wrote, in an unusual outburst of humility, "but it is my opinion . . ." and he goes on to say that, when our government refused to take Dean's machine seriously, it was tantamount to our giving away the solar system.

> Oh, the Dean machine, the Dean machine,
> Why, you put it right in a submarine,
> And it flies so high that it can't be seen,
> The wonderful, wonderful Dean machine!

wrote Damon Knight in a science-fiction fan magazine. (The rest of his poem pokes fun at other Campbell enthusiasms.) Campbell's loyal readers may still feel that scientific orthodoxy is just too inflexible to accept the fact that Newton's laws of motion have to be revised, but Campbell himself has been so quiet lately about the Dean drive that one suspects even he has become convinced it doesn't work.[1] The current lull in *Analog* is ominous. What will John latch onto next?

In the medical field, the most newsworthy recent event, relevant to our topic, is the Food and Drug Administration's discovery that Krebiozen, the much publicized cancer drug, is nothing more than creatine. Creatine is an inexpensive chemical that costs 30 cents a gram. Most of the 5,000 patients who have taken K, as it is called, during the past 13 years made a "donation" of $9.50 per dose, each dose containing one one-hundred-thousandth of a gram. A few years back, the backers of K quoted the government a price of $170,000 per gram. A normal man has about 120 grams of creatine in his body, and previous research has shown that the chemical has no effect on cancer cells. Backers of K continue to insist that K is *not* creatine, but the government's case seems airtight. "Dr." Carlton Fredericks, the well-known radio commentator on diet, went all-out for K early in 1963. I never listen to him, so I am unable to report on how he responded to the FDA's discovery. (I put the "Dr." in quotes because most of Frederick's followers think he has a degree in medicine, or at least in nutrition. It is a doctor of philosophy degree, from the New York University School of Education, for a thesis on how his female listeners responded to his own radio broadcasts.[2])

In book publishing, the great scandal of recent years has been Simon and Schuster's promotion of Dr. Herman Taller's worthless best-seller,

Calories Don't Count. The editors at S & S, seeing in the manuscript a chance to make a financial killing, were careful not to send the manuscript to a single expert for evaluation (a practice normally followed on books of a scientific nature). It was skillfully rewritten by Roger Kahn, a free-lance sports writer. Worse than that, an assistant to one of the editors was asked to insert into the manuscript references to safflower oil capsules, and to mention that they could be purchased from Cove Pharmaceuticals, a firm in New York. Two vice presidents of S & S bought stock in this firm.

The capsules were branded worthless by the FDA. Because the tie-in of book and capsules amounted to a mislabeling of a product, the FDA seized copies of the books along with capsule supplies. S & S has since removed the reference to Cove Pharmaceuticals, but the book continues to sell widely as a paperback, deluding tens of thousands of overweight readers into thinking they can reduce without cutting their calorie intake. Dr. Taller, a gynecologist, made a fortune on the book, so you can imagine what the take was for S & S. It was their most profitable book in 1962.

The latest display of scientific immaturity has been in the pages of *Harper's Magazine.* It was in *Harper's,* January, 1950, that Eric Larrabee, then one of the editors, first revealed to the public the bizarre cosmological fantasies of Immanuel Velikovsky. Velikovsky's first book, *Worlds in Collision,* was published by Macmillan with much fanfare. Later, under pressure from scientists, Macmillan passed the book over to Doubleday. Thirteen years later, a full-page advertisement in the *New York Times Book Review* (July 28, 1963) announced that Larrabee was at it again. In an article in the August issue he argued that, although "orthodox" scientists (those stubborn, biased souls who refuse to read Velikovsky with an open mind) have not yet acknowledged Velikovsky's theory, here and there reputable scientists are beginning to take parts of it seriously. The article is a masterpiece of evasive argument.

The heart of Velikovsky's theory is that within recorded history a gigantic comet erupted from Jupiter. It finally settled down to become the planet Venus, but before doing so it passed close to earth, causing the earth's rotation to slow down or possibly stop. (In this way Velikovsky finds scientific explanations for such biblical tales as Joshua's successful effort to make the sun and moon stop moving through the sky.) Naturally, such a break in the continuity of the earth's spin would cause enormous inertial effects on the entire surface of the earth. Velikovsky thinks that at that time the earth also reversed magnetic poles.

Geologists have never denied that all sorts of catastrophic events have occurred in the earth's geological past, including many reversals of magnetic poles. What makes Velikovsky's view so preposterous is its insistence, absolutely vital to the theory, that such catastrophes occurred as late as 1500 B.C. I will cite only one example of Larrabee's technique. He quotes a paragraph from an article on earth magnetism that appeared in *Scientific American,* September, 1955. The author, S. K. Rankin, writes that the earth's magnetic poles indeed have undergone reversals in the past. What Larrabee slyly conceals from the reader is that Rankin is considering events during the earth's Tertiary period, which he dates between 60 and 1 *million* years ago!

That a worldwide seismic catastrophe, on the scale required by Velikovsky's theory, could have occurred about 1500 B.C. is so completely ruled out by geologic evidence that not a single reputable geologist has taken it seriously. Yet here is *Harper's,* a magazine presumably edited by and read by well-educated Americans, defending a geological theory so far out that not even John Campbell has been able to work up enthusiasm for it.

The editors of *Harper's* were kind enough to permit Donald Menzel, a Harvard astronomer, to explain in their December issue why astronomers ignore Velikovsky, but Larrabee is given the last word in a page of rejoinders. The same issue contains an article by Upton Sinclair called "My Anti-Headache Diet." It seems that for fifty years Sinclair was plagued by headaches. He tried to stop them by becoming first a vegetarian, then a raw food enthusiast. But a book by someone named Salisbury convinced him that raw food was making a "yeastpot" (whatever that is) of his stomach; so he tried Salisbury's miracle diet of fresh, ground-up beef. No miracle. He next turned to a "fasting cure," living for 11 days on water and orange juice. That didn't work either. He read somewhere that soldiers of King Cyrus were forced to perspire every day, so he tried a "sweat cure." Finally, at the age of 76, his headaches stopped. Naturally, Sinclair (whose knowledge of nutrition is on a par with Larrabee's knowledge of geology) attributes the "cure" to the fad he was trying at the time: a diet of brown rice and fresh fruit. He describes this wonder-working diet, recommending it to all readers of *Harper's* who suffer from headaches.

Books on occultism and psychic phenomena continue to be published by the country's largest, most respectable houses. In recent years there has been a marked trend toward interest in occultism on the part of American psychoanalysts. Freud himself believed in mental telepathy, and for many years took seriously the numerological views of his dearest friend, Dr. Wilhelm Fliess. Jung's occult beliefs are now well known. The Summer, 1963, issue of the *Psychoanalytic Review* contains an article by occultist Nandor Fodor (he is a member of the magazine's editorial board) in which he argues that Jung was a medium capable, at times, of inducing poltergeist phenomena. In England, Mark Hansel, a professor of psychology at the University of Manchester, has been investigating classic tests of ESP, tests supervised by J. B. Rhine in the United States, and in England by S. G. Soal.[3] Hansel's studies have uncovered strong presumptive evidence of deliberate fraud in many cases in which fraud had not previously been suspected. One would expect such disclosures to be widely reported in U.S. newspapers and magazines. Why haven't they been? The answer is simple: negative evidence for ESP lacks the dramatic news value of positive evidence.

So it goes. It will be a long time until the average citizen is well enough informed about science to make the promotion of popularly written pseudoscientific books unprofitable. And as long as they are profitable, you can be sure they will be written and printed.

NOTES

1. Bibliographies of articles relating to the Dean drive can be found in "Detesters, Phasers and Dean Drives," by G. Henry Stine, in *Analog Science Fiction,* June, 1976, and "In Search of the Bootstrap Effect," by Russell E. Adams, Jr., ibid., April, 1978. The drive surfaced again in an article by Stine in *Destinies,* October, 1979. See also an amusing discussion of the drive by Milton Rothman in "On Designing an Interstellar Spaceship," *Isaac Asimov's Science Fiction Magazine,* September, 1980.

2. A few years ago I accidentally heard a portion of Fredericks's radio program. He was gung ho for Laetrile as a cancer remedy. To my astonishment he brought up his earlier defense of Krebiozen and said that defending Laetrile against the medical establishment gave him a strong feeling of *déjà vu.* It certainly gave me that feeling.

3. Leading parapsychologists were outraged by Hansel's evidence, revealed in several articles, that Soal had fudged his data. Not until Betty Markwick reported her sensational findings (see Chapters 10 and 19) did they admit that all of Soal's work is now suspect.

Postscript

This was written at the request of a friend, Ricky Jay, now a top professional magician and the author of a very funny book, *Cards as Weapons* (Darien House, 1977). At the time I wrote, Ricky was editor of *The Quid,* a publication of his high school. I have here restored the last two paragraphs, which were cut for lack of space, and have added the footnotes.

5

Manifesto of the Institute
of General Eclectics

1. Membership in the Institute shall be restricted to anal erotics who have strong compulsions to collect, preserve, and classify philosophical ideas.

2. The *Basic Axiom* of the Institute shall be that all philosophical systems are in fundamental agreement, apparent differences arising from variable verbal formulations and/or variable emphases. Beneath superficial differences in color, shape, texture, odor, the systems are manifestations of one unified, simple, basic stuff.

3. Members shall be required to purchase and read *Science and Sanitation,* by the late Count Aulayore Beeyemski, founder of General Eclectics, to acquire a thorough understanding of the Count's countless reasons for opposing the contemporary positivist attempt to purge philosophy of "obsolete" metaphysical systems.

4. The symbol of the Institute, which shall appear on all letterheads, publications, etc., shall be the *Orphic Egg*—primordial form of the cosmos and the unity toward which all creation strives. As a common object, easy to grasp, smooth, colorless, perfect in symmetry yet not uninteresting in shape like a sphere, the egg symbolizes the accessibility, unity, clarity, simplicity, and symmetry of eclectic philosophical thought.

5. Because different philosophic systems are merely different verbal formations of the same Orphic Egg, all metaphysical debate shall be considered as pointless as the war in *Gulliver's Travels,* fought over the question of how to break an egg.

Reprinted with permission from *Stranger Than Fact,* Summer 1964.

6. The Basic Axiom shall be symbolized by a pedagogic device called *The Structural Similarium,* consisting of a series of plastic, egg-shaped beads of differing sizes and colors, strung on cord. The device shall also play indispensable roles in several of the Institute's secret sex rituals.

7. Male members shall be required to shave their heads daily, the egg-shaped appearance of their skulls serving as a reminder of the basic unity of all egg-headed speculation.

8. Members of the Institute are forbidden to belong to any religious organization. Each Sunday they shall attend services of a different sect until all accessible sects have been exhausted, at which point the visits shall be repeated and this practice shall be continued throughout the life of the member.

9. Members of the Institute are forbidden to hold partisan political views. During election campaigns members shall wear buttons of all political parties and on election day cast an eclectic vote for all candidates. In time of war, members shall remain neutral. If eligible for draft, they shall register as conscientious objectors.

10. Members of the Institute are forbidden to make "value judgments" concerning any moral or aesthetic matter. All cultures shall be regarded as equally good, all artistic works as equally beautiful. All possible variations of cocktails shall be served at social functions of the Institute. A clearing house shall be established through which members may periodically exchange such possessions as neckties, hats, silverware, jewelry, books, recordings, paintings, children, spouses, etc.

11. Chief project of the Institute shall be the publication of dictionaries by which assertations in one philosophical language shall be translated into assertations of another: i.e., Plato-Aristotle, Kant-James, Russell-Dewey, Freud-Barth, Carnap-Sartre, etc. Funds shall be set aside for research on the development of electronic machines by which philosophical systems may be elaborated, new systems devised, and translations from system to system effected rapidly and accurately.

12. Funds shall also be set aside for the publication in approximately one thousand volumes of a *Summa Summa Dialectica,* which will combine Aristotle's passion for analysis and classification with Plato's passion for mixture and synthesis, as outlined in Count Beeyemski's earlier work, *Prolegomena to a Future Summa Summa.* The *Summa Summa Dialectica* will explore dialectically all possible expressions of all possible systems of all possible metaphysical ideas.

13. The Institute shall publish a *Museum of Identica* in which striking identities of philosophic-literary expressions shall be dramatized for public enlightenment. Examples: George Herbert Mead's discussion of his concept of "taking the role of the other" and the popular song "You're Nobody if Nobody Loves You"; D. H. Lawrence's novel *Women*

in Love and the song, "I only Want a Buddy, not a Sweetheart"; the collected works of Santayana and Dr. Seuss's book, *Mulberry Street.*

14. The Institute shall maintain a monthly periodical titled *But,* the title derived from the terminal lines of Humpty Dumpty's song in Chapter 6 of *Through the Looking Glass:*

> And when I found the door was shut,
> I tried to turn the handle, *but*—

Members of the Institute shall cultivate the habit of ending sentences with "but" to suggest that all expressions have an obverse, rear side, as true, beautiful, and good as the front.

15. Additional funds shall be allocated for more intensive investigation into the cathartic effects of General Eclectic training on neurotics and psychotics whose rigid dogmatisms and inability to cope with alternative attitudes are major causes of their mental constipation.

16. The Institute's Secretary shall be a woman named *Barbara,* after the traditional name for the syllogism: All A is B, all B is C, therefore all A is C.

17. When members of the Institute are accused of being anal erotics, their response shall be, "Oh yeah? Anal erotic, my but!"

Postscript

This light-hearted piece of satire, written for a science fiction fanzine, is directed partly toward Count Alfred Korzybski and his Institute of General Semantics and partly toward the eclectic views of a philosopher I prefer to leave nameless.

6

Dermo-optical Perception:
A Peek Down the Nose

Science reporting in United States newspapers and mass-circulation magazines is more accurate and freer of sensationalism than ever before, with pseudoscience confined largely to books. A reverse situation holds in the Soviet Union. Except for the books that defended Lysenko's theories, Soviet books are singularly free of pseudoscience, and now that Lysenko is out of power, Western genetics is rapidly entering the new Russian biology textbooks. Meanwhile, Russian newspapers and popular magazines are sensationalizing science much as our Sunday supplements did in the 1920s. The Soviet citizen has recently been presented with accounts of fish brought back to life after having been frozen 5,000 years, of deep-sea monsters that leave giant tracks across the ocean floor, of absurd perpetual-motion devices, of extraterrestrial scientists who have used a laser beam to blast an enormous crater in Siberia, and scores of similar stories.

By and large, the press in the United States has not taken this genre of Soviet science writing seriously. But in 1963 and 1964 it gave serious attention to a sudden revival, in Russia's popular press, of ancient claims that certain persons are gifted with the ability to "see" with their fingers.

The revival began with a report, in the summer of 1962, in the Sverdlovsk newspaper *Uralsky Rabochy*. Isaac Goldberg, of First City Hospital in Lower Tagil, had discovered that an epileptic patient, a 22-year-old woman named Rosa Kuleshova, could read print simply by moving a fingertip

Reprinted with permission from *Science,* February 11, 1966. Copyright 1966 by the American Association for the Advancement of Science.

over the lines. Rosa went to Moscow for more testing, and sensational articles about her abilities appeared in *Izvestia* and other newspapers and popular magazines. The first report in the United States was in *Time*, 25 January 1963.

When I first saw *Time's* photograph of Goldberg watching Rosa, who was blindfolded, glide her middle finger over a newspaper page, I broke into a loud guffaw. To explain that laugh, I must back up a bit. For 30 years my principal hobby has been magic. I contribute to conjuring journals, write treatises on card manipulation, invent tricks, and, in brief, am conversant with all branches of this curious art of deception, including a branch called "mentalism."

For half a century professional mentalists—performers, such as Joseph Dunninger, who claim unusual mental powers—have been entertaining audiences with "eyeless vision" acts. Usually the mentalist first has a committee from the audience seal his eyes shut with adhesive tape. Over each eye is taped something opaque, such as a powder puff or a silver dollar. Then a large black cloth is pulled around the eyes to form a tight blindfold. Kuda Bux, a Mohammedan who comes from Kashmir, is perhaps the best known of today's entertainers who feature such an act. He has both eyes covered with large globs of dough, then many yards of cloth are wound like a turban to cover his entire face from the top of his forehead to the tip of his chin. Yet Kuda Bux is able to read books, solve mathematical problems on a blackboard, and describe objects held in front of him.

Now I do not wish to endanger my standing in the magic fraternity by revealing too much, but let me say that Kuda Bux and other mentalists who feature eyeless vision do obtain, by trickery, a way of seeing. Many ingenious methods have been devised, but the oldest and simplest, surprisingly little understood except by magicians, is known in the trade as the "nose peek." If the reader will pause at this point and ask someone to blindfold him, he may be surprised to discover that it is impossible, without injury to his eyes, to prepare a blindfold that does not permit a tiny aperture, on each side of the nose, through which light can enter each eye. By turning the eyes downward one can see, with either eye, a small area beneath the nose and extending forward at an angle of 30 to 40 degrees from the vertical. A sleep-mask blindfold is no better; it does not fit snugly enough around the nose. Besides, slight pressure on the top of the mask, under the pretense of rubbing the forehead, levers out the lower edge to permit even wider peeks.

The great French magician Robert Houdin (from whom Houdini took his name), in his memoirs,[1] tells of watching another conjuror perform a certain card trick while blindfolded. The blindfold, Robert Houdin writes, "was a useless precaution . . . for whatever care may be taken to deprive a person of sight in this way, the projection of the nose always leaves a vacuum sufficient to see clearly." Pushing wads of cotton or cloth into the two apertures accomplishes nothing. One can always, while pretending to adjust the blindfold, secretly insert his thumb and form a tiny space under the wadding. The wadding can actually be an asset in maintaining a wider aperture than there would be without it. I will not go into more subtle methods currently used by mentalists for overcoming such apparent obstacles as adhesive tape criss-crossed over the eyelids, balls of dough, and so on.

If the mentalist is obtaining information by a nose peek (there are other methods), he must carefully guard against what has been called the "sniff" posture. When the head of a blindfolded person is in a normal position, the view down the nose covers anything placed on the near edge of a table at which the person is seated. But to extend the peek farther forward it is necessary to raise the nose slightly, as though one is sniffing. Practiced performers avoid the sniff posture by tilting the head slightly under cover of some gesture, such as nodding in reply to a question, scratching the neck, and other common gestures.

One of the great secrets of successful blindfold work is to obtain a peek in advance, covered by a gesture, quickly memorize whatever information is in view, then later—perhaps many minutes later—to exploit this information under the pretense that it is just then being obtained. Who could expect observers to remember exactly what happened five minutes earlier? Indeed, only a trained mentalist, serving as an observer, would know exactly what to look for.

Concealing the "sniff" demands much cleverness and experience. In 1964, on a television show in the United States, a girl who claimed powers of eyeless vision was asked to describe, while blindfolded, the appearance of a stranger standing before her. She began with his shoes, then went on to his trousers, shirt, and necktie. As her description moved upward, so did her nose. The photograph in *Time* showed Rosa wearing a conventional blindfold. She is seated, one hand on a newspaper, and sniffing. The entire newspaper page is comfortably within the range of a simple nose peek.

After the publicity about Rosa, Russian women of all sorts turned up, performing even more sensational feats of eyeless vision. The most publicized of these was Ninel Sergyeyevna Kulagina. The Leningrad newspaper *Smena*, 16 January 1964, reported on her remarkable platform demonstration at the Psychoneurological Department of the Lenin-Kirovsk District. The committee who examined Ninel's blindfold included S. G. Fajnberg (Ninel's discoverer), A. T. Alexandrov, rector of the University of Leningrad, and Leonid Vasiliev, whose laboratory at the University is the center of parapsychology research in Russia. No magicians were present, of course. While "securely blindfolded," Ninel read from a magazine and performed other sensational feats. Vasiliev was reported as having described her demonstration as "a great scientific event."

There were dozens of other DOP claimants. The magazine *USSR* (now *Soviet Life*), published here in English, devoted four pages to some of them in its February 1964 issue.[2] Experiments on Rosa, this article said, made it unmistakably clear that her fingers were reacting to ordinary light and not to infrared heat rays. Filters were used which could block either light or heat. Rosa was unable to "see" when the light (but not heat) was blocked off. She "saw" clearly when the heat rays (but not light) were blocked off. "The fingers have a retina," biophysicist Mikhail Smirnov is quoted as saying. "The fingers 'see' light."

Accounts of the women also appeared in scientific publications. Goldberg contributed a report on his work with Rosa to *Voprossy Psikhologii* in 1963.[3] Biophysicist N. D. Nyuberg wrote an article about Rosa for *Priroda,* May 1963.[4] Nyuberg reports that Rosa's fingers, just like the human eye, are sensitive to three color modes and that, after special training at the neurological institute, she "succeeded in training her toes to distinguish between black and white." Other discussions of Rosa's exploits appeared in Soviet journals of philosophy and psychology.

Not only did Rosa read print with her fingers, she also described pictures in magazines, on cigarette packages, and on postage stamps. A *Life* correspondent reported that she read his business card by touching it with her elbow. She read print placed under glass and cellophane. In one test, when she was "securely blindfolded," scientists placed a green book in front of her, then flooded it with red light. Exclaimed Rosa: "The book has changed color!" The professors were dumbfounded. Rosa's appearance on a TV program called "Relay" flushed out new rivals. *Nedelya,* the supplement of *Izvestia,* found a 9-year-old Kharkov girl, Lena Bliznova, who staggered a group of scientists by reading print ("securely blindfolded") with fingers held a few inches *off* the page. Moreover, Lena read print just as easily with her toes and shoulders. She separated the black from the white chess pieces without a single error. She described a picture covered by a thick stack of books (see my remarks about exploiting previously memorized information).

In the United States, *Life* (12 June 1964) published a long uncritical article by Albert Rosenfeld,[5] the writer whose card Rosa had read with her elbow. The Russian work is summarized and hailed as a major scientific breakthrough. Colored symbols are printed on one page so the reader can give himself a DOP test. Gregory Razran, who heads the psychology department at Queens College, New York, is quoted as saying that perhaps "some entirely new kind of force or radiation" has been detected. Razran expected to see "an explosive outburst of research in this field. . . . To see without the eyes—imagine what that can mean to a blind man!"

Let us hope that Razran, in his research, will seek the aid of knowledgeable mentalists. In a photograph of one of his DOP tests, shown in the *Life* article, the subject wears a conventional sleep-mask, with the usual apertures. She is reaching through a cloth hole in the center of an opaque partition to feel one of two differently colored plates. But there is nothing to prevent her from reaching out with her other hand, opening the cloth a bit around her wrist, then taking a nose peek through the opening.

The most amusing thing about such experimental designs is that there is a simple, but never used, way to make sure all visual clues are eliminated. A blindfold, in any form, is totally useless, but one can build a light-weight aluminum box that fits over the subject's head and rests on padded shoulders. It can have holes at the top and back for breathing, but the solid metal must cover the face and sides, and go completely under the chin to fit snugly around the front of the neck. Such a box eliminates at one stroke the need for a blindfold, the cumbersome screen with arm holes, various bib devices that go under the chin, and other clumsy pieces of apparatus designed by psychologists unfamiliar with the methods of mentalism. No test made without such a box over the head is worth taking seriously. It is the only way known to me by which all visual clues can be ruled out. There remain, of course, other methods of cheating, but they are more complicated and not likely to be known outside the circles of professional mentalism.

In its 1964 story *Life* did not remind its readers of the three pages it had devoted, in 1937, to Pat Marquis, "the boy with the X-ray eyes".[6] Pat was then 13 and living in Glendale, California. A local physcian, Cecil Reynolds, discovered that Pat could "see" after his eyes had been taped shut and covered with a blindfold. Pat was carefully tested by reporters and professors, said *Life,* who could find no trickery. There are photographs of Pat, "securely blindfolded," playing ping-pong, pool, and performing similar feats. Naturally he could read. Reynolds is quoted as saying that he believed that the boy "saw" with light receptors in his forehead. Pat's powers were widely publicized at the time by other magazines and by the wire services. He finally agreed to being tested by J. B. Rhine, of Duke University, who caught him nose peeking.[7]

The truth is that claims of eyeless vision turn up with about the same regularity as tales of sea serpents. In 1898 A. N. Khovrin, a Russian psychiatrist, published a paper on "A rare form of hyperaesthesia of the higher sense organs,"[8] in which he described the DOP feats of a Russian woman named Sophia. There are many earlier reports of blind persons who could tell colors with their fingers, but "blindness" is a relative term, and there is no way now to be sure how blind those claimants really were. It is significant that there are no recent cases of persons known to be totally blind who claim the power to read ordinary print, or even to detect colors, with their fingers, although it would seem that the blind would be the first to discover and develop such talents if they were possible.

Shortly after World War I the French novelist Jules Romains, interested in what he called "paroptic vision," made an extensive series of tests with French women who could read while blindfolded. His book, *Vision Extra-Rétinienne* should be read carefully by every psychologist tempted to take the Russian claims seriously,[9] for it describes test after test exactly like those that have been given to today's Russians. There are the same lack of controls, the same ignorance of the methods of mentalism, the same speculations about the opening of new scientific frontiers, the same unguarded predictions about how the blind may someday learn to "see," the same scorn for those who remain skeptical. Romains found that DOP was strongest in the fingers, but also present in the skin at any part of the body. Like today's Russian defenders of DOP, Romains is convinced that the human skin contains organs sensitive to ordinary light. His subjects performed poorly in dim light and could not see at all in total darkness. Romains thought that the mucous lining of the nose is especially sensitive to colors, because in dim light, when colors were hard to see, his subjects had a marked tendency to "sniff spontaneously."

The blindfolding techniques Romains used are similar to those used by the more recent investigators. Adhesive tape is crossed over the closed eyes, then folded rectangles of black silk, then the blindfold. At times cottom wool is pushed into the space alongside the nose; at times a projecting bib is placed under the chin. (Never a box over the head.) Anatole France witnessed and commented favorably on some of Romains' work. One can sympathize with the novelist when he complained to a U.S. reporter that both Russian and American psychologists had ignored his findings and had simply "repeated one-twentieth of the discoveries I made and reported."[10]

It was Romains' book that probably aroused magicians in the United States to devise acts of eyeless vision. Harlan Tarbell, of Chicago, worked out a remarkable act of this type which he performed frequently.[11] Stanley Jaks, a professional mentalist from Switzerland, later developed his method of copying a stranger's signature, upside down and backward, after powder puffs had been taped over his eyes and a blindfold added."[12] Kuda Bux uses still other techniques.[13] At the moment, amateurs everywhere are capitalizing on the new wave of interest in DOP. In my files is a report on Ronald Coyne, an Oklahoma boy who lost his right eye in an accident at the age of seven. When his left eye is "securely blindfolded," his empty right eye socket reads print without hesitation. Young Coyne has been appearing at revival meetings to demonstrate his miraculous power. "For thirteen years he has had continuous vision where there is no eye," reads an advertisement in a Miami newspaper for an Assembly of God meeting. "Truly you must say 'Mine eyes have seen the glory of God.'"

The most publicized DOP claimant in the United States is Patricia Stanley. Richard P. Youtz, of the psychology department at Barnard

College, was discussing the Soviet DOP work at a faculty lunch one day. Someone who had taught high school in Owensboro, Kentucky, recalled that Patricia, then a student, had astounded everyone by her ability to identify objects and colors while blindfolded. Youtz traced Patricia to Flint, Michigan, and in 1963 he made several visits to Flint, tested her for about sixty hours, and obtained sensational results. These results were widely reported by the press and by such magazines of the occult as *Fate*.[14] The soberest account, by science writer Robert K. Plumb, appeared in the *New York Times*, 8 January 1964.[15] Mrs. Stanley did not read print, but she seemed able to identify the colors of test cards and pieces of cloth by rubbing them with her fingers. Youtz's work, together with the Russian, provided the springboard for Leonard Wallace Robinson's article "We have more than five senses" in the *New York Times Magazine*, Sunday, 15 March 1964.

Youtz's first round of tests, in my opinion, were so poorly designed to eliminate visual clues that they cannot be taken seriously. Mrs. Stanley wore a conventional sleep-mask. No attempt was made to plug the inevitable apertures. Her hands were placed through black velvet sleeves, with elastic around the wrists, into a lightproof box constructed of plywood and painted black. The box could be opened at the other side to permit test material to be inserted. There was nothing to prevent Mrs. Stanley from picking up a test card or piece of colored cloth, pushing a corner under the elastic of one sleeve, and viewing the exposed corner with a simple nose peek. Youtz did have a double sleeve arrangement that might have made this difficult, but his account of his first round of tests, on which Mrs. Stanley performed best, indicate that it was attached only on the rare occasions when a photo-multiplier tube was used.[16] Such precautions as the double sleeve, or continuous and careful observation from behind, seemed unnecessary because Mrs. Stanley was securely blindfolded. Moreover, there was nothing to prevent Mrs. Stanley from observing, by nose peeks, the test material as it was being placed into the light-tight box.

Here is a description of Mrs. Stanley's performance by the *New York Times* reporter who observed her: "Mrs. Stanley concentrates hard during the experiments. . . . Sometimes she takes three minutes to make up her mind. . . . She rests her forehead under the blindfold against the black box as though she were studying intently. Her jaw muscles work as she concentrates."[17] While concentrating, she keeps up a steady flow of conversation with the observers, asking for hints on how she is doing.

Youtz returned to Flint in late January 1964 for a second round of tests, armed with more knowledge of how blindfolds can be evaded (we exchanged several letters about it) and plans for tighter controls.[18] I had been unsuccessful in persuading him to adopt a box over the head, but even without this precaution, results of the second round were not above chance expectation. These negative results were reported by the *New York Times,*[17] but not by any other newspaper or news magazine that had publicized the positive results of the first round of tests. Youtz was disappointed, but he attributed the failure to cold weather.[19]

A third series of tests was made on 20 April for an observing committee of four scientists. Results were again negative. In the warm weather of June, Youtz tested Mrs. Stanley a fourth time, over a three-day period. Again, performance was at chance level. Youtz attributes this last failure to Mrs. Stanley's fatigue.[19] He remains convinced that she does have the ability to detect colors with her fingers and suspects that she does this by sensing delicate differences in temperature.[20] Although Russian investigators had eliminated this as an explanation of Rosa's powers, Youtz believes that his work with Mrs. Stanley, and later with less skillful Barnard students, will eventually confirm this hypothesis. He strongly objects to calling the phenomenon "vision." None of his subjects has displayed the slightest ability to read with their fingers.

In Russia, better-controlled testing of Rosa has strongly indicated nose peeking. Several articles have suggested this, notably those by L. Teplov, author of a well-known book on cybernetics, in the 1–7 March 1964 issue of *Nedelya,* and in the 25 May issue of the Moscow *Literaturnaya Gazeta.* Ninel Kulagina, Rosa's chief rival, was carefully tested at the Bekhterev Psychoneurological Scientific Research Institute in Leningrad. B. Lebedev, the institute's head, and his associates summarize their findings as follows:[21]

> In essence, Kulagina was given the same tasks as before, but under conditions of stricter control and in accordance with a plan prepared beforehand. And this was the plan: to alternate experiments in which the woman could possibly peek and eavesdrop with experiments where peeking would be impossible. The woman of course did not know this. As was to be expected, phenomenal ability was shown in the first instance only. In the second instance [under controls] Kulagina could distinguish neither the color nor the form. . . .
>
> Thus the careful checking fully exposed the sensational "miracle." There were no miracles whatever. There was ordinary hoax.

In a letter to *Science,* Joseph Zubin, a biometrics researcher at the New York State Department of Mental Hygiene, reported the negative results of his testing of an adolescent who "read fluently" after blindfolds had been secured around the edges with adhesive tape.[22] Previous testing by several scientists had shown no evidence of visual clues. It became apparent, however, that the subject tensed muscles in the blindfolded area until "a very tiny, inconspicuous chink appeared at the edge. Placing an opaque disk in front of the chink prevented reading, but not immediately. The subject had excellent memory and usually continued for a sentence or two after blocking of the reading material." Applying zinc ointment to the edges of the adhesive proved only temporarily effective, because muscle tensing produced new chinks (made easier to detect by the white ointment). A professional magician, Zubin reports, participated in the investigations. Zubin doesn't name him, but he was James Randi.

The majority of psychologists, both here and in the Soviet Union, have remained unimpressed by the latest revival of interest in DOP. In view of the failures of subjects to demonstrate DOP when careful precautions were taken to rule out peeks through minute apertures, and in view of the lack of adequate precautions in tests that yielded positive results, this prevailing skepticism appears to be strongly justified.

NOTES

1. J. E. Robert-Houdin, *Confidences d'un Prestidigitateur* (Blois, 1858), chap. 5; English translation, *Memoirs of Robert-Houdin: Ambassador, Author, and Conjuror* (London, 1859); reprinted as *Memoirs of Robert-Houdin: King of the Conjurers* (Dover, New York, 1964).

2. *USSR 89*, 32 (1964).

3. For English translation, see I. Goldberg, *Soviet Psychol. Psychiat. 2*, 19 (1963).

4. For English translation, see N. D. Nyuberg, *Federation Proc. 22*, T701 (1964).

5. A. Rosenfeld, "Seeing color with the fingers," *Life 1964*, 102–13 (12 June 1964).

6. "Pat Marquis of California can see without his eyes." *Life 1937*, 57–59 (19 Apr. 1937).

7. J. B. Rhine, *Parapsychol. Bull. 66*, 2–4 (Aug. 1963).

8. A. N. Khovrin, in *Contributions to Neuropsychic Medicine* (Moscow, 1898).

9. J. Romains, *Vision Extra-Rétinienne* (Paris, 1919); English translation, *Eyeless Vision*, C. K. Ogden, transl. (Putnam, New York, 1924).

10. J. Davy, *Observer*, 2 Feb. 1964.

11. See H. Tarbell, "X-ray eyes and blindfold effects" in *The Tarbell Course in Magic* (Tannen, New York, 1954), vol. 6, pp. 251–261. Tarbell speaks of his own work in this field as a direct result of his interest in Romains' work, and briefly describes an eyeless vision act by a woman who performed under the stage name of Shireen in the early 1920s.

12. See. M. Gardner, *Sphinx 12*, 334–337 (Feb. 1949); *Linking Ring 34*, 23–25 (Oct. 1954); also, G. Groth, "He writes with your hand," in *Fate 5*, 39–43 (Oct. 1952).

13. A description of an early eyeless vision act by Kuda Bux will be found in H. Price, *Confessions of a Ghost-Hunter* (Putnam, New York, 1936), chap. 19.

14. P. Saltzman, *Fate 17*, 38–48 (May 1964).

15. R. K. Plumb, "Woman who tells color by touch mystifies psychologist," in *New York Times*, 8 Jan. 1964; see also Plumb's follow-up article, "6th Sense is hinted in ability to 'see' with fingers," *ibid.*, 26 Jan. 1964. The *Times* also published an editorial, "Can fingers 'see'?" 6 Feb. 1964.

16. R. P. Youtz, "Aphotic Digital Color Sensing: A Case under Study," photocopied for the Bryn Mawr meeting of the Psychonomic Society, 29 Aug. 1963.

17. "Housewife is unable to repeat color 'readings' with fingers," *New York Times*, 2 Feb. 1964.

18. For an exchange of published letters, see M. Gardner, *New York Times Magazine*, 5 Apr. 1964, and R. P. Youtz, *ibid.*, 26 Apr. 1964.

19. R. P. Youtz, "The Case for Skin Sensitivity to Color: with a Testable Explanatory Hypothesis," photocopied for the Psychonomic Society, Niagara Falls, Ontario, 9 Oct. 1964.

20. See R. P. Youtz, letter, *Sci. Amer. 212,* 8–10 (June 1965).

21. B. Lebedev, *Leningradskaya Pravda,* 15 Mar. 1964; translated for me by Albert Parry, department of Russian studies, Colgate University.

22. J. Zubin, *Science 147,* 985 (1965).

Postscript

Dermo-optical perception is still taken seriously by yahoos of the paranormal both here and abroad. *Fate* continues to run articles promoting it (see issues of September 1967, May 1976, and July 1976). The Parapsychology Foundation of New York published *The Paranormal Perception of Color,* by Yvonne Duplessis, in 1975.

In the USSR, Nina Kulagina has gone on to more sensational demonstrations of her psi abilities and is now Russia's most publicized psychic. Rosa Kuleshova, after performing for a while with a circus, died in 1978, her fame eclipsed by Nina. I had lunch with Albert Rosenfeld a few years ago when he was science editor of the *Saturday Review,* and although we did not talk about his *Life* article, I found him firmly convinced of Uri Geller's ability to bend and break metal by psychokinesis. Gregory Razran died in 1972.

Ronald Coyne, the one-eyed evangelist from Tulsa who sees with his eyeless socket, is still going strong on Pentecostal revival circuits. You can read more about him in a 77-page booklet, *When God Smiled on Ronald Coyne,* by his mother, Mrs. R. R. Coyne. It was first published in Tulsa in 1952. And there is a long-playing record on which Coyne tells his story. In 1966 both book and record were obtainable from the Centre of Evangelism, Box 640, Cloverdale, British Columbia, Canada, and from Ronald Coyne Revivals, Tulsa, Oklahoma.

Youtz replied to my *Science* paper with a long letter that appeared in *Science,* May 20, 1966, page 1108. Two years later he was still defending fingertip vision in an article, "Can Fingers See Color?" in *Psychology Today,* February 1968.

A UPI news report of February 15, 1980, reported an outbreak of eyeless vision in China. "Using today's scientific knowledge," said the Chinese magazine *Nature,* "we still cannot explain this kind of phenomenon." The article spoke of the ability of two sisters in Peking, Wang Bingn, 11, and Wang Qiang, 13, to identify Chinese letters on paper by tucking the paper under their armpits. In Shanghai, the same article reported, there are children able to read writing on paper placed in their ears.

The Israeli psychic, Uri Geller, used to perform what magicians call the "blindfold drive"—driving a car while blindfolded—and he still uses a crude nose-peek during a blindfold bit in his standard magic act. (See *Further Confessions of a Psychic,* by Uriah Fuller, 1980.) In an interview published in the Autumn 1980 issue of *MetaScience Quarterly,* Uri made his usual denial that he ever uses magician's tricks. Asked about his blindfold drives, Uri responded: "It was a form of telepathy. . . . I stopped doing this because it can be so easily duplicated by magic." Carl Sagan and I are singled out as "dirty negative people."

The Reverend Richard Ireland, of Tucson, Arizona, has been doing a crude eyeless-vision act for decades, complete with the usual tape over the eyes and a "secure" blindfold. Only the *National Enquirer* paid any attention to him until December 1980, when he suddenly emerged from the woodwork as the "psychic" who helped Dallas oilman Jerry Conser find two whopping oil fields. Mr. Conser, who believes that today's upsurge of psi power heralds the Second Coming of Christ, has set up the Millennium Foundation, based in San Francisco, to provide a million dollars for parapsychological research.

7

Targ's ESP Teaching Machine

In modern extrasensory-perception (ESP) experiments probability and statistics play indispensable roles in determining if ESP events have indeed occurred. Targets are set up, subjects make a large number of guesses, and then the results are analyzed to see if there are significant deviations from chance. The results are usually recorded by hand, which has given rise to a persistent criticism. Because those who record ESP data are almost always firm believers in ESP, often with a large personal stake in a favorable outcome, the possibility of belief biasing the results looms very large.

The biasing can, of course, be entirely unconscious. Over and over again it has been demonstrated that people with a strong belief make unwitting recording errors that tend to favor their belief. In tests of psychokinesis (PK), for example, when subjects try to influence the fall of dice, secret cameras have shown that handwritten records kept by "sheep" (believers) display significant errors favoring PK, whereas similar records kept by "goats" (skeptics) display an equal bias in the other direction.

With the rise of electronic and computer technology it naturally occurred to many workers in the field of ESP research that one simple way to guard against unconscious recording errors is to make the process as automatic as possible. Let the machine, incorporating an efficient randomizer, select the targets, and design the machine so that it makes a

This article, which appeared in *Scientific American,* October 1975, and the letters to *Scientific American* quoted in the Postscript are reprinted with permission of the publisher. Copyright 1975–1976 by Scientific American.

permanent, unalterable record of both targets and trial guesses. It is true that such machines are not fraudproof; witness the scandal last year when Walter J. Levy, Jr., director of J. B. Rhine's Institute for Parapsychology, resigned after it was found that he had been tinkering with the apparatus to improve scores. Apart from cases of outright chicanery, however, an electronic apparatus is an excellent way to eliminate unconscious bias.

Several crude devices for testing ESP were used on rare occasions from the late 1930s on, but the first major tests with an electronic machine were made in 1962. They were done with a system called VERITAC, which had been designed and built by a worker at the Air Force Cambridge Research Laboratories. The system randomly selects digits from 0 through 9. It prints a record of the chosen digit, the subject's guess as to what the digit is, the time of each trial and the time interval between selection of the target and the guess. Counters on the control console provide instant feedback of results, but the counters can be disconnected if it is wished. After a trial run VERITAC goes into a locked condition and remains locked until a teletypewriter prints out the data.

The machine can be set for one of three modes. In the clairvoyant mode the subject guesses the digit after it has been selected. In the precognitive mode the guess precedes the selection. And in the general extrasensory perception (GESP) mode the target is observed by someone who acts as a telepathic sender to a subject in another room. Hence a hit can be the result of telepathy, clairvoyance, or both.

In the 1962 experiment each of 37 subjects completed five runs of 100 trials each for each of the three modes, making a total of 55,500 trials. When results were analyzed, using the familiar chi-square test for statistical significance, there was no deviation from chance either for the entire group or for any individual. Nor were there significant differences in the scores of sheep and goats.

C. E. M. Hansel, discussing this historic experiment in his book *ESP: A Scientific Evaluation* (Scribner's, 1966), pointed out that VERITAC's instant feedback of results to the subject made it an ideal teaching machine. "With the VERITAC machine, subjects could be given long practice sessions so that any ESP ability that might be present could be strengthened. Thus parapsychologists would have both a testing and a training machine. It could also be modified to provide a reward after each hit and punishment, such as a mild electric shock, after each miss. It would then constitute a conditioning machine. . . ."

"If 12 months' research on VERITAC can establish the existence of ESP," Hansel wrote on the final page of his book, "the past research will not have been in vain. If ESP is not established, much further effort could be spared and the energies of many young scientists could be directed to more worthwhile research."

Most parapsychologists did not look favorably on that kind of ESP testing. One exception was Russell Targ, then a physicist with Sylvania Electric Products specializing in laser and plasma research. In 1966, the year Hansel's book appeared, a short note in *Electronics* (December 26, page 36) reported that Targ was working on an ESP teaching box designed by David B. Hurt, an engineer at Fairchild Camera and Instrument. The subject tries to guess which of four buttons will light up, the notice said, and the box "reinforces by punishment as well as by reward."

Five years later, with a grant from the Parapsychology Foundation (founded by the well-known spiritualistic medium Eileen J. Garrett), Targ and Hurt designed and built a more advanced ESP teaching device. In 1972 Targ was hired by the Electronics and Bioengineering Laboratory of the Stanford Research Institute (SRI). Since then he and his associate Harold E. Puthoff, a physicist and Scientologist who had joined the SRI staff a year earlier, have been engaged in parapsychological research. The two men have become best known for their testing of Uri Geller, the Israeli magician who professes to have paranormal powers. Here, however, we shall be concerned only with their ESP-teaching-machine experiment. It marks the second milestone in the attempt with an electronic apparatus to establish the existence of ESP abilities in man.

The research was made possible by a grant of $80,000 from the National Aeronautics and Space Administration, with the Jet Propulsion Laboratory of the California Institute of Technology serving as the administrator. The final 61-page report was published by SRI in August of last year with the title "Development of Techniques to Enhance Man/Machine Communication." The authors are Targ, Phyllis Cole, and Puthoff. Since Targ was the senior investigator, I shall henceforth use his name only.

The report is not classified. Its cover states that it was prepared for distribution "in the interest of information exchange." Since the work was financed by public funds, anyone interested in it is entitled to ask SRI for a copy. The address is Menlo Park, Calif. 94025. I have been told that only 50 copies of the report were printed and that all have long since been distributed but that SRI has permission to reprint the report any time it wants to do so. It is to be hoped that this will be done, because it is an important report that every serious student of parapsychology should have access to.

Let us have a look at Targ's machine [*see illustration on following page*]. Models of it are manufactured by Aquarius Electronics, Albion, Calif. (Similar and more compact models are now being made by other companies.) There are four square panels, each of which can display a colored transparency. Before any picture is displayed, however, a randomizer in the machine selects one of the four pictures as the target. The subject tries to guess the target, indicating his choice by pressing the

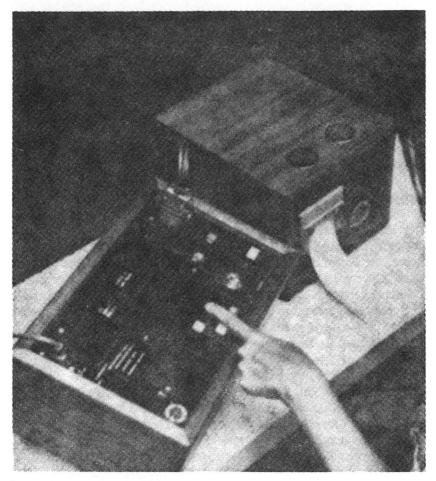

A subject works with Targ's ESP Teaching Machine.

square button nearest that panel. As soon as the subject indicates his choice, a light goes on behind the correct target picture to provide feedback and reinforcement. When there is a hit, a bell sounds. A counter to the right of the panels displays the number of the trial (from 1 to 25). A second counter displays the number of hits.

If a subject feels that he does not "know" the right button, he can press a "pass" button below the panels, and no guess will be recorded. Another button to the right of the pass button resets the counters to zero. Above the panels are five "encouragement lights" to provide additional reinforcement. The first legend, "Good beginning," lights up as soon as there are six hits, and it goes out if the hits rise to eight. The second legend, "ESP ability present," lights up at eight hits. "Useful at Las Vegas" appears at 10 hits. Twelve hits light up "Outstanding ESP ability," and 14 hits "Psychic, medium, oracle."

To the left of the panels is a rotary switch. Throughout the NASA project the machine was set for clairvoyance. The switch can also be set for precognition and telepathy. For telepathic testing it is necessary to plug a "telepathy adapter" into the model's output jack. This accessory box, connected to the teaching machine by a 25-foot cable, displays the targets to a telepathic sender in another room, who sees them before the subject makes his choice. A report on the use of an earlier model of the machine in its precognitive mode is given in a paper by Puthoff and Targ, published as Chapter 22 of astronaut Edgar D. Mitchell and John White's anthology *Psychic Exploration* (Putnam's, 1974). The two authors present their theory that events send out waves that propagate backward in time but decay rapidly. The closer the event to the precognition, the stronger the precognition; therefore the machine is designed to select its target with a quarter-second to one-second delay after a choice has been made. The authors think that the "familiar *déjà vu* phenomenon is the most common form of precognition," not (as some parapsychologists have argued) a hazy recollection of an experience in a previous incarnation. They are also convinced that awaking just before an alarm clock rings is another familiar instance of precognition. Since that is a "large, timely and unpleasant event," its backward wave in time makes a strong impression on the sleeping mind.

Back to clairvoyance. The first phase of Targ's NASA project was to test two individuals under informal conditions. Subject A1, identified only as the son of an SRI scientist, worked at home with the machine, with his father recording the data on score sheets. Subject A2, identified only as a scientist not at SRI, worked in the laboratory but kept his own handwritten records. Subject A1 made 9,600 trials, obtaining a mean score of 26.06 hits per 100. The rising slope of his learning curve was .077. Subject A2 had a mean score of 30.50 over 1,400 trials and a learning slope of .714. These were encouraging results, but because of the lack of controls it was called Phase 0 and represented only a pilot study.

Phase 1 tightened the controls a bit by hooking a printer up to the machine. A typical printout is shown in the illustration on page 80. The printer counts the number of trials from 1 through 25 (holding the count for a "pass"), records the machine's choice of target (0 through 3), records the guess, and keeps a running total of hits.

Of the 145 volunteer subjects for Phase 1, 100 were "employees, relatives, and friends" of SRI (79 adults and 21 children under 15). All did their work alone in an SRI laboratory, keeping their own records. Each worked on two or more machines in different locations. Twenty-two subjects of junior-high-school age or younger were at a private school where an experimenter was in attendance. The remaining 23 subjects were

```
0 0 0 0 0 4 —— RESET
2 5 0 7 2 0
2 4 0 7 1 2
2 3 0 7 0 3
2 2 0 7 3 0
2 1 0 7 0 0 —— HIT
2 0 0 6 0 3
1 9 0 6 0 1
1 8 0 6 1 0
1 7 0 6 2 2 —— HIT
1 6 0 5 0 3
1 5 0 5 2 0
1 4 0 5 2 0
1 3 0 5 2 7 —— PASS
1 3 0 5 2 3
1 2 0 5 3 3 —— HIT
1 1 0 4 3 0
1 0 0 4 2 1
0 9 0 4 2 2 —— HIT
0 8 0 3 1 3
0 7 0 3 0 0 —— HIT
0 6 0 2 3 7 —— PASS
0 6 0 2 2 7 —— PASS
0 6 0 2 0 3
0 5 0 2 2 2 —— HIT
0 4 0 1 0 3
0 3 0 1 2 0
0 2 0 1 1 1 —— HIT
0 1 0 0 2 1
```

TRIAL   SCORE   MACHINE'S   SUBJECT'S
                  CHOICE   CHOICE

Paper-tape record of an ESP machine trial.

junior-high-school students at a public school where the tests were supervised by teachers.

The overall scores of the 145 subjects were at the chance level for both ESP and learning. A questionnaire given to the private-school students revealed that 15 of the 22 students actually tried to get low scores. "This tendency to experiment with different modes of interacting with the machine," Targ writes, "was not taken into account in recording or analyzing data." Nine of the 145 subjects had slight upward learning slopes, and 11 showed significant ESP. None showed a significant learning decline.

Targ was well aware that the Phase 1 controls were much too loose to justify the NASA investment in the study. Although he does not spell it out, it is easy to see how bias could have crept in. In the first place, the printer keeps no record of the total number of trials by any subject. There was no supervision of the 100 SRI employees, relatives, and friends. One may assume that a high proportion of them were sheep. On the further assumption that no one consciously cheated, how could their unconscious bias operate?

The most obvious way is by decisions as to when a run of 25, recorded on paper tape, is to be preserved or discarded. Suppose there is a sudden disturbance: someone comes into the room, a fire engine goes by outside, or a telephone rings. If the run is low in hits, there could be a strong feeling that the noise disturbed ESP and that the run should therefore not be counted. Or there might be an internal disturbance to justify tearing off and throwing away a run. The subject's foot falls asleep, his head starts to ache, disturbing thoughts cross his mind, and so on. His finger could fumble and give him the impression that he had pressed the wrong button. Imagine yourself acting as an unsupervised subject. Now suppose any of the above disturbances, which could be a plausible basis for discarding a run, occurs. You note, however, that the run is high in hits. Would you then discard it?

Suppose you decide, but only in a vague way, to make a practice run. As you watch hits accumulate on the counter, would it not be easy to fool yourself into believing a practice run had not been intended after all? You keep the run. If the hits had been low, you would have discarded it.

All of this obviously also applies to the students. At the private school how carefully did the experimenter supervise the subjects? Did he keep watch at all times, or did he occasionally read a book or leave the room? And would the experimenter have strenuously objected if a student explained why he did not want to save a run?

At the public school how well did the teachers supervise the subjects? Targ tells us that many subjects "complained of the noise and confusion inherent in the location." And again: "Several dozen Phase 1 participants had complained that the clatter of the printer was a distraction." I am not guessing when I say that the paper-tape records of Phase 1 were turned in to Targ in disconnected bits and pieces.

Targ clearly perceives the weakness of his experimental design for this phase. It is the design of a physicist trained to investigate physical laws — laws that do not exhibit psychological quirks. An experimental psychologist would have constructed a printer that kept an unalterable record of all trials. Subjects would have been required to start at Trial 1, continue to a predetermined limit agreed to by both the experimenter and the subject and then turn in an unbroken tape. VERITAC was carefully designed to forestall bias by the simple expedient of keeping a time record of all trials. In Targ's defense it should be said that he regarded Phase 1 as being no more than a loose screening process designed to pick out high scorers in preparation for the crucial Phase 2, during which all psychological bias would be eliminated.

To eliminate it, a Model 33 Teletype was plugged into the system so that in addition to the paper printout a record of all trials was kept on punched tape. The punched tape was necessary not only for keeping an unalterable total record but also for ease in computer reading and analysis. The punched tape was fed to a computer on a trial-by-trial basis. The computer analyzed the data while the choices were being made.

Only the best subjects from Phase 0 and Phase 1 were used. There were 12 in all. This included subject A2 from Phase 0. (Subject A1, a student, had left the area to return to college.) Eleven subjects were chosen from Phase 1. Because of complaints about the noise of the printer during Phase 1, the printer was kept in another room with the teletype. Indeed, both the printer and the teletype were in the experimenter's own office, where they were inaccessible to the subjects.

The final outcome of Phase 2 must have disappointed Targ. No subject did better than chance on hits. No subject showed a significant learning curve. In short, the experiment was a failure.

One feature of Phase 2 is of unusual interest. Subject A13, who had "demonstrated some paranormal ability in other tests conducted at SRI," was offered the following rewards: $1 for 10 hits in 25, $2 for 11 hits, $5 for 12, $10 for 13, and $20 for 14.

Rhine has continually expressed his conviction that monetary motivation (among others) greatly increases ESP. "Subject motivation to score high," he wrote in 1964, "has long stood out as the mental variable that seems most closely related to the *amount* of psi effect shown in test results." To support his hypothesis Rhine invariably recalls that famous occasion in 1932 when Hubert Pearce, his star performer, made a run of 25 correct guesses of ESP cards—a miracle by anybody's statistical criteria. Rhine motivated Pearce by offering him $100 for each correct hit. The final sum of $2,500 was so large that Rhine had to tell Pearce he did not really mean it. That has always struck me as a dirty trick to play on Pearce, who was poor in those Depression days.

Since then Rhine has reported many other instances of perfect runs of 25, usually after some kind of motivation. The most notable was in 1936, when one of Rhine's assistants was testing Lillian, a nine-year-old who had been the highest scorer among a group of children. As Rhine told it in 1944:

"One day, after an unusually high score, she looked out the window, a happy little smile on her face.

"'Don't say anything,' she said. 'I'm going to try something.'

"The alert experimenter cut the cards again, just to be safe."

Continuing the description from Rhine's 1964 account: "Before the run the child paused, laid the cards down, and, with eyes closed, moved her lips as if talking to herself. She played off the cards without appearing to be focusing on the backs."

Apparently Rhine had learned the hazards of offers of cash. In this case Lillian had been earlier promised 50 cents if she made a perfect score. One presumes that the payoff was made. The following week in Rhine's laboratory Lillian's powers failed when she tried to guess ESP cards in sealed envelopes. On one of these tests, however, she made 24 misses. Rhine considers that a "nearly significant" instance of "psi missing," or ESP avoidance of the target.

There are other instances of subjects obtaining perfect scores in which strong motivation, Rhine believes, commonly played a major role.

Not once does he entertain the possibility that strong motivation also motivates clever subjects to cheat. Magicians can give you 20 easy ways to obtain perfect runs with ESP cards. They range from seeing the large ESP symbols reflected in the experimenter's glasses (under certain conditions they can even be seen reflected by the corneas of people not wearing glasses) to almost imperceptible fingernail nicks along the edges of cards and even subtler methods that I would prefer not to disclose (but that can easily be reinvented by an ingenious child). It is noteworthy that not a single run of 25 has been obtained under controlled conditions, in spite of Rhine's persistent belief that in the uncontrolled cases the chances of success were genuinely $1/5^{25}$, or one in 298,023,223,876,953,125.

There is little doubt that Targ's money offer gave subject A13 a strong incentive. As Targ puts it, he "was highly motivated to generate trials." Out of more than 20,000 trials about 13,500 were made under the payment agreement. Alas, the offers had no effect. Subject A13's scores, faithfully recorded on the unalterable punched tape, remained at the chance level.

The results confirmed numerous tests by goats that have shown no correlation between motivation and ESP. Such failures are generally not reported to the public because they are not newsworthy. Richard C. Sprinthall and Barry S. Lubetkin gave the results of such a test in the *Journal of Psychology* (vol. 60, 1965, pages 313–318). Twenty-five volunteer students were asked to guess ESP cards for a run of 25, and 25 other volunteer students were given an identical test except that a firm offer was made of $100 to anyone who got 20 hits. The highest score among the 50 was 10. Overall scores were at the chance level. The mean score for the unmotivated group, 5.56, was slightly higher than the 5.40 for the motivated group.

Whenever a major experiment, such as the SRI test of Targ's ESP machine, is a conspicuous failure, parapsychologists themselves become strongly motivated to give reasons for the failure. If the test is supervised by a skeptical psychologist, or even if a goat is a mere observer, the favorite excuse is to invoke a kind of Catch 22: Skepticism destroys the subtle operation of psi. It is a catch unique to parapsychology. In other sciences failure by a doubting scientist to replicate an experiment is counted as disconfirming evidence. Because psi powers are said to be adversely influenced by doubt, however, parapsychologists are not impressed by replication failures unless they are obtained by sheep. In this case no goat was present, so that Targ turned to Catch 23.

Catch 23 asserts that psi powers are negatively influenced by complexity. As Rhine once phrased it, ". . . elaborate precautions take their toll. Experimenters who have worked long in this field have observed that the scoring rate is hampered as the experiment is made complicated, heavy, and slow-moving. Precautionary measures are usually distracting in themselves." Catch 23 achieves a truly remarkable result. It makes it impossible to establish psi powers by tests that are convincing to the goats who are the vast majority of professional psychologists. As long as testing is informal and under sloppy controls, you get results. If you tighten controls, the experiment inevitably gets complicated and scores fall.

Let us look at the way Targ puts it. "First, the subjects were definitely aware that they were in a test situation, despite attempts to provide a quiet, pleasant, nonthreatening atmosphere. All knew they were selected to participate because they performed well during the screening process, and this knowledge created varying degrees of tension. . . .

"The subjects in this experiment have uniformly complained about the new experimental conditions in that 'It all feels different, being connected to a computer' despite the fact that the new working conditions were much quieter and more congenial than those in the pilot studies. We have spent considerable time interviewing the more articulate of our previous high-scoring subjects. From these conversations we have determined that they have lower levels of confidence when working with the printer connected to the teaching machine than they do when working with an experimenter watching them. And they have the least comfort when working with the teletypewriter punch operating. We have not done blind studies to confirm these perceptions, but they are certainly supported by the decline in scoring rate observed as we have progressed through the three recording techniques used in the program."

At no time does Targ consider the hypothesis that Phase 2 disconfirmed the existence of clairvoyant powers. "Based on arguments by E. P. Wigner . . . ," Targ goes on, "we may hypothesize that increasing the complexity of the observation system for an event makes the event increasingly sensitive to 'observer' effects. Therefore, we may have a situation wherein the more complex configuration for observing a subject's performance causes greater perturbation of his perceptual channel."

That, of course, is the voice of a physicist familiar with observer effects in quantum mechanics. It is not the voice of a psychologist. In this case the "observer" is not even a person. It is a computer in another room!

What is Targ to do? With an important experiment showing no results, it would be understandable if he sought a way to "rehabilitate" (his word) some of the good scorers in the two pilot phases. What better way to do this than to drop the "complex" controls of Phase 2 and go back to the absence of controls in Phase 0.

Eight subjects were used for Phase 3. Seven made no printouts of any kind; they were merely observed by an experimenter. Their results were at the chance level. The eighth subject, A3 in Phase 1, asked to use the printer and work without an observer. He was allowed to make as many practice runs as he liked. Indeed, he made 4,500 practice guesses as against 2,500 "real" guesses. His score was at the chance level on the practice runs. His "real" runs showed slight ESP and a moderate upward learning slope.

This partly rehabilitated subject is the only subject identified in the report. He is Duane Elgin, a policy-research analyst at SRI. The report closes with an appendix in which Elgin states his firm belief in ESP and discusses his reaction to the failure of Phase 2. The thing that disturbed him most, he writes, was his constant confusion over whether he was in a clairvoyant state of mind or a precognitive one. When he made a guess, was he guessing the picture just selected or was his mind aiming at the picture that would be chosen next? Targ makes much of this confusion, which he hopes can be minimized in future experiments. One wonders why Elgin did not also worry about the possibility that his PK powers might be causing the randomizer to select the picture he intended to choose next. (Such a possibility had not been overlooked by Targ. In 1972 he had announced that this PK hypothesis "will be the subject of a future investigation.") Elgin ends by stating his belief that the tests were valuable exercises for his "psychic muscles." He feels he is much better now in "other situations where I might use ESP abilities, in particular, telepathy, precognition, and clairvoyance."

The story of the failure of this expensive experiment is almost a paradigm of what has happened numerous other times in ESP research. High-scoring subjects are first identified by loosely controlled screening, then as their testing proceeds, under better (that is, more complex) controls, their psi powers mysteriously fade. In addition to Catch 22 and Catch 23, parapsychologists have a string of other good ones. Catch 24 says that, for reasons nobody understands, high-scoring subjects tend to lose their powers. Subject A3's ability did return. In spite of that, and regardless of Targ's desire to repeat the tests in some manner that would avoid the awful complexity of a computer's "observing" the trials from a distant location, NASA decided to provide no additional funds.

Postscript

My *Scientific American* article was understandably not well received by the true believers. Gertrude Schmeidler wrote the friendliest letter, mainly chiding me for not mentioning the use of machine recording by Helmut Schmidt and Charles Tart. Tart was miffed mainly because I hadn't mentioned his own work with machines, especially with his 10-choice trainer (see Chapters 18 and 30), and because I had called Puthoff a Scientologist.

"Does that mean," Tart asked, "we should be suspicious of any scientist who has been associated with views outside of scientific orthodoxy?

If so, we shall have to stop reading *Scientific American* and most scientific journals, for I have it on good authority that many of the authors in these publications are Christians, Jews, etc., and thus believers in all sorts of scientifically preposterous things!" Tart, by the way, is a devout Lutheran, but I have never held that against him. Scientology clearly is a different matter. Unlike traditional religions it demands a strong belief in psi powers and holds many scientific views that are absurd. To say that P is a Scientologist is not at all like saying Tart is a Lutheran; it is more like saying that a physicist is a member of a flat-earth society.

The anticipated letter from P and T appeared in *Scientific American,* January 1976:

Sirs:

We should like to take this opportunity to reply to the comments of Martin Gardner in his critique of our NASA research report.

The research pertains to the use of an automatic, solid-state machine that randomly selects from among four hidden targets while a subject tries to choose which target was selected. The machine provides immediate feedback as to the machine state, and rings a bell for correct subject responses, to allow him to try to use this feedback and reinforcement to improve his scores. Of the 147 volunteer subjects, six were identified whose learning performance was significant at the 0.01 level or better; the binomial probability of this occurring by chance is less than 0.004. At the other extreme, no subject had a negative learning slope of equal significance. In our report we took these preliminary findings to indicate that "there is evidence for paranormal functioning from our work with the ESP teaching machine." This evidence includes one subject who achieved scores at the $p < 10^{-6}$ level of significance in his 2,800 trials.

Gardner's major criticism of the experiments is based on an error in fact, namely his misconception of the manner in which data were collected. Subjects made runs of 25 trials. These trials were automatically printed on continuous fanfold paper tape, which carries a permanent record of every trial, machine state and trial number. After a series of 8 to 10 runs the subject would bring the continuous fanfold to one of the experimenters for entry in the experimental log. The tapes were always delivered to us intact with all runs recorded. They were never torn into "disconnected bits and pieces," as Gardner asserts (implying that an individual could *post hoc* select which runs he turned in). Since we were interested in evidence of learning within each day's session, it was of particular importance to us to have the complete intact tape.

Russell Targ
Harold Puthoff

The same issue ran a letter from Phyllis Cole, the third member of the research team:

Sirs:

As the scientist who actually conducted the research, I should like to respond to Martin Gardner's critique of ESP teaching-machine research. . . .

Gardner's attack on the research was based on a number of misconceptions. First, data were not collected from tapes torn at the whim of subject and/or experimenter into "disconnected bits and pieces." Fanfold paper data tapes were torn off only at the end of a daily session; each session record was delivered to me intact. Second, Phase 3 research with Mr. Elgin was conducted as for all other subjects in Phase 3: I sat beside the subject and recorded scores by reading the total number of hits per run of 25 trials from the digital display on the face of the machine. Third, prior to the initiation of each Phase 3 session Mr. Elgin specified how many "practice" runs of 25 trials he wished to do. In no instance was such a decision capriciously based on the outcome of a run. Mr. Elgin's average score over the 2,500 experimental trials in Phase 3 had a probability of 5×10^{-4}, hardly "slight ESP" as Gardner asserts. We consider this satisfactory replication of his earlier work with a printer as a recording device, in which $p < 10^{-6}$. . . .

Phyllis M. Cole

The final ellipsis in Cole's letter is for a paragraph in which she quotes a passage from my 1952 *Fads and Fallacies* often quoted by parapsychologists, in which I said that I shared "to a certain degree" the "irrational prejudice" of American psychologists against ESP. By this I meant nothing more than the kind of prejudice parapsychologists have for the reality of ESP. They think the evidence warrants belief, I think it does not. My use of the word "irrational" was a poor choice of adjective. I hereby withdraw it. I have a rational prejudice, just like P, T, and C.

My reply to P, T, and C followed their letters, but before it was written I had sent a letter to P, with copies to his superior, Earle Jones, and to Ron Deutsch, then head of SRI's public relations office:

Dear Dr. Puthoff:

I have just read with interest the letter that you and Dr. Targ sent to *Scientific American*. If I read it correctly, you are asserting that all the data of Phase 1 is based on complete, unbroken fanfold tape, and that sufficient precautions were taken to insure that all runs made by all 147 participants were turned in and incorporated in the published data.

I understand, of course, that this was the experimental design. But your description of how it worked out in actual practice is in sharp conflict with information I received from a source I consider completely reliable. It was this source that emboldened me to write, "I am not guessing when . . ."

The two descriptions of the state of the records are so conflicting that I would like to make the following request. If we hired a statistician in the area, acceptable to all of us, and asked him to examine the original tapes and send us a detailed report, could this be arranged?

Martin Gardner

This letter was never answered. My published reply to P, C, and T follows:

Sirs:

The account by Targ and Puthoff of how data were collected for Phase 1 of their ill-fated test of Targ's ESP teaching machine differs radically from information provided by a reliable source who had access to the original records. Accordingly I wrote to Dr. Puthoff on October 16, asking that he allow *Scientific American* to arrange for a statistician acceptable to all parties to inspect the original tapes. There has been no reply.

Even in cases where a subject turned in an unbroken tape there is no way to determine if other runs have been omitted, because the machine's printer does not count *runs*. It counts only guesses from 1 to 25 within a run. I have no doubt that Phase 1 was designed with the expectation that complete tapes of all runs would be turned in by all 147 participants. I do question that it worked out that way. I am told that some participants, annoyed by the printer's clatter, stopped using it and turned in handwritten records. Until the original tapes can be checked by an outside expert I stand by my statement that they are in "bits and pieces."

Ms. Cole corrects an important error. I had said that in Phase 3, when Duane Elgin was "rehabilitated," his scores were printer-recorded. The facts are: Elgin scored significantly in Phase 1, when the printer was used. In Phase 2, when the computer provided safeguards against unconscious bias, his score dropped to chance. In Phase 3 no printed records of *any kind* were made. Ms. Cole sat beside him and kept a handwritten tally of his 7,000 guesses, of which 4,500 were decided (in advance) to be practice runs that were not counted.

We have here a classic instance of one firm believer being observed by another firm believer, both with a strong vested interest in a favorable outcome. The test was not even single-blind. None of the three investigators has made any comment on the central point of my article: that the results of Phase 2, the only phase with adequate safeguards, were entirely negative.

Martin Gardner

Other interesting letters were received. Dale Ann Kagan discovered a curious anomaly in the illustration showing a run of 25 choices that I had reproduced from the official SRI report (see p. 80). Reading from bottom up, note that the machine's third target, 2, matches the subject's choice two steps later. Continuing upward, every fifth choice by the machine matches the subject's choice two steps later. Five such hits on what parapsychologists call − 2 targets are not significant in a run of 25 choices, but spaced at equal intervals? Was the subject following a guessing strategy or is this pure coincidence? I believe it to be the latter, but it does show how easy it is to find striking patterns in *post hoc* analysis of ESP data.

Richard B. Hoppe, a psychologist at Kenyon College, reminded me that the *p*-value, so often cited by parapsychologists, is not an indication of the strength of the relationship being tested. This value depends on many variables, of which only one is the difference between the expected score and the actual score. "The number of trials, which in ESP work is generally huge by psychological standards, is an important determinant of the *p*-value." When Cole denies that Elgin's score in his final test showed "slight ESP," she supports her claim with a *p*-value based on 7,000 guesses, of which 2,500 were deemed (in advance) to be nonpractice runs by Elgin and Cole. No third observer was present to make sure that 4,500 runs were declared in advance to be practice, a marked flaw in the experimental design, but we will let this pass, trusting in the honesty of both participants. But even with 2,500 legitimate guesses, only an occasional recording error could run the *p*-value to the asserted height. If there were such errors, the *p*-value would be no more than a measure of the experiment's bias.

Both Tart and Schmeidler lectured me by mail on the method by which parapsychologists routinely exclude recording errors by the simple expedient of not allowing the recorder to see the target selections. In this way, recording errors are random and have little effect on the overall results. Schmeidler went so far as to add: "I doubt that either of the two major American parapsychology journals would accept any research where this rule wasn't followed."

Apparently both Tart and Schmeidler were misled by my mistake in saying that Elgin's final test was automatically recorded. Elgin, too, wrote to protest this error, without realizing that had I not made it the implication was far more damaging to P, T, and C. Picture the situation. Phase 2 of the experiment, when controls were adequate, produced no results. P, T, and C desperately needed to justify NASA's funding. Their "rehabilitation" of Elgin consisted of allowing him to make 2,500 guesses, all hand recorded by C, one of the researchers, sitting beside Elgin and observing both targets and guesses! The least P and T could have done would have been to have the recording done by an outsider with no emotional interest in the outcome.

I do not know what Elgin is currently up to, but in its December 27, 1976, issue *New York Magazine* printed a sensational interview with him by William K. Stukey, entitled "Psychic Power: The Next Super-weapon?" Elgin was then a 33-year-old "futurologist" at SRI. The article deals mainly with a lengthy report by Elgin in which he predicts that in the 1990s a major civil war may develop between the nonpsychic majority in the military-industrial complex and a small band of psychic radicals who will destroy computers, weapon systems, and communication networks, perhaps even the minds of their opponents, by using their mighty PK powers. The only hope, he believes, is for Americans to accept the reality of psi and use it for peaceful purposes and for curing their medical ills. Elgin firmly believes in his own psi powers. Readers are urged to look up this incredible interview and also to read Elgin's "Powers of Mind: The Promise and the Threat," in *New Realities,* vol. 1, no. 1, 1977.

In their book *Mind-Reach,* reviewed in Chapter 30, P and T devote a chapter to the "Loyal Opposition" in which I am heavily castigated for my *Scientific American* piece. Their section on this (pages 180–181) is a masterpiece of deceptive writing. They vigorously deny that their NASA-funded experiment was a failure, asserting that Elgin achieved scores at odds of better than a million to one! There is not a word about the test's total failure during the only phase with adequate controls; not a word about the shameless manner in which Elgin's "rehabilitation" was achieved.

P and T wrongly assert that I accused subjects of fraud when I clearly pointed out how easily unconscious bias could operate. They reprint their letter to *Scientific American,* with no mention of my reply or my earlier request for the opportunity to have an outsider examine the original tapes. Until those tapes are inspected by someone I respect, I cannot accept the claim that the tapes support the data of Phase 1. Even if they do, and my informant is wrong, it is only the failure of Phase 2 that matters and that remains the core of this unfortunate experiment.

8

Magic and Paraphysics

The purpose of conjuring, at least most of the time, is to entertain audiences by pretending to violate natural laws. In a curious way this has something in common with how the universe behaves. When a person is mystified by a good magic trick it is because he can't figure out how the magician did it. When a physicist is mystified by an unexpected observation it is because he can't figure out how the universe did it.

The big difference, of course, is that the universe plays fair. Its tricks may operate by principles of incredible subtlety, and we may never discover all of them, but it keeps performing its illusions over and over again, always by the same methods. Or so it seems. If a scientist tries to discover one of the methods, the universe, so far as anyone can tell, doesn't go out of its way to flimflam him. "God may be subtle but He is not malicious," Einstein is often quoted as having said. Or, as he put it in a letter, "Nature hides her secrets through her intrinsic grandeur but not through deception."

The magician, by contrast, is a consummate liar. His principles, borrowed in part from physics and psychology (but mostly they are *sui generis*), are soaked through and through with deliberate falsification of the most reprehensible sort. It is not so much what a magician says as what he implies. He will show the queen of hearts, turn it face down on top of the deck, and apparently deal it to the table. He may even say, "And we'll place the queen over here," knowing full well that the card he is putting

This article, which appeared in *Technology Review*, June 1976, and the letters quoted in the Postscript are reprinted with permission of the publisher.

there is no longer the queen. But most of the time it is what the magician does, not what he says, that is deceptive. He may tap an object to prove it solid when only the spot he taps is solid. He may casually show the palm of his hand to prove he has nothing concealed when something is on the back of his hand.

Any magician will tell you that scientists are the easiest persons in the world to fool. It is not hard to understand why. In their laboratories the equipment is just what it seems. There are no hidden mirrors or secret compartments or concealed magnets. If an assistant puts chemical *A* in a beaker he doesn't (usually) surreptitiously switch it for chemical *B*. The thinking of a scientist is rational, based on a lifetime of experience with a rational world. But the methods of magic are irrational and totally outside a scientist's experience.

The general public has never understood this. Most people assume that if a man has a brilliant mind he is qualified to detect fraud. This is untrue. Unless he has been thoroughly trained in the underground art of magic, and knows its peculiar principles, he is easier to deceive than a child.

Some physicists also have not understood this. In the late nineteenth and early twentieth centuries, a number of prominent scientists (Oliver Lodge, William Crookes, John Rayleigh, Charles Richet, Alfred R. Wallace, and others) were firmly persuaded that mediums, aided by discarnate "controls," could levitate tables, materialize objects, and call up audible and even photographable spirits from the vasty deep. An Austrian astrophysicist, Johann Zöllner, wrote a book called *Transcendental Physics* about an American medium, Henry Slade, who specialized in producing insipid chalked messages from the dead on slates, and knots in closed loops of cord.[1] Zöllner believed that Slade could move the cord in and out of four-space. It was as impossible for anyone to convince Zöllner that so charming a man as Slade could be a magician as it was impossible for Houdini to persuade Conan Doyle that he (Houdini) did not perform his escapes by dematerializing his body.

And now the wretched story is happening all over again, with Uri Geller in the center of a cyclone of irrationalism that is churning over the Western world. Geller is a young, personable Israeli who began his spectacular career by performing what magicians call a "mental act" in Israeli night spots. An American parapsychologist, Andrija Puharich, discovered him, introduced him to Edgar Mitchell, the astronaut who once walked the moon and who now runs his own organization devoted to investigating the paranormal. Mitchell financed Geller's trip to the United States and arranged for him to be tested at the Stanford Research Institute by Harold Puthoff and Russell Targ, two former laser physicists now engaged in full-time psychic research. After a series of poorly designed experiments with Geller, Puthoff and Targ published their favorable findings in *Nature*.[2]

Although Puthoff and Targ are personally convinced of Geller's ability to bend metal by PK (psychokinesis) and to perform even more remarkable miracles, their *Nature* report was limited to Geller's power of ESP (extrasensory perception). His most sensational feat was guessing correctly, eight times in a row, the number on a die that had been shaken in a metal file box by "one of the experimenters." It later turned out that Geller had been allowed to handle the box, and that many prior trial runs had been made. Because the experimenters always shook the box before Geller was permitted to touch it, Geller's handling seemed irrelevant, so it was not mentioned in the *Nature* report. This seemingly trivial detail gave Geller a splendid chance to obtain information by a technique known to conjurors.[3] Had Puthoff and Targ been aware of this technique it would have been easy to take steps to preclude it. The fact that they did not makes the dice experiment worthless.

Puthoff and Targ are prominent among a small group of well-trained physicists, some with doctorates, who like to call themselves "paraphysicists." Most of them are active in psychic research and in publishing popular books and articles. All are convinced that psi phenomena have been firmly demonstrated by parapsychologists. That the overwhelming majority of psychologists deny this is dismissed as the stubborn prejudice of Establishment science against what Thomas Kuhn calls a new "paradigm." Paraphysicists look upon themselves as in the vanguard of a new scientific revolution that will be more shattering of old paradigms than the Copernican revolution. After all, they say, did not the establishment persecute Galileo?

In England the best known of this new breed of physicist is John Taylor, a mathematical physicist at King's College, London. When the *New Scientist* conducted a poll of its readers in 1975 they found that Taylor was regarded as among the world's top 20 scientists! The reasons for this esteem are Taylor's frequent appearances on BBC television, his popular books (including one on black holes), and his loud espousal of the "Geller effect." His latest book, *Superminds,* not only argues that Geller can bend spoons, keys, and metal bars by the power of his mind, but that hundreds of British superkids, teen-age and younger, can do the same thing.[4]

Oddly, Taylor never actually *sees* anything bend, nor has he been able to capture the actual bending on video. He calls this the "shyness effect." Bending usually occurs only when nobody is looking. He has given his children crudely sealed tubes with a straight metal bar inside. They take the tubes home and come back with the bar bent. For some reason, which Taylor is unable to fathom, the children are successful only when the tubes are inadequately sealed.

At Bath University, two psychologists designed a simple test for six young spoonbenders. The observer was told to relax his vigilance after

20 minutes. Rods and spoons Gellerized all over the place while the unsuspecting children were being secretly videotaped through one-way mirrors. In every case where something bent, the children were seen doing the bending by "palpably normal means." One little girl had to put a rod under her feet to bend it. Others held a spoon below a table and used two hands.[5] Taylor had not thought it worth-while to design such a test because he had already decided that all his children were honest.

Taylor is not sure what the mysterious force is that produces the Geller effect. In *Superminds* he considers many possibilities—gravity, the weak force, neutrinos, tachyons, intermediate bosons, magnetic monopoles, quarks. Some of these have been proposed by other paraphysicists as the source of psi phenomena. Taylor finally opts for electromagnetism. This is scoffed at by parapsychologists because, following J. B. Rhine, they believe the psi force to be unknown to science. The possibility that the Geller effect may be caused by deception is, of course, ruled out by Taylor on the grounds that he personally witnessed it.

A notion of how gullible physicists can be if they have a strong compulsion to believe in paranormal events can be gained by considering a dramatic occasion at Birkbeck College, London, on June 21, 1974. Uri Geller was demonstrating his powers for a small group of physicists. The most distinguished man present was David Bohm, a world-renowned expert on quantum mechanics. Also present were paraphysicists John Hasted, Keith Birkinshaw, Ted Bastin, and Jack Sarfatt (who has since restored his family name, Sarfatti), and psychic researcher Brendan O'Regan, who had arranged the demonstration.

Geller's outstanding achievement was producing a "very strong burst from a Geiger counter tube that he held in his hand. The creation of the burst happened almost simultaneously with Geller's expressed intention to create it. . . . The creation of the burst was correlated with strong breathing and signs of great physical exertion on Geller's part." I quote from a stirring press release sent out by Sarfatti.[6] Geller repeated the Geiger counter bit on the following day for the writer Arthur Koestler and others. "Koestler reported a strong sensation simultaneous with the Geiger tube burst," says Sarfatti, and was "visibly shaken for several minutes." Science-fiction writer Arthur C. Clarke, also there, said it was time for the magicians to "put up or shut up."

"My personal professional judgment as a Ph.D. physicist," Sarfatti concludes, "is that Geller demonstrated genuine psychoenergetic ability at Birkbeck, which is beyond the doubt of any reasonable man, under relatively well controlled and repeatable experimental conditions."[7]

Note the clear implication in Sarfatti's release that having his doctorate in physics made him specially qualified to rule out deception. Well, how would a magician with no Ph.D., but with a knowledge not possessed by Sarfatti, have reacted had he been present? Although I am not much of a performer, magic has been my principal hobby for fifty years. When I read Sarfatti's account, the first thing that occurred to me was that Geller could have had a piece of harmless radioactive substance concealed on his person. While twisting about in simulated physical stress he could have simply brought the tube close to his beta source. It could have been in the tip of a shoe, above his knee, in his mouth, behind an ear, under a collar, taped to his chest, in back of his belt. It is not hard to obtain a beta source. A luminous watch dial produces excellent crackling in a Geiger counter. When Philip Morrison once asked Sarfatti if anyone had examined Geller for a beta source, Sarfatti replied that no one had thought of such a possibility, and that it was an "ingenious idea." Magicians find this response hilarious.

Was it a "repeatable" experiment as Sarfatti's release states? Perhaps repeatable in front of Ph.D. paraphysicists, but not in front of knowledgeable magicians.[8] Indeed, Geller's methods are both old-fashioned and well known. The interested reader can learn most of them by reading the references cited in notes 2 and 3.

The publication of Geller's methods has had, so far, little effect on the mind-sets of paraphysicists. Their reaction is exactly the same as the reaction of their counterparts around 1900 when confronted with obvious fraud by a physical medium. First, they say, the fact that a magician can duplicate a psychic event does not mean the psychic does it that way. Second, even if he *does* sometimes do it that way it doesn't mean that he does it that way *all the time*. Every paraphysicist now concedes that Geller occasionally cheats. After all, is not the poor lad under terrible pressure to produce results, especially on television? How can you blame him for using a little prestidigitation when the psi power is not available? If Geller is caught cheating, as he has been many times, so what? *Then* he cheated. But when nobody catches him, what he does is genuine. This is one of those old heads-I-win-tails-you-lose arguments which paraphysicists seem to find satisfying.

At the moment, Sarfatti is director of what he calls the Physics/Consciousness Research Group, a tax-exempt, nonprofit organization in San Francisco. The group is funded by Werner Erhard, a former Scientologist (the church expelled him in 1971), now running a movement of his own called est (Erhard Seminars Training), designed to raise the consciousness of anyone willing to pay for its bizarre processing.[9]

Sarfatti is a leading theoretician among U.S. paraphysicists. According to his autobiography, written in 1975 for Ken Kesey's new magazine, *Spit in the Ocean,* he was born in Brooklyn in 1939 and obtained his doctorate at San Diego State College. About ten years ago he began to have "odd subliminal experiences" which he attributes to "communication with other modes of consciousness." He found he could practice "automatic writing" (writing without conscious control of the hand) and several of his published scientific papers, he says, were written that way.

He discovered he had "a kind of collective mind experience" with his associate and "comrade in psychic exploration," physicist Fred Wolf. As his consciousness continued to soar, he was greatly aided by the est processing. In a 1975 letter to the magician James Randi, he says that he has "serious information indicating a high probability that extraterrestrial contacts are being made." This information comes "from sober and scientifically trained people with no axe to grind," such as J. Allen Hynek, the astronomer who is now so active in UFO research. Sarfatti and Wolf are co-authors with Bob Toben of a wild paperback, *Space-Time and Beyond* (Dutton, 1975). The text plus the cartoons will tell you all about the theoretical underpinnings of such things as time travel, levitation, precognition, and the ability to bend a spoon.

Sarfatti's theory of psi stems from David Bohm's attempt to escape from the terrible discontinuities of quantum mechanics. In quantum theory one event seems to influence another without any transmission of energy through space or time. These bedrock discontinuities are utterly unlike those that occur on the macroscopic level. Rotate a beam of light and a spot will move across a wall, over an edge, then leap quickly to a more distant wall. But this leap is easily explained by considering the beam's source. Discontinuities on the quantum level cannot be explained this way. An event happens at point A, another event at point B, but nothing propagates in between. We can do no better than "explain" the event by invoking statistical laws. Conventional quantum theory leaves no room for "hidden variables"—something else about nature, as yet unknown to us, that would restore causality.

Einstein did not like this vision of "God playing dice" (as he called it) and Bohm, although he posits no deity, is equally disturbed. In Bohm's vision the universe has infinite levels of structure, like a vast sea without a bottom. On each level there are discontinuities that vanish when we perceive the patterns on the next level below. Beneath the entities of quantum mechanics are subquantum entities we do not yet understand. The entities we know are like tips of icebergs. We relate them by statistical laws only because we do not yet see how they are causally related on a deeper level. When we finally comprehend the deeper level it, too, will have its discontinuities, magic violations of cause and effect, that can be explained only by going still deeper. For all we know, these levels may go on forever. Our poor quantum mechanics is no more than an upper layer of what Milton called the world's "dark unbottom'd infinite Abyss."

Sarfatti elaborates. If we could but see the universe in what Bohm has called its "unbroken wholeness" we might see that every particle is connected with every other particle, every event with every other event, no matter how far away in space or time. Shake your finger, you shake the universe. Wiggle something here, everything wiggles. Where Sarfatti departs from other physicists is in the language he chooses for describing the sub-quantum level. Sarfatti calls it "consciousness." He is thus in the classic tradition of philosophical idealism. Behind the crazy, paradoxical world of everyday experience, behind the even crazier world of microphysics, is Mind.

Into this ancient vision Sarfatti stirs several modern ingredients: the "Einstein-Rosen-Podolsky" paradox, John Wheeler's wormholes and superspace, Hugh Everett's "many worlds" interpretation of quantum mechanics, and especially the implications of an argument advanced by John S. Bell.[10] Bell has shown that no *local* hidden variable theory (that is, a set of equations to describe the properties at a point in space and time) is consistent with quantum mechanics. This leaves open, however, the possibility that a nonlocal hidden variable theory—one that applies to the entire universe—may be consistent with quantum mechanics. There is not the slightest evidence for such a theory, but the logical possibility of such a theory—one in which God does *not* play dice—allows Sarfatti and others to posit it.

Think of the world as an immense, intricate puppet show. Everything has a "string" attached to it, and the strings are all held by the Great Puppeteer. It looks as though puppet *A* throws a particle to puppet *B,* but this is an illusion. The Puppeteer moves *A*'s arm, then carries the particle over to *B* and moves *B*'s arm to catch it. No matter how random and acausal events seem on the microlevel, causality is restored by positing the Great Puppeteer. Jung called it "synchronicity." Leibniz called it "pre-established harmony." Whatever you call the Puppeteer—God, Being, the Tao, Brahmin, the Absolute—its infinite strings provide a connectedness that permits the transfer of information at instantaneous speeds through space and time.

Such a transfer must not be thought of as violating relativity's dictum that signals cannot go faster than light. There are no instant signals in the sense that energy is transferred. Nothing "moves." No time "elapses." It is what Sarfatti calls "instant superluminal transfer" of information by means of "hyperdynamical connection." In our metaphor, a puppet tugs a string here, the Great Puppeteer instantly tugs a string there.

The concept is simple, but the paraphysicist makes it sound scientific by hoking it up with technical jargon. The "measure of information," according to Sarfatti, is "the degree of order in the energy already existing at a particular place. This kind of information is coded directly to the superluminal de Broglie quantum matter waves." It is Sarfatti's persuasion that the human mind has natural detectors of de Broglie waves on the quantum molecular level. "The introduction of this kind of direct quantum information into waking consciousness often appears as 'paranormal' or 'psychic.' Certain kinds of altered states of consciousness . . . seem to facilitate awareness of direct quantum information. . . ."[11]

There is more. Not only can the psychic pick up information instantly from any part of the universe, he can also transmit it instantly. He simply uses his PK powers to wiggle a wave function here—possibly by altering the spin-state of a quantum system—and presumably wiggling it with some kind of code. Since this wiggles everything, the receiving psychic can pick it up instantly. If an establishment physicist doesn't buy this, Sarfatti and his friends consider him hopelessly mired in "electromagnetic chauvinism." (Poor John Taylor. Even though a brother paraphysicist, he, too, is mired in electromagnetic chauvinism.)

Here, at long last, is a truly marvelous theory for explaining all the mind-blowing, spoon-bending wonders of psi. Information can be conveyed instantly into the skull of anyone, especially if he is psychic, from any part of the universe, from anywhere in the past, present or future. Telepathy, clairvoyance, psychokinesis, poltergeists, precognition, psychic healing, out-of-body experiences, and so on, no longer suffer from the lack of a physical theory. Our consciousness, according to Sarfatti and others, can instantly perceive and influence any part of the universe. It can leave the body and roam, faster than a photon, through endless reaches of space and time. If a superintelligence in some distant galaxy wishes to communicate with Uri Geller and give him the power to bend a spoon, there is no reason why it can't. Indeed, this is just what Puharich claims is the source of Geller's power.[12] Geller himself has validated this in TV talk-shows and in an autobiography ghosted by the journalist John G. Fuller.[13]

This is not the place to go into more details about Sarfatti's great theory of "superluminal information transfer." I wish to consider a much humbler matter. Before the paraphysicist develops elaborate theories to explain how Geller can bend a spoon, would it not be wise to make sure first that Geller actually *can* bend a spoon? By PK, that is. Now I do not wish to get into trouble with magician friends by exposing methods used by honest charlatans, but perhaps they will forgive me if I consider in detail Uri's most publicized feat. How does Geller bend a car key?

First, it is important to understand that there is no single method. There are dozens of ways to bend car keys, some of them developed by magicians after Geller made the trick popular, and which Geller never uses because they are too complicated and not adaptable to his casual, impromptu brand of magic. But Geller himself has many ways of bending keys, depending on the circumstances. If he is performing for one person, say a reporter or a Gellerite who has asked for a private demonstration, he will do it one way. If he is in front of a large audience he adopts other procedures. The method he uses depends on who is watching, how many are watching, and how closely they are watching. If he suspects a magician is watching, he won't bend the key at all.

Here is a typical scenario based on the observation of many friends, some of them magicians whom Geller did not know were present and who actually saw his exact "moves." Let's assume Geller is in an office with a group of scientists gathered to witness his awesome powers. Some of them believe Geller has those powers. Others are skeptical but curious. None knows much about magic.

In performing for such an audience Geller has one overwhelming psychological advantage over every magician: he comes on as a psychic. A magician is expected to perform his miracles rapidly and cleanly, without fail, while everyone watches like a hawk to catch the trickery. No magician, when he gets up to perform, dares say, "I'm sorry, ladies and gentlemen. I intended to show you my great trick of floating a burning light bulb across the room, but unfortunately I don't feel like doing it. There are skeptics in the audience. The vibes are unfavorable."

The psychic, on the other hand, is under no obligation to do anything, and Geller plays this rôle with superb skill. He begins by saying that he is very nervous, being in such distinguished company, and he doesn't know whether anything will happen or not. All he can do is try. Things are more likely to happen, he says, if everyone wants them to happen. The power he has is not peculiar to himself. Everybody has it. So—if everyone will try their best to make things happen, maybe they will. But don't be disappointed if they don't.

This little speech has the effect of discouraging skeptics from voicing doubts. It also gets Geller off the hook if he finds that conditions do not permit him to do much. Most of all it allows him plenty of time to perform the most trivial of tricks. No magician could possibly get away with taking half an hour to make a key bend, but this often happens with Geller. He will borrow a car key, stroke it, nothing happens. He will put it aside and try later. Again nothing happens. Perhaps on the third or fourth try it will bend.

The reason for this delay is that Geller cannot bend the key until he obtains strong enough misdirection to bend it secretly. The secret bending takes only an instant. Most car keys bend easily, especially if they are long and have a low-cut notch. Geller prides himself on his strength (he works out with bar bells, Puharich tells us). If you have strong fingers you can bend most car keys simply by resting the key crosswise on the fingers and pressing firmly with the thumb. Stronger keys require pressing the tip against the side of a table, the table leg, the side of a chair, or whatever firm surface is handiest. In any case, the bending can be done in a split second. Of course it must be done at a moment when no one is looking.

To obtain the necessary misdirection Geller creates a maximum amount of chaos by moving around the room and going quickly from one experiment to another. Here are some of the ploys he has used to get the needed misdirection.

1. Geller has tried twice to bend a key but without success. He tries a third time, letting someone hold the base of the key while he gently strokes

it with a finger. Again nothing happens. Geller acts disappointed. Everyone is disappointed. He starts to put the key aside once more. No one is paying much attention because the trick has failed. At that instant someone in the room makes a funny remark. Everyone turns toward him and laughs. It is the moment Geller has been waiting for. His hand drops to the side of the chair while he himself is laughing. Who except a trained magician would be watching his hand at that instant? Geller immediately puts the key aside, carefully placing it in a spot where it 'is partly concealed so that no one can see the bend. He may not try the key again for another ten minutes.

2. Geller is performing for one person. Both are seated in chairs. The key fails to bend. Perhaps, Geller says, they are sitting too far apart. To move his chair closer, Geller's hands drop to the sides of the chair. As he moves the chair, the tip of the key is pressed against the chair's leg.

3. Geller is in his own apartment entertaining a guest. He sits on a sofa behind his glass-topped coffee table. There seems to be nothing near him he can use for a pressure bend. Who would guess that the thick glass of the coffee table will serve admirably? As soon as a bit of misdirection occurs, and the spectator's attention is diverted, the key is bent against the edge of the glass.

4. Geller is entertaining a group of people in an office. They are watching too closely for him to obtain the misdirection he needs. Geller is apologetic. Sometimes it helps the metal bend, he says, if there is a lot of metal nearby. He points across the room and asks, "Is that a metal file cabinet?" If he is in a living room he points to a radiator. Every head turns. In that instant his hand lowers and puts in the work. If the key is weak he bends it in his hand. All he has to do now is hold the key at one end, concealing the bend, walk to the file cabinet, let someone hold an end of the key, then miraculously bend it.

5. On many occasions Geller finds it necessary to leave the room to obtain strong misdirection. In 1974 when he was performing for a group of people in Ottawa, a friend of mine in the audience told me that, after many failures to bend a key, Geller asked if there was an elevator in the hallway. The large amounts of metal in the shaft, he said, might help. Geller then dashed into the hallway, his spectators trailing. Sure enough, in front of the elevator door the key bent.

6. Another one of Geller's favorite excuses for leaving the room is to say that running water helps a key bend. He actually used this preposterous excuse on the paraphysicists at Birkbeck College. Let me quote the relevant passage from Sarfatti's ecstatic press release:

"Geller then succeeded in bending several pieces of metal by psychoenergetic action. These objects included the blade of a knife and a key belonging to Bohm. The flow of water from a tap onto the metal seemed to make the bending occur more easily."

To a magician this means that the paraphysicists had been watching too closely. Geller suggested flowing water. Everybody moved to a spot where the key could be held under a tap. In the process of getting there, Geller obtained the needed misdirection. He could have bent the key in his hand, against the side of a doorway as he passed through, or in a dozen other ways. The point is: no one is watching on the way to the sink.

It is important to realize that Geller puts the bend in the key before, sometimes long before, he pretends that the actual bending takes place. Let's suppose he finds a chance to bend the key after a second failure. The key has been put aside, but behind something or partly under something so the bend is not visible. Ten minutes later, when he picks up the key again, he holds it so that only half the key projects from his fingers. Because the visible half is straight, everyone assumes the entire key is straight. Sometimes he rubs the bent key back and forth across a table top. The action and sound strengthen the impression that the key is flat. The key is then given to someone to hold at one end while Geller's fingers surround the bend.

While Geller is gently massaging the key he usually asks if the key is beginning to feel warmer. Since it is being handled, it *is* getting warmer, but most people respond readily to suggestion and imagine that the key is feeling much warmer. Geller continues to rub the key. Slowly he lowers his fingers and allows the bend to come into view. It really looks as if the key is bending at that moment, especially if that is what you are convinced is happening. Geller is a master at creating this illusion. He will see to it that the flat side of the key is toward the audience as he lowers his fingers. Then he will twist the key gradually to bring the bend slowly into view. At the same time he will shout excitedly, "Look! It's starting to bend!" All this combines to create a strong illusion. Many people will swear later that they saw the key slowly bend, the way a match bends while it is burning.

Sometimes Geller will hand a key, already bent after an earlier "failure," to a spectator. If no one else is present, Geller may ask him to hold the key by its tip but above his head where he can't see it is already bent. Geller will then announce that he is going to attempt something he seldom does. He is going to move across the room, ten feet away, and try to bend the key without even touching it.

The person holding the key naturally assumes it is unbent. Geller, ten feet away, tries hard to make the key bend. He walks forward, examines the bent key, and acts tremendously disappointed. Nothing has happened! Before the spectator has looked at the key—why should he examine it since the key clearly failed to bend?—Geller is anxious to try once more. The key goes back in the person's hand, the hand is raised. Geller moves twenty feet away. Now he feels the power surging through him! He breathes heavily and seems to be undergoing considerable stress.

Yes—he *knows* the key is bending! "Do you feel it bending?" If the spectator is suggestible he imagines he does. Geller tells him to look at the key. *Mirabile dictu!* It is bent 30 degrees! To his dying day the spectator will insist that Geller was 20 feet away when he caused the key to bend. Moreover, he will insist, Geller *never touched the key.* Over and over again, reporters whose keys Geller has bent in this way have written that Geller bent their keys without touching them. What they mean is that Geller was not touching the keys at the times they assumed they were bending. The fact that Geller handled the keys many times before the great miracles occurred seems totally irrelevant. Indeed, they forget this entirely.

These are a few of the dodges Geller uses for just one of his little miracles. I have not mentioned all of his key-bending techniques. For example, I have said nothing about how Geller can be secretly aided by his friend, Shipi Shtrang, who is often with him, sometimes disguised as one of the innocent spectators. And Geller has other close friends who occasionally "stooge" for him. The use of stooges, a term magicians use for secret assistants, is a branch of magic in itself. And I have omitted other methods, not using stooges, because they are being employed by magician friends who are now more skillful at key bending than Geller.

Magicians are, of course, under the enormous disadvantage of being known as magicians. As a result, they are expected to bend a key under conditions far more stringent than those demanded for Geller. A Gellerite will approach the Amazing Randi with his fist closed over a key. "You claim you can do anything Geller does," he will say. "Okay, I have my car key inside my first. Now let's see you make it bend."

"May I inspect the key?" Randi asks.

"You may *not,*" says the angry Gellerite. "The key is in my fist. Make it bend without touching it. That's exactly what Geller did when he bent my key."

What can Randi do? He may feebly protest that those were *not* the conditions under which Geller bent the key, but what Gellerite is going to believe him?

There are two traits that characterize the true Gellerite. He is a person of enormous gullibility, the gullibility strengthened by an enormous compulsion to believe. And he is a person incapable of recognizing the absurd. (If he is a physicist, I would add a third trait: the egotism of believing he is competent to detect fraud.) In the heyday of spiritualism, the physicists who were convinced that spirits could float trumpets through the air were not in the least amused by the singular fact that spirits could do this only in total darkness. Why should the dear departed operate only in the dark? To a magician the answer is obvious.

If Geller has the power to bend metals, why is it necessary to bend them only under the conditions of a magic performance? If Geller possesses paranormal powers, why do they manifest themselves in such picayune ways as bending a spoon? If Geller can bend a metal bar by PK, why can't he straighten it again?

The feats performed by Geller's chief rivals are even funnier. Ted Serios, a Chicago bellhop, persuaded such leading parapsychologists as Thelma Moss, Charles Tart, Gertrude Schmeidler, William Cox, and Jule Eisenbud that he could cause his memory of old photographs to register on Polaroid film merely by looking into the lens through a roll of paper that he held in front of the lens. When two magicians explained how easily Serios could have faked it, Serios lost his power and faded from the psi scene.[14] Yet not one of the parapsychologists mentioned above has altered his or her opinion of the genuineness of Serios's power.

The leading PK performer in Russia is Nina Kulagina, who makes objects move across a table and ping-pong balls float in space. American magicians, who have seen her only on film, are enormously unimpressed.[15] Dean Kraft, a Brooklyn boy, made a stir two years ago when he was written up in the *Village Voice*.[16] His specialty was making a pen follow him across a rug and pieces of candy hop out of a bowl. He soon gave this up (too much of a strain, he says) to become a psychic healer. Curing the sick seems to be much easier for him than moving pens. He now treats, he claims, about 30 patients a day. His healing is all "free," but patients donate to a fund set up for him by Judy Skutch, one of the early financial backers of Geller.

Another recent PK wonder in the United States is Felicia Parise, a former medical assistant at Maimonides Hospital in Brooklyn. After seeing a film about Kulagina, Parise discovered that she, too, could make corks, tinfoil, and bottles move and roll across a table. She was discovered by Charles Honorton, then director of parapsychology research at the Maimonides Medical Center. According to Honorton, Parise put a small plastic bottle on the formica top of her kitchen counter, concentrated, and the bottle moved two inches away from her. "I then examined the counter," this distinguished parapsychologist told a reporter. "I virtually took it apart to ensure that there was no mechanical aid she could have used. . . . But I found none." Has Parise discovered the invisible nylon threads magicians sometimes use? Who knows? Honorton made no effort to have her observed by a qualified magician.[17]

"It seems to me," wrote Conan Doyle in *The Coming of the Fairies,* "that with fuller knowledge and with fresh means of vision, these people are destined to become just as solid and real as the Eskimos." By "people" Doyle meant tiny creatures with filmy wings that two girls had photographed in the woods of Yorkshire.[18]

Substitute "psi energies" for "people," and "laws of present-day physics" for "Eskimos," and you have the heart of what the paraphysicists are trying to tell us.[19]

NOTES

1. Zöllner's book, must reading for all Uri Geller watchers, was first published in Germany in 1879. An English translation by C. C. Massey (1880) had many British and U.S. editions. Zöllner's investigations of Slade were assisted and endorsed by physicists William E. Weber and Gustave Fechner, and mathematician W. Scheibner. Alfred Wallace and Lord Rayleigh were firmly convinced of Slade's powers. For a defense of Slade, see Conan Doyle, *History of Spiritualism* (1926). For Slade's methods, consult the 1887 report of the Seybert Commission, which caught Slade in outright fraud; J. W. Truesdell, *Bottom Facts of Spiritualism* (1883); Walter Prince, "A Survey of American Slate-Writing Mediumship," in Section 2, *Proceedings of the American Society for Psychical Research, Inc.*, vol. 15 (1921); Harry Houdini, *A Magician Among the Spirits* (1924); and John Mulholland, *Beware Familiar Spirits* (1938).

2. "Information Transmission under Conditions of Sensory Shielding," by Russell Targ and Harold Puthoff, *Nature*, vol. 251, October 18, 1974, pp. 602–607. For mild criticism of this paper see the editorial on p. 559 of the same issue. For strong criticism, see "Uri Geller and Science," by Joseph Hanlon, *The New Scientist*, vol. 64, October 17, 1974, pp. 170–186, and the first chapter of *Mediums, Mystics and the Occult*, by Milbourne Christopher, T. Y. Crowell (1975). For still stronger criticism, see *The Magic of Uri Geller*, by James Randi, Ballantine Books (1975).

3. This technique was explained by Houdini in a rare pamphlet that Randi reprints in his book. It is also explained in *Confessions of a Psychic*, an anonymous booklet published by Karl Fulves (1975) for sale to the magic trade. The booklet pretends to be the secret diary of Geller's chief rival, a mythical Uriah Fuller, but it contains the most detailed explanations to date of Geller's methods.

4. *Superminds* was published in 1975, in England by Macmillan, in the U.S. by Viking. See also Taylor's article, "The Spoon Benders," in *Psychic*, vol. 6, December (1975) pp. 8–12, and my review of *Superminds* in the *New York Review of Books*, October 30, 1975, pp. 14–15.

5. See "Spoon Bending: An Experimental Approach," by Brian R. Pamplin and Harry Collins, *Nature*, vol. 257, September 4, 1975, p. 8.

6. Sarfatti's release was reprinted in psychic journals around the world, and in *Science News*, vol. 106, July 20, 1974, p. 46.

7. Since I wrote this article, Sarfatti had lunch with magician James Randi, who fractured a spoon and moved the hands of a watch in a way that Sarfatti found indistinguishable from his observations of Geller. This prompted Sarfatti to reverse his opinion and fire off another press release (dated November 19, 1975) which begins: "On the basis of further experience in the art of conjuring I wish to publicly retract my endorsement of Uri Geller's psycho-energetic authenticity." This release appeared as a letter in *Science News*, December 6, 1975, p. 355. "I do not think," Sarfatti writes, "that Geller can be of any serious interest to scientists who are currently investigating paraphysical phenomena." Sarfatti does not doubt that PK powers exist. He merely doubts now that Geller has them.

8. One must add "knowledgeable" because there are no doctorates in magic, and obviously anyone, no matter how puerile his magic background, can pose as an authority. Several self-styled experts on conjuring, considered eminently unknowledgeable by other magicians, have watched Geller perform and pronounced him genuine: notably William E. Cox, an associate of J. B. Rhine.

9. On Scientology, see *Cults of Unreason*, by Christopher Evans, Farrar, Straus & Giroux (1974). On est, see "The New Narcissism," by Peter Marin, *Harper's Magazine*, October, 1975; "We're Gonna Tear You Down and Put You Back Together," by Mark Brewer, *Psychology Today*, August 1975; "The Führer Over EST," by Jesse Kornbluth, *New Times*, March 19, 1976, pp. 36–52.

10. For an excellent semi-technical explanation of the EPR paradox and Bell's theorem, see "Quantum Theory and Reality" by Bernard d'Espagnat, in *Scientific American,* October 1979.

11. My quotations from Sarfatti are taken from releases distributed by his Physics/ Consciousness Research Group, and from his essay, "The Physical Roots of Consciousness," in *The Roots of Consciousness,* by Jeffrey Mishlove. Mishlove is a graduate student in philosophy at the University of California, Berkeley. His big, lavishly illustrated book (with color plates showing how psychic surgeons remove diseased body tissues without slicing the skin) was published by Random House in 1975. It is an incredible Mishlovemash of every crazy aspect of the current psi scene.

12. See *Uri,* by Andrija Puharich, Doubleday (1974).

13. *Uri Geller: My Story,* by Uri Geller, Praeger (1975). John G. Fuller, the actual author, should not be confused with Curtis Fuller, editor of the psychic pulp magazine, *Fate*; with Willard Fuller, the psychic dentist of Jacksonville, Florida (he fills cavities without touching the teeth); or Uriah Fuller, the legendary rival of Uri Geller. John G. Fuller is the author of many books, some of them on UFOs and on the occult. His book *Arigo: Surgeon of the Rusty Knife,* T. Y. Crowell (1974), is an all-out defense of a famous Brazilian medical quack whose surgical procedures were guided by instructions whispered in his left ear by a dead German doctor.

14. See "An Amazing Weekend with the Amazing Ted Serios," by Charles Reynolds and David B. Eisendrath, Jr., *Popular Photography,* October, 1967, pp. 81f.

15. Parapsychologists take Nina Kulagina very seriously. She began her career as one of several Russian ladies who claimed to be able to read *Pravda* with their fingertips. After a short term in prison (for black marketeering) she emerged as the Soviet Union's number one psychic. There is a good section about her in *Psychic Discoveries Behind the Iron Curtain,* by Sheila Ostrander and Lynn Shroeder, Prentice-Hall (1970). J. Gaither Pratt, a former associate of Rhine, is one of Mrs. Kulagina's strongest supporters. See *The Psychic Realm,* by Pratt and Naomi A. Hintze, Random House (1975).

16. "The Brooklyn Healer," by Brian Van der Horst, *Village Voice,* December 23, 1974; reprinted in *Cosmopolitan,* August 1975.

17. Honorton announced his discovery in his "Report on the Psychokinesis of Felicia Parise" at a 1973 convention of the Parapsychological Association, an affiliate (since 1969) of the American Association for the Advancement of Science. His claims were backed by Graham Watkins, then working for Rhine. Watkins reported on Parise's ability to make a compass needle move, and to fog unopened film near the compass—all, of course, under "strict" laboratory controls, no magicians present.

A film documenting Parise's powers was shown at the convention by physicist Edwin May. "Physics does not have any idea how these phenomena work," May told a reporter. "Now we're trying to find out where her power comes from." See "Amazing U.S. Woman Moves Objects with Mind Power," by Paul Bannister, *National Enquirer,* December 30, 1975, p. 4; and "Apparent Psychokinesis on Static Objects by a 'Gifted' Subject: A Laboratory Demonstration," by Graham K. Watkins and Anita M. Watkins, *Parapsychology Research 1973,* pp. 132-134.

18. Doyle's preposterous book was reprinted in 1972 by Samuel Weiser, Inc., Manhattan's leading occult bookstore.

19. Compare Doyle's statement with the following remark by Wilbur Franklin, professor of physics at Kent State University and one of the nation's leading Gellerites. "I'm convinced there is nothing mysterious about Geller's or any other psychic's feats. Once we understand the natural laws that govern such things, we will also understand psychic phenomena as clearly as we understand such natural laws as gravity." (*The Star,* December 30, 1975, p. 17.)

Postscript

The following letter from Harold Puthoff and Russell Targ was printed in *Technology Review,* October/November 1976:

> In Martin Gardner's article on "Magic and Paraphysics" some references were made to the work with Uri Geller at Stanford Research Institute. Unfortunately, Gardner's statements concerning what happened at S.R.I. and what we published are grossly in error. We therefore wish to inform your readers of the facts involved, all of which can be independently verified on the basis of information available in the public domain.
>
> To begin, Gardner states that "Although Puthoff and Targ are personally convinced of Geller's ability to bend metal by PK (psychokinesis) and to perform even more remarkable miracles, their *Nature* report was limited to Geller's power of ESP (extrasensory perception)."
>
> Gardner is wrong on both counts. We are in fact *not* convinced of Geller's ability to bend metal, and our negative findings *were* reported in the very *Nature* article to which Gardner refers (R. Targ and H. Puthoff, vol. 252, No. 5476, p. 604): "It has been widely reported that Geller has demonstrated the ability to bend metal by paranormal means. Although metal bending by Geller has been observed in our laboratory, we have not been able to combine such observations with adequately controlled experiments to obtain data sufficient to support the paranormal hypothesis." A more detailed statement is found in the S.R.I. film, *Experiments with Uri Geller,* the text of which was released accompanying a March 6, 1973, presentation at a Columbia University physics colloquium. With regard to metal bending, the text states: "One of Geller's main attributes that had been reported to us was that he was able to bend metal. . . . In the laboratory we did not find him able to do so. . . . [it] becomes clear in watching this film that simple photo interpretation is insufficient to determine whether the metal is bent by normal or paranormal means. . . . It is not clear whether the spoon is being bent because he has extraordinarily strong fingers and good control of micro-manipulatory movements, or whether in fact the spoon 'turns to plastic' in his hands, as he claims."
>
> In discussing our dice-box experiment, Gardner goes on to claim that the reported run of correct guesses as to which die face was uppermost was selected out of a longer run which included "many prior trial runs." That is completely false. The facts are exactly as reported in the *Nature* paper and in the S.R.I. film. The experiment was performed ten times, with Uri passing twice and giving a correct response eight times. These ten trials were the *only* ten; they were not selected out of a longer run—there were *no* prior trials nor follow-up trials, as Gardner claims.
>
> Gardner's errors appear to be due to his taking at face value the erroneous speculations of Geller's self-appointed debunker, the Amazing Randi.

My reply to the letter above appeared in the same issue:

I have enormous admiration for the expertness of Puthoff and Targ in one field—verbal obfuscation. Let me take each of their two counts in turn:

Count 1. When I say that Puthoff and Targ are "personally convinced" of Geller's ability to bend metal, I use the phrase in the ordinary language sense, as when an astronomer says he is personally convinced that quasars are not within our galaxy. It is a probability estimate, as are all scientific beliefs. When Puthoff and Targ deny they are convinced Geller can bend metal, they mean they are not "convinced" because they have not proved such ability in their laboratory. Privately, in letters and conversation, they have expressed their personal beliefs that Geller has such ability. If Puthoff and Targ wish to make precise their present beliefs about Geller's PK powers, let them give it as a probability estimate: How do they *now* rate the probability that Geller has PK ability? If they rate it low, it means they have changed their minds.

In my article I said that Jack Sarfatti, the paraphysicist who first staked his reputation on the PK powers of Uri Geller, but who recently branded Geller a fraud, "did not doubt" that others have PK ability. Sarfatti telephoned me to say that this is not true. He *does* doubt the existence of PK. He then added that he doubts "everything" except the existence of himself. This surprised me because I doubt even my own existence—for all I know I may be just a figment in the Red King's dream (consult Lewis Carroll's *Through The Looking Glass*). When I said Sarfatti does not doubt PK, I meant it only in the ordinary language sense, as when a physicist says he does not doubt electromagnetism.

Count 2. I did *not* say in my article that in the dice-box test Puthoff and Targ selected ten guesses out of a longer run. I said that many "prior trial runs" had been made. A trial run, in ordinary language, is a practice run. Yet every time someone points out that practice runs were made with the dice-box, Puthoff and Targ obfuscate by denying that the ten guesses were "selected" from previous trials. They were not so selected.

But that is not the point. The point is that a very large number of practice runs were made during which Geller was allowed to handle the box. This gave him all the time he needed to devise a method of cheating when the final test of ten trials was made. That was all I said and all I meant.

Anyone reading the letter from Puthoff and Targ would assume that there were no practice runs. Without drawing upon private information, I content myself with the following published data:

—In the July, 1973, issue of *Psychic* there are two photographs of Geller performing the dice-box test. In the first picture we see Geller recording his guess of the die's face, the closed box about four inches from his hand, while Targ watches. In the second, we see Geller opening the box to check on his guess. We assume it is the box used in the famous test because we see S.R.I. printed on top. In the obfuscatory Puthoffian-Targian dialect, this is not a "follow-up trial" because it is not part of the test they reported.

—In John Wilhelm's carefully researched book, *The Search for Superman*, just published by Pocket Books, there is a report on dice-box tests conducted in Geller's motel room. Geller did all the shaking, although Puthoff insists it was vigorous enough to "ensure an honest shake." Commented

Targ: "He's like a kid in that he had something that made a lot of noise and he just shook and shook it." Targ also told Wilhelm that in the famous run of ten Geller was allowed to place his hands on the box in a dowsing fashion.

Targ told Wilhelm that S.R.I. has a good-quality videotape of another dice test in which Geller, five times in a row, correctly wrote down the die's number *before* the box was shaken. Targ first shook the box, then Geller took it and dumped out the die. Since magicians familiar with dice cheating know a variety of ways to control a fair die when it is dumped out of a box, how about letting magicians see this valuable videotape? Why keep it top secret? Targ was so impressed by *this* test that he told Wilhelm he suspects that, even in the run of ten that they reported, Geller probably used precognition, not clairvoyance, to guess the number he later "would see when he opened the box." Note: Targ said it was Geller who opened the box!

In the sound track of the S.R.I. film from which Puthoff and Targ quote in their letter, the following occurs: "Here is another double-blind experiment in which a die is placed in a metal file box. . . . The box is shaken up with neither the experimenter nor Geller knowing where the die is or which face is up. This is a live experiment that you see—in this case, Geller guessed that a four was showing but first he passed because he was not confident. You will note he was correct and he was quite pleased to have guessed correctly, but this particular test does not enter into our statistics."

Now Puthoff and Targ have not, so far as I know, revealed whether this single trial, which was recorded on videotape, was part of *the* test of ten trials. If it was not, then surely it was in a practice run. If it was, then presumably the entire dice-box test was recorded on videotape. In the interest of scientific truth, Puthoff and Targ should make this entire videotape available to inspection by magicians. It then could be determined unequivocally whether the theory suggested by James Randi, as to how Geller might have cheated, is a viable one. Come, gentlemen, let us see the entire tape! If we are wrong, we will humbly apologize.

Martin Gardner

Much has happened since. Taylor has reversed his opinion of metal bending (see Chapter 16 of this book). P and T are doing their best to forget about Uri as they devote their energies to research on clairvoyance, or what they prefer to call "remote viewing" (see Chapter 30). Incidentally, Puthoff recently disclosed to a friend of mine some startling new information about the famous ten-shake die test. It did not take place at one time, said P, but was extended over a period of several days!

This answers the third of eleven questions that psychologist Richard Kammann (doing his best to obtain information for *The Psychology of the Psychic,* Prometheus Books, 1980, a book he wrote with David Marks) sent to Puthoff in 1978. Some of the other questions were:

Which trials were actually videotaped and which other trials were filmed? Which trials were neither taped nor filmed, and which were both?

Were the trials all conducted in one room at SRI, or were some done in other rooms? Were some done in Uri's motel room?

Were the videotapes or films run continuously during each trial, or were they at some time switched off?

Is there any reason why you chose a "pass" trial to include in the SRI film on Geller? Was this one of the two pass trials reported in the *Nature* article?

Is it possible to learn the time that Geller took on each trial after the die was shaken?

This letter was never answered. This was typical of a long history of evasions and nonresponses that Kammann and Marks experienced with P and T, including their persistent refusal to supply any transcripts of remote-viewing tests. We will probably never know exactly what happened during the die test unless Uri some day decides to tell all.

Sarfatti responded to my article with a long letter to *Technology Review* (March/April 1977):

I wish to comment on Martin Gardner's article, "Magic and Paraphysics," which contains extensive descriptions of my research, opinions, and speculations.

1. My name was not changed from Sarfatt to Sarfatti. Rather, my original name was officially restored to me. My father's name was Hyman Sarfatti.

2. My Ph.D. in physics is not from San Diego State College, which does not grant that degree. It is from the University of California at Riverside. (I was an assistant professor of physics at San Diego State College.) I have also studied with David Bohm at the University of London and with Abdus Salam at the International Center for Theoretical Physics in Trieste.

3. I am open to the possibility of extra-terrestrial communication from advanced civilizations. However, I do not uncritically or dogmatically believe that such communications with higher intelligence actually are occurring. I maintain a healthy skepticism of the increasing flow of subjective and circumstantial evidence reported by people in responsible positions that allude to extra-terrestrial contact by means other than electromagnetic signals. It is definitely a misrepresentation to identify my position with that of Andrija Puharich and Uri Geller.

Contrast my view of the extra-terrestrial question with that of Dr. Frank Drake, director of the National Astronomy and Ionosphere Center at Cornell. In his article, "On Hands and Knees in Search of Elysium" (*Technology Review,* June, 1976) we find statements such as: "All the S.E.T.I. [Search for Extraterrestrial Life] people have been tantalized by the belief that there are alien radio signals passing through our offices and homes which could be detected now with existing equipment if we but knew in which direction and on which frequency to listen" (p. 25).

Dr. Drake's position is much more certain about the possibility of extra-terrestrial communication than even I am!

Both S.E.T.I. and my organization, the Physics/Consciousness Research Group (PCRG), fundamentally agree that attempts to communicate with extra-terrestrial intelligence are sane and desirable. S.E.T.I. thinks extra-terrestrials almost certainly exist, that they will communicate by electromagnetic signals and that no earthperson has yet received such signals.

PCRG entertains the possibility that extra-terrestrials exist and that they may choose to use another means apparently allowable within the formal structure of quantum mechanics, namely superluminal quantum communication. An experiment being conducted by Aspect (its design is reported in *Physics Letters*, 54A, August 25, 1975) is capable of disproving the superluminal quantum communication possibility. If Aspect's experimental results are negative and are confirmed, then PCRG will no longer entertain the superluminal possibility.

4. I deny Mr. Gardner's statement, "Sarfatti does not doubt that PK powers exist." I do doubt that PK powers—and superluminal information transfer—exist. However, I am open to the possibility of their existence since quantum mechanics apparently has room for them, e.g., John A. Wheeler's "participatory" and Eugene Wigner's "consciousness" interpretations of quantum measurement.

The recent paper by Henry P. Stapp, "Are Superluminal Connections Necessary?" (*Lawrence Berkeley Laboratory* 5559, November 8, 1976) provides strong support for the superluminal information transfer interpretation of the quantum theory. Thus, Professor Stapp writes: "If the statistical predictions of quantum theory are true in general and if the macroscopic world is not radically different from what is observed, then what happens macroscopically in one space-time region must in some cases depend on variables that are controlled by experimenters in far-away, space-like-separated regions. . . . The central mystery of quantum theory is 'how does information get around so quick?' . . . How does the information about what is happening everywhere else get collected to determine what is likely to happen here? . . . Quantum phenomena provides *prima facie* evidence that information gets around in ways that do not conform to classical ideas. Thus the idea that information is transferred superluminally is, a priori, not unreasonable."

Superluminal information transfer, if it exists, would surely be used for communications by an advanced extra-terrestrial civilization that did not wish to wait years between information transmissions. Accordingly, why not spend a few million to really check out the quantum superluminal possibility before spending billions of dollars on electromagnetic searches such as Project Cyclops? S.E.T.I., in ignorance of the possible quantum alternative, seems to want to settle the issue of rival forms of extra-terrestrial communication without debate. PCRG does not intend to let that happen. If Cyclops would be of value for purely scientific reasons, independent of the extra-terrestrial issue, then perhaps it should be supported. But the creators of the project have not made that clear to the public.

5. Mr. Gardner writes: ". . . the concept [of superluminal information transfer] is simple, but the paraphysicist makes it sound scientific by hoking

it up with technical jargon. The 'measure of information,' according to Sarfatti, is 'the degree of order in the energy already existing at a particular place. . . .'"

This is more than hoking with technical jargon. For example, David Hawkins writes: ". . . the physical concept of work, as distinguished from energy, has itself an informational aspect. The performance of 'useful work' . . . is to produce a situation having a certain order or information. It is to inform a physical system in some way, to transfer order or information to it. To say that free energy is energy available for external work is to say that order cannot come into existence *ex nihilo,* but only by transfer" (*Language of Nature,* p. 216).

Since Lord Kelvin in the nineteenth century, the notion of order as the thermodynamic quality of energy has been well established. The new idea I present is of a nonlocal, superluminal (i.e., space-like) transfer of the quality of energy, not the amount of energy. I suggest that the essence of quantum theory is the *nondynamical* transfer of order (i.e., quality of energy) without the dynamical transfer of energy through space. As Schrodinger wrote of the Einstein-Rosen-Podolsky effect: "It is rather discomforting that [quantum] theory should allow a system to be steered or piloted into one or another type of state at the experimenter's mercy in spite of his having no access to it" (*Proceedings of the Cambridge Philosophical Society,* 31, 555 [1935]).

6. Werner Erhard's support of PCRG does not imply that Mr. Erhard or est (Erhard Seminars Training) fully endorse every action and policy of PCRG, or vice versa. PCRG is not a part of est. Mr. Erhard and I share a warm personal relationship that enriches both our lives independent of ideological and scientific beliefs.

Jack Sarfatti

I once mentioned to a high official of est that Sarfatti had ended a letter by saying that he and Werner Erhard shared a "warm personal relationship." The official laughed and said, "Werner doesn't have a warm personal relationship with anybody." In any case, after est turned down Sarfatti's request for new funding for his PCRG, Sarfatti became a bitter foe of est, mailing out endless press releases accusing Erhard of being a native fascist. In recent years Sarfatti has been promoting an invention, for which he has pending patents, designed to transmit coded information faster than light. I know of no other physicist who thinks it will work. If it does, Sarfatti will become one of the greatest physicists of all time.

As I write, Bantam Books is about to reprint *Space-Time and Beyond.* I know of no more knuckleheaded volume than this. Even Sarfatti now agrees. He has broken with his two former collaborators and has refused to allow his name on the new edition, as revised by Toben. The agent, by the way, who sold the original book to Dutton was none other than Ira Einhorn, now a fugitive from justice, having jumped bail in 1981 while awaiting trial in Philadelphia for the murder of his girl-friend, Holly Maddux.

I wrote to John Macrae III, president of Dutton, to get confirmation of his plans to allow Bantam to reprint. His reply contains a classic statement that deserves airing. "Whether the book is or is not sound, or, as you say, valueless, is beyond my capacity to judge." No doubt. But are not responsible publishing houses obligated to send science books, which they are not competent to judge, to professionals for evaluation before they publish? Morally obligated, perhaps, but this practice is seldom followed when a moronic manuscript has great potential for meeting the public's hunger for scientific hogwash.

Wilbur Franklin, the Kent State paraphysicist mentioned in my last note, died unexpectedly in the spring of 1978. His most sensational psi claim had been in an article reprinted in *The Geller Papers,* edited by Charles Panati. Using an electron microscope, he had examined a fracture that Uri had produced in a platinum ring and concluded that the unusual fracture surfaces could only have been produced by paranormal means. Not until 1977 did he discover his error. A reexamination of the fracture disclosed, to Franklin's vast dismay, that the crack had occurred at an incomplete braze where the jeweler had attached the ring's shank to the part of the ring holding the gemstone. Franklin conceded that mechanical pressure could easily have caused the break, though he continued to believe that Geller had broken the ring paranormally. His sad letter about this mistake appeared in the *Humanist,* September/October 1977.

That same year the *Chicago Journal* obtained a spoon broken by Uri in front of an audience. Uri had selected the spoon from a pile of some 30 spoons on a table. A test spoon of similar type was then broken by bending it back and forth until the fatigued metal snapped. The two spoons were sent to a consulting firm for examination by electron microscope. The examination failed to show any difference between the two fractures. Details were reported in the *Journal*'s January 19, 1977, issue. Uri's usual method of bending spoons on stage is to have one of his assistants, disguised as a spectator, bring up one or more spoons that have been fatigued to the extent that stroking will cause the spoon to bend and break at the weakened spot. No self-respecting magician would stoop to performing such kindergarten feats, but gullible Gellerites think they are witnessing miracles.

9

The Irrelevance of Conan Doyle

There are some trees, Watson, which grow to a certain height and then suddenly develop some unsightly eccentricity. You will see it often in humans.

Conan Doyle, *The Adventure of the Empty House*

What has that eminent Spiritualist . . . to do with Sherlock Holmes?

T. S. Eliot

Questions similar to Eliot's can be asked about many another famous scrivener whose name has been associated with allegedly fictional characters. What has that sixteenth-century Spanish drifter and one-armed soldier to do with Don Quixote and Sancho Panza? What has that kinky-haired, pie-faced French mulatto, lecher, spendthrift, and literary hack to do with Athos, Porthos, Aramis, and d'Artagnan?

The answer, of course, is "Nothing." The case of Cervantes is particularly instructive, because it has so much in common with that of Conan Doyle. The two books about the adventures of the Knight of La Mancha tell the story of a long friendship between a dreamer—yet man of action—and his faithful down-to-earth companion. We now know, thanks to the recent efforts of Spanish scholars, that these adventures were written, not by Cervantes, but by Sancho Panza. After the death of his master,

Reprinted with permission from *Beyond Baker Street,* ed. by Michael Harrison (Bobbs-Merrill, New York, 1976). © 1976 by Bobbs-Merrill.

Sancho sold his memoirs to Cervantes, who, scoundrel that he was, kept them hidden until Sancho, too, had died.

One should have suspected this long before the truth came out. Cervantes had little interest in Don Quixote. It was his poetry and plays, all written in a careful, classical style, of which he was proud. Only because he was seriously in debt did he allow his name to appear on Sancho's sprawling, carelessly written work.

Sancho was, of course, a much greater writer than Cervantes. He was far from the slow-witted person he made himself out to be, but like James Boswell and John Watson he modestly underplayed himself to pay greater homage to his friend. Unfortunately, his stories about the Don were written in his old age, when his memory was starting to fade, and they are filled with lapses that Cervantes would never have allowed to remain in the manuscript had he troubled to go over it carefully. Cervantes had so little interest in the first half of Sancho's memoirs that it was not until ten years later, when he desperately needed money again, that he issued the sequel. This time he edited more carefully, adding passages in which he tried to explain the contradictions he had failed to catch in the earlier volume.

There are many reasons for believing that Doyle had as little to do with Watson's manuscripts as Cervantes with Sancho Panza's. Like Cervantes, Doyle had no interest in—indeed, he had contempt for—the stories he pretended were his own. But as soon as they became great popular successes, bringing him an income needed for other projects, he let them continue to appear under his own name, touching them up here and there, but editing them so hastily that many of Watson's contradictions, like those of Sancho's, were allowed to remain.

The strongest internal evidence that neither Cervantes nor Doyle wrote the stories for which they became famous is simply the enormous contrast between the mentality and philosophical outlook of supposed author and hero. Don Quixote was a man of firm Roman Catholic faith and high moral principles, with a passion for chivalry. Cervantes hated chivalry. He allowed his name to appear on Sancho's books because he mistakenly supposed them to be attacks on faith and chivalry. His infidelities to his wife, the episodes involving his mistresses, the affairs of his daughter—all were so sordid that early biographers of Cervantes fell back on Latin when they supplied details.

The equally great contrast between the minds of Holmes and Doyle has often been noted. Was Gilbert Chesterton the first to point out how much more Doyle had in common with Dr. Watson? It is true that Doyle and Watson were both medical men, slow thinkers, good writers, and sensitive to the poetry of London; yet there was one overwhelming difference between the two men that has not, I believe, been sufficiently recognized. I refer to Watson's abiding respect for rationality, science, and common sense.

It has many times been pointed out that Holmes's so-called deductions were actually inductions. Like the scientist trying to solve a mystery of nature, Holmes first gathered all the evidence he could that was relevant to his problem. At times he performed experiments to obtain fresh data. He then surveyed the total evidence in the light of his vast knowledge of crime, and of sciences relevant to crime, to arrive at the most probable hypothesis. Deductions were made from the hypothesis; then the theory was further tested against new evidence, revised if need be, until finally the truth emerged with a probability close to certainty.

Although Watson seldom played a role in this complex process, with enormous respect he watched it unfold. Frequently mystified by the speed and efficiency of Holmes's method, he never failed to admire it, to accept its final results, and on one occasion, after the procedure had been explained to him, to exclaim, "How absuraly simple!"

Nothing could be more remote from the mind-set of Watson's alleged creator. Doyle spent the last twelve years of his life in a tireless crusade against science and rationality. It is a period usually glossed over quickly in biographies of Doyle, but, in view of today's explosion of interest in spiritualism and all things occult, it is good to review it as an object lesson. Above all, it provides overwhelming evidence that Doyle had almost nothing to do with either Holmes or Watson.

It has been said that Doyle's conversion to spiritualism, like the recent case of Bishop James Pike, was an emotional reaction to the death of his son. Not so. Even when he was a young Irish-Catholic, Doyle had a strong interest in psychic phenomena. His crusade for spiritualism began in 1916, two years before his son died. Although several British scientists were caught up in the craze, notably Oliver Lodge and William Crookes, Doyle rapidly became the movement's most infuential fugleman. He lectured and debated everywhere. His literary labors for the cause were prodigious. In addition to innumerable pamphlets, magazine articles, introductions to books by others, letters, and book reviews, the following volumes of spiritualist apologetics flowed from his pen: *The New Revelation, The Vital Message, The Wanderings of a Spiritualist, Our American Adventure, Our Second American Adventure, The Case for Spirit Photography, Psychic Experiences, The Mystery of Spiritualism, The Land of Mist,*[1] *The Edge of the Unknown,* and, not least, a monumental two-volume *History of Spiritualism.*

It is no good to say that Doyle had become senile. Clearly he had not. His final years were remarkably vigorous and productive. His last book, *The Edge of the Unknown,* published in 1930, the year that he died at age seventy-one, is a model of lucid, beautifully structured prose. Thousands of people were deeply influenced by his books and lectures. Dr. Joseph B. Rhine, the eminent parapsychologist, is on record as saying that it was a speech by Doyle that first inspired him to turn from botany, in which he had been trained, to the study of psychic phenomena.

In *Memories and Adventures* (pp. 392–94), Doyle gives a dramatic summary of why he believes in spiritualism. He had seen his dead mother and nephew so plainly that he could have counted the wrinkles on one, the freckles on the other. He had conversed at length with spirit voices. He had smelled the "peculiar ozonelike smell of ectoplasm." Prophecies he heard were swiftly fulfilled. He had "seen the dead glimmer up upon a photographic plate" untouched by any hand but his own. His wife, a medium whose writing fingers would be seized by a spirit control, had produced "notebooks full of information . . . utterly beyond her ken." He had seen heavy objects "swimming in the air, untouched by human hand." He had seen "spirits walk around the room in fair light and join in the talk of the company." On his wall was a painting done by a woman with no artistic training, but who had been possessed by an artistic spirit.

He had read books written by unlettered mediums who transmitted the work of dead writers, and he had recognized the writer's style, "which no parodist could have copied, and which was written in his own hand-writing." He had heard "singing beyond earthly power, and whistling done with no pause for the intake of breath." He had seen objects "from a distance projected into a room with closed doors and windows." Why, Doyle concludes, should a man who has experienced all this "heed the chatter of irresponsible journalists, or the head-shaking of inexperienced men of science? They are babies in this matter, and should be sitting at his feet."

Those are the strong words of a profoundly sincere man. They are also the words of a man with a temperament utterly alien to that of both Holmes and Watson. The bitter truth is that Doyle was an incompetent observer of supposed psychic events. He was ignorant of even the rudiments of magic and deception, hopelessly naïve, capable of believing anything, no matter how flimsy the evidence. Over and over again the great mediums of the day who produced psychical phenomena were caught in fraud by the Holmeses and Watsons of science. Over and over again Doyle refused to recognize even the possibility of fraud except in a few cases where it was so patently obvious that everyone in the spiritualist movement recognized it. Even in such rare cases Doyle was quick to explain deception away as a temporary aberration on the part of genuine psychics. Were they not pressured into cheating by the incessant demands of skeptics for phenomena that could not always be produced at will?

In many cases Doyle flatly refused to believe fraudulent mediums even when they made full confessions and explained in detail exactly how they cheated. The most sensational of such confessions was by Margaret Fox, one of the Fox sisters of upper New York State whose ability to produce spirit raps by cracking the first joint of a big toe had started the modern spiritualist craze. Margaret Fox's remarkable confession, made in 1888 when she was eighty-one, appeared in the New York *World,* October 21, and you can read it in Harry Houdini's *A Magician Among the Spirits.* That night, on the stage of New York's Academy of Music, under the close scrutiny of three physicians, Margaret took off a shoe, put one foot on a stool and demonstrated her toe-cracking technique to a hushed audience.

How did Doyle react to her confession? Like other prominent spiritualists, he refused to believe it. Nor did he believe Houdini when the magician tried to persuade him that prominent conjurors of the day who were capitalizing on the spiritualist movement by claiming supernormal powers were not genuine psychics. The Davenport brothers, for example, were friends of Houdini. He knew their methods well but was unable to convince Doyle that they were tricksters. Julius Zancig, another magician and friend of Houdini, had perfected a secret code by which he could transmit information quickly to his wife. Just as some magicians today pretend to be genuine mind readers because it enhances their reputation and increases their earnings, so the Zancigs found they could make more money posing as psychics than by doing straight magic. Doyle never doubted the authenticity of their telepathic abilities. Magicians found this as hilarious then as they do today whenever a famous writer or scientist goes on record as believing that some magician-turned-psychic has supernormal powers.

Indeed, Doyle even refused to believe Houdini's repeated denials that he, Houdini, was not psychic. Doyle's essay, "The Riddle of Houdini,"[2] is one of the most absurd documents in the history of parapsychology. Here is Doyle, the supposed creator of Sherlock Holmes, arguing soberly that his friend Houdini was in reality a medium who performed his escapes by dematerializing his body!

Houdini's protests fell on uncomprehending ears. Doyle readily admitted that Houdini was a skilled conjuror, but he argued that the magician's escapes were on such an "utterly different plane" from that of other magicians that it was an "outrage of common sense to think otherwise." Why, if Houdini was a genuine psychic, did he deny his singular powers? "Is it not perfectly evident," Doyle answered himself, "that if he did not deny them his occupation would have been gone forever? What would his brother-magicians have to say to a man who admitted that half his tricks were done by what they would regard as illicit powers? It would have been 'exit Houdini.'"[3]

There is scarcely a page in any of Doyle's books on the occult that does not reveal him to be the antithesis of Holmes. His gullibility was boundless. His comprehension of what constitutes scientific evidence was on a level with that of members of London's flat-earth society. Consider, for example, the story he tells in *The Coming of the Fairies.*[4]

In 1917, in the Yorkshire village of Cottingley, a sixteen-year-old girl, Elsie Wright, was being visited by her ten-year-old cousin, Frances Griffiths. Elsie was a dreamy little girl who for years had loved to draw pictures of fairies. She had a fair amount of artistic talent, had done some designing for a jeweler, and once worked a few months for a photographer. The two girls loved to spend hours in a glen back of the cottage where, they told Mr. and Mrs. Wright, they often played with fairies.

One day the girls borrowed Mr. Wright's camera, and Elsie snapped a picture of Frances in the woods. When Mr. Wright developed the plate he was astonished to see four scantily clad Tinkerbells, with large butterfly wings, prancing merrily in the air under Frances's chin. Two months later Frances took a picture of Elsie that showed her beckoning a tiny gnome wearing black tights and pointed hat (a bright red hat, the girls recalled) to step into her lap.

The two photos reached Doyle by way of Edward L. Gardner, a theosophist and occult journalist. Doyle wrote to Houdini in great excitement: "I have something . . . precious, two photos, one of a goblin, the other of four fairies in a Yorkshire wood. A fake! you will say. No, sir, I think not. However, all inquiry will be made. These I am not allowed to send. The fairies are about eight inches high. In one there is a single goblin dancing. In the other, four beautiful, luminous creatures. Yes, it is a revelation."

In the December 1920 issue of the *Strand Magazine,* the monthly that had printed so many of Watson's marvelous tales, Doyle and Gardner collaborated on "An Epoch-Making Event—Fairies Photographed." The article blew up a storm. Several newspapers attacked the pictures as fakes, but hundreds of readers wrote to Doyle about the fairies that they, too, had seen in their gardens. Three years after the first two fairy pictures had been taken, Gardner brought the two cousins together again at the same cottage (the girls insisted the fairies would not "come out" unless both were there) and let them borrow his camera. Eventually the girls succeeded in obtaining three more fairy photos. Gardner was not present during any of this picture-taking. Why? Because the girls convinced him the fairies were extremely shy and would not come out for a stranger!

The three new pictures appeared in the *Strand* in 1921, and the following year Doyle reproduced all five in his book *The Coming of the Fairies.* Of the three new photos, one shows a fairy with yellow wings (the girls always supplied details about the colors) offering a posy of "etheric harebells" to Elsie. A second shows an almost nude young lady, with lavender wings, leaping toward Elsie's nose.

Neither girl is in the third photograph. A winged fairy is on the left, another on the right. Both are either partly hidden behind twigs, or the twigs are showing through their transparent bodies. The girls recalled seeing the two creatures, but said they had noticed only a misty glow between them. On the photograph this glow proved to be something that looks like nothing more than a piece of silk hanging on some branches. According to Doyle's caption in the book's first British edition, it is "a magnetic bath, woven very quickly by the fairies, and used after dull weather and in the autumn especially. The sun's rays through the sheath appear to magnetize the interior and thus provide a 'bath' that restores vitality and vigour."

Doyle was now firmly persuaded that the fairies were not "thought forms" projected into the camera by the girls, like the photographs which Jule Eisenbud argues, in his *World of Ted Serios* (William Morrow, 1967), were projected onto Polaroid film by a Chicago bellhop. Doyle believed that the fairies belonged to "a population which may be as numerous as the human race . . . and which is only separated from ourselves by some difference of vibrations."

Moreover, Doyle was convinced that a revelation of the existence of these little people would go far toward combatting the materialism that dominated modern science, and so paved the way for an acceptance of the greater revelation of spiritualism. In 1920 he wrote to Gardner:

I am proud to have been associated with you in this epoch-making incident. We have had continued messages at séances for some time that a visible sign was coming through—and perhaps this was what is meant. The human race does not deserve fresh evidence. . . . However, our friends beyond are very long-suffering and more charitable than I, for I will confess that my soul is filled with a cold contempt for the muddleheaded indifference and the moral cowardice which I see around me.

Doyle noticed that one of the four fairies, in the first picture taken by the girls, is playing a double pipe. A similar pipe is held by the gnome in the second picture. Is not this the traditional pipe of Pan? According to the girls, it made a "tiny little tinkle" that could barely be heard when all was still. And if the fairies have pipes, why not other belongings? "Does it not suggest a complete range of utensils and instruments . . . ?" Doyle asks. "It seems to me that with fuller knowledge and with fresh means of vision, these people are destined to become just as solid and real as the Eskimos."

One of the funniest (and saddest) aspects of Doyle's preposterous book is that the five pictures he so proudly displays are not even clever fakes. The lack of modeling on the fairy figures, and their sharp outlines, indicate that Elsie had simply drawn them on stiff paper, then the girls had cut them out and stuck them in the grass or supported them in the air with invisible wires or threads. (The pictures could have been faked in other ways, but this seems the most likely.) The little ladies have hairdos that were fashionable at the time. There is not the slightest blurring of their fluttering wings. In every picture the fairies look as flat as paper dolls.

Unlike Dr. Watson, Doyle could never bring himself to exclaim, "How absurdly simple!" Not once did he doubt the genuineness of the fairy photos, although he did own that proof of their authenticity was less "overwhelming" than for the authenticity of photographs of discarnates on the "other side." The two girls never obtained another fairy picture. Doyle reports on a visit in 1921 to the Cottingley glen by a clairvoyant named Geoffrey Hodson. He was accompanied by the two girls. The place swarmed with elves, gnomes, fairies, brownies, goblins, water nymphs, and other elusive creatures, all seen and vividly described by Hodson and the girls, but the little people refused to appear on any camera plates.[5]

In 1971 both Elsie and Frances were interviewed by the BBC. The two elderly ladies insisted that their father had not faked the pictures. When Elsie was asked point blank if she or Frances had faked them, she was unwilling to deny it. "I've told you they're photographs of figments of our imagination," she said, "and that's what I'm sticking to." The same question was put to Frances, who was interviewed separately. Frances asked how Elsie had answered it. When told, she said she had nothing to add.[6]

Well, what is one to make of an eminent writer who believed that Houdini dematerialized his body to perform his escapes and that the glens of England teem with wee folk who now and then allow themselves to be seen and photographed by us mortals? However you answer, one thing is certain. Such a man could never have constructed, as figments of *his* imagination, the coldly rational Holmes or his admiring Dr. Watson.

It was not, I think, Doyle who made this pair immortal. It was the other way around. Holmes and Watson, intent on guarding their privacy, permitted Sir Arthur to take credit for inventing them. In doing so, they conferred upon him that earthly immortality that his authentic but undistinguished writings could never have provided.

NOTES

1. *The Land of Mist* is actually a novel, but one that rattles with spiritualist drumbeating. Doyle's fictional scientist, George Edward Challenger (of *Lost World* fame), now a widower, is converted to spiritualism when he gets a message from his discarnate wife. Before the *Strand* serialized the novel, Doyle called it *The Psychic Adventures of Edward Malone*.

One of the strongest indications that Holmes was not Doyle's creation is that Holmes, unlike Professor Challenger, never became a spiritualist. True, he once remarked (in "The Adventure of the Veiled Lodger"), echoing one of Doyle's favorite themes: "The ways of fate are indeed hard to understand. If there is not some compensation hereafter, then the world is a cruel jest." But had Doyle actually written this story at the time he claimed, when his interest in spiritualism was at its zenith, Holmes surely would have said more than that.

2. This essay was first published as a pamphlet and serialized in the *Strand Magazine* as "Houdini the Enigma," vol. 74, August and September 1927. It is reprinted in Doyle's *The Edge of the Unknown* (1930), currently in print as a Berkley paperback.

3. On Doyle's relationship with Houdini, see *Houdini and Conan Doyle: The Story of a Strange Friendship,* by Bernard M. L. Ernst and Hereward Carrington (New York: Albert and Charles Boni, 1932). Consult also the chapter on Doyle in Houdini's *A Magician Among the Spirits* (New York: Harper and Brothers, 1924), and the many references to Doyle in *Houdini, the Untold Story,* by Milbourne Christopher (New York: Thomas Y. Crowell, 1969).

4. *The Coming of the Fairies* was first published in 1922: in London by Hodder and Stoughton, in New York by George H. Doran. An enlarged edition, to which Doyle added more fairy photographs from England and other lands, was issued in London in 1928 by Psychic Press. Samuel Weiser, New York, reprinted the Doran edition in paper covers in 1972.

5. Hodson gives a full account of this in his book, *Fairies at Work and Play,* published in London by the Theosophical Society Publishing House, 1921. The same house, in 1945, published Edward L. Gardner's *Fairies: The Cottingley Photographs and Their Sequel,* a book containing the best reproductions of the five photos. The fourth revised edition appeared in 1966. Both books are still in print.

The latest retelling of the story of Doyle and the fairy pictures is "Exploring Fairy Folklore," a two-part article by Jerome Clark, *Fate* magazine, September and October 1974.

6. Tapes of the BBC interviews with Elsie and Frances are owned by Leslie Gardner (son of Edward L. Gardner), who also owns much unpublished material on his father's investigation of the fairy pictures. For comments on the BBC interviews, see Robert H. Ashby's letter in *Fate,* January 1975, pp. 129-30, and "The Cottingley Fairy Photographs: A Re-Appraisal of the Evidence," by Stewart F. Sanderson, in *Folklore,* vol. 84, Summer, 1973, pp. 89-103. The latter article is a presidential address given by Sanderson at the Folklore Society's annual meeting in London, March 1973. It is an excellent summary, by a skeptic, of the history of the fairy photos.

Postscript

Jerome Clark, in his flatulent *Fate* articles (cited above in Note 5), defended the genuineness of the fairy photos. In a later piece ("The Cottingley Fairies: The Last Word," *Fate,* November 1978) Clark ate crow. The reason: a discovery reported in *Ghosts in Photographs* (Harmony Books, 1978), by occult journalist Fred Gittings.

Gittings found a children's book, *Princess Mary's Gift Book,* published in England in 1915 by Hodder and Stoughton, the same house (ironically) that later published Doyle's treatise on the fairies. In the gift book is a poem by Alfred Noyes called "A Spell for a Fairy" that tells how to conjure up the wee creatures. The poem's final illustration shows three dancing fairies. When you compare them with the three in the first Cottingley photograph, you see at once they are line-for-line copies. One of the girls obviously had drawn them on cardboard, added wings, then cut them out, and the girls had stuck them in the grass just as the skeptics had always said.

This revelation convinced Clark that the photos were indeed a hoax. Will it convince Jule Eisenbud? I doubt it. Many thought-photographs of Ted Serios were found to correspond line for line with published photographs. This did not disturb Eisenbud in the least. He still firmly believes that Ted saw those pictures in magazines, imprinted them on his mind, then years later projected them by his psychic powers onto Polaroid film. One can similarly argue that the two girls saw the fairy pictures in the gift book, remembered them, and later psychically projected three of the prancing ladies onto the camera plates. Paranormal hypotheses never die. They just momentarily fade, only to bloom again in full strength.

Robert Sheaffer's article, "The Cottingley Fairies: A Hoax?" (*Fate,* June 1978) was written before Gittings's discovery, but it presents additional strong evidence of fakery. Sheaffer had earlier pulled off a delightful hoax of his own. His "Cottingley Photographs: Winged Astronauts?" suggested that the fairies could have been creatures from UFOs. It was printed in *Official UFO Magazine,* October 1977, by editors too stupid to realize that Sheaffer had his tongue in his cheek.

Frances and Elsie were interviewed again, their remarks appearing in the October 25, 1975, issue of the British magazine *Woman.* The ladies stuck by their psychic guns.

For reproductions of the picture copied by the girls, see Gittings's book, Clark's apologetic article, James Randi's recent *Flim-Flam!* (Lippincott & Crowell, 1980).

10

Great Fakes of Science

Politicians, real-estate agents, used-car salesmen, and advertising copy-writers are expected to stretch facts in self-serving directions, but scientists who falsify their results are regarded by their peers as committing an inexcusable crime. Yet the sad fact is that the history of science swarms with cases of outright fakery and instances of scientists who unconsciously distorted their work by seeing it through lenses of passionately held beliefs.

Gregor Johann Mendel, whose experiments with garden peas first revealed the basic laws of heredity, was such a hero of modern science that scientists in the thirties were shocked to learn that this pious monk probably doctored his data. R. A. Fisher, a famous British statistician, checked Mendel's reports carefully. The odds, he concluded, are about 10,000 to 1 that Mendel gave an inaccurate account of his experiments.[1]

Brother Mendel was a Roman Catholic priest who lived in an abbey in Brünn, now part of Czechoslovakia. More than a century ago, working alone in a monastery garden, he found that his plants were breeding according to precise laws of probability. Later, these laws were explained by the theory of genes (now known to be sections along a helical DNA molecule), but it was Brother Mendel who laid the foundations for what later was called Mendelian genetics. His great work was totally ignored by the botanists of his time, and he died without knowing he would become famous.

Most of the monk's work was with garden peas. Seeds from dwarf pea plants always grow into dwarfs, but tall pea plants are of two kinds. Seeds from one kind produce only talls. Seeds from the other kind produce both talls and dwarfs. Mendel found that when he crossed true-breeding talls with dwarfs he got only talls. When he self-pollinated these tall hybrids he got a mixture of ¼ true-breeding talls, ¼ dwarfs, and ½ talls that did not breed true.

Today one says that tallness in garden peas is dominant, dwarfness is recessive. Mendel's breeding experiment is like shaking an even mixture of red and blue beads in a hat, then taking out a pair. The probability is ¼ you will get red-red, ¼ you will get blue-blue, and ½ you will get red-blue. These, however, are "long-run" probabilities. Make such a test just once, with (say) 200 evenly mixed beads, and the odds are strongly against your getting *exactly* 25 red pairs, 25 blue, and 50 mixed. Statisticians would be deeply suspicious if you reported results that precise.

Mendel's figures are suspect for just this reason. They are too good to be true. Did the priest consciously fudge his data? Let us be charitable. Perhaps he was guilty only of "wishful seeing" when he classified and counted his talls and dwarfs.

Geologists find strange things in the ground, but none so strange as the "fossils" unearthed by Johann Beringer, a learned professor of science at the University of Würzburg. German Protestants of the early eighteenth century, like so many American fundamentalists today, could not believe that fossils were the relics of life that flourished millions of years ago. Professor Beringer had an unusual theory. Some fossils, he admitted, might be the remains of life that perished in the great flood of Noah, but most of them were "peculiar stones" carved by God himself as he experimented with the kinds of life he intended to create.

Beringer was ecstatic when his teen-age helpers began to dig up hundreds of stones that supported his hypothesis. They bore images of the bodies of strange insects, birds, and fishes never seen on earth. One bird had a fish's head—an idea God had apparently discarded. Other stones showed the sun, moon, five-pointed stars, and comets with blazing tails. He began to find stones with Hebrew letters. One had "Jehovah" carved on it.

In 1726 Beringer published a huge treatise on these marvelous discoveries. It was written in Latin and impressively illustrated with engraved plates. Colleagues tried to convince Beringer he was being bamboozled, but he dismissed this as "vicious raillery" by stubborn, establishment enemies.

No one knows what finally changed the professor's mind. It was said that he found a stone with his own name on it! An inquiry was held. One of his assistants confessed. It turned out that the peculiar stones had been carved by two peculiar colleagues, one the university's librarian, the other a professor of geography.

Poor trusting, stupid Beringer, his career shattered, spent his life's savings buying up copies of his idiotic book and burning them. But the work became such a famous monument to geological gullery that twenty-seven years after Beringer's death a new edition was published in Germany. In 1963 a handsome translation was issued by the University of California Press. Beringer has become immortal only as the victim of a cruel hoax.[2]

Was Paul Kammerer the victim of a similar hoax, or was he himself the perpetrator? In any case, when someone applied India ink to (or perhaps injected it into) the feet of several of Kammerer's frogs, the career of one of the most respected of Viennese biologists was brought to an inglorious end.

Kammerer was the last great champion of a theory of evolution called Lamarckism. In this view, named for the French naturalist Jean Lamarck, acquired traits are somehow passed on to descendants: when giraffes stretched their necks to nibble high leaves, their offspring were born with longer necks. Darwin himself was a Lamarckian. Modern genetics discards this theory, replacing it with the Mendelian view that natural selection operates on variations produced by random mutations.

In 1910 Lamarckism was still the "establishment" view, but the new Mendelian theory was rapidly gaining ground. Eager to defend the older theory (he had written a book about it called *The Inheritance of Acquired Characteristics*), Kammerer devised a simple experiment with a species of frog known as the "midwife toad."

Most toads mate in water. To keep a firm grip on the female's slippery body, the male toad develops dark "nuptial pads" on his feet. The male midwife toad, which mates on land, lacks such pads. Kammerer's scheme was to force midwife toads to copulate under water for several generations, then see if they develop nuptial pads. It was a stupid experiment, because, had it succeeded, Mendelians would have explained it as no more than a revival of a genetic blueprint. Nothing so complicated as a nuptial pad could have developed in just a few generations.

But Kammerer went ahead with his plan and soon reported it to be a huge success. The black pads had indeed appeared. The news was sensational, especially in Russia where Lamarckism then completely dominated biology. Russian scientists were so impressed that they offered Kammerer a post at the University of Moscow.

No sooner had Kammerer accepted this offer than it was discovered that his toad specimens had been crudely faked. It was the biggest science scandal of the decade. Kammerer blamed it all on an assistant, but nobody believed him. In 1926, at age 46, he took a pistol and shot himself through the head.

Kammerer continued to be a great hero in the Soviet Union throughout the period when Joseph Stalin and the plant-breeder Trofim Lysenko, both enthusiastic Lamarckians, saw to it that Mendelian geneticists were banished to Siberia. Now that Lysenko is dead and Soviet genetics has gone Mendelian, it is hard to find a biologist anywhere in the world who takes Lamarckism seriously.

Arthur Koestler's *The Case of the Midwife Toad* (Random House, 1971) argues that Kammerer was probably innocent of fraud and that Lamarckism is still a viable theory. Of course Koestler could be right, though at the moment it seems unlikely. For two reports on recent developments that have Lamarckian overtones, see "Lamarck Lives—In the Immune System," by Colin Tudge, in *New Scientist* (February 19, 1981), and "Fighting Lamarck's Shadow," by Susan West, in *Science News* (March 14, 1981).

Are live eggs psychic? Can a fertilized egg use its powers of PK (psychokinesis) to influence electronic machines? Leading parapsychologists around the world thought so until the great 1974 scandal, at Dr. J. B. Rhine's famed Institute of Parapsychology, in Durham, North Carolina, threw these eggstraordinary results into the ash can.

When *Time* did a cover story called "The Psychics" in 1974, the most persuasive letter to the science editor, Leon Jaroff, protesting *Time*'s "unfair" charges of fraud in ESP research, was from Dr. Walter J. Levy, Jr. Levy was then the 26-year-old director of Rhine's institute. Three months later Levy resigned in disgrace.

Rhine had been extremely proud of the sensational results that his brilliant, boyish-looking protégé had obtained by using computers to record and evaluate data. Levy's most impressive experiment involved fertilized chicken eggs beneath an incubator that was turned off and on at chance intervals by a randomizing device. Laws of probability dictated that the heat would be on half the time, but computer records showed it to be on more often than off. Clearly the eggs had influenced the randomizer by PK. When Levy used hard-boiled eggs, no PK. His findings appeared in Rhine's journal in an abstract titled "Possible PK by Chicken Embryos to Obtain Warmth" (*Journal of Parapsychology,* vol. 34, 1970, p. 303).

Dr. Levy's undoing was a later PK test he made with rats. Electrodes had been implanted in the rodents' brains so that when a randomizer turned on the current the rats got intense pleasure shocks. Levy found that the rats used their PK powers to get their kicks 55 percent of the time. This bolstered his earlier animal-psi results on the PK abilities of gerbils and hamsters to avoid *unpleasant* shocks.

Three of Levy's older co-workers began to smell a rat. One of them, peeping from a concealed spot, saw Levy repeatedly pull a plug from a recorder that caused it to register only hits. A secret set of instruments, installed without Levy's knowledge, recorded the expected dull score of 50 percent.

Confronted by this evidence Levy admitted everything. He blamed his downfall on overwork and the pressure to get results. The young parapsychologist hasn't been heard from since he got caught, as one science writer put it, with his fingers in the circuitry.[3]

How do bugs of opposite sex find one another, even in the night? The answer is that one of them emits a strong sexual odor called a pheromone. We can't smell it, but the bug can.

In 1976 a chemist at Pennsylvania State University reported that the sex odors of certain insects depend on what they eat. The university trumpeted this amazing result, which threatened to revolutionize pest control, until other chemists at the same university repeated the experiment and got negative results.

This happens over and over again at top universities and research centers. A scientist, overeager for recognition and greedy for research grants, rushes into print with a staggering claim. Colleagues try to replicate the experiment. They fail. The great claim fades into oblivion.

The most notorious of recent instances, in 1973, quickly grew into a medical Watergate. William T. Summerlin, chief of transplant immunology at the world-renowned Sloan-Kettering Institute for Cancer Research in New York City, announced a stupendous breakthrough. He had, he claimed, grafted a patch of skin from a black mouse to the back of a white mouse, and the white mouse had not rejected it. The skin had been carefully prepared by special techniques. If true, medical benefits would be enormous—not only for grafts and transplants, but also for cancer control.

Summerlin had the backing of his boss, Robert Good, president of Sloan-Kettering and co-author of many of Summerlin's papers. The great discovery hit a brick wall when it was found that the black patches were as phony as Kammerer's nuptial pads. Summerlin had painted them on the white mice to convince his colleagues of what he firmly believed to be true. It was a black and white case. The patches easily washed off with alcohol. Good could hardly believe it. "I trusted him," he said. "He came as a respected scientist."

Summerlin was given a leave of absence for psychiatric treatment. He never went back. Today he is practicing medicine in a small Louisiana town. Even if his grafting methods should later prove to be sound, his career as a research scientist is finished.[4]

It is easy to understand how scientists can fake or fudge results, but surely this sort of thing is impossible in pure mathematics. Wrong! A celebrated case involves an Italian mathematician known only as Lazzarini. He is famous for having reported in 1901 an experimental calculation of the decimals in pi, the most ubiquitous of all irrational numbers.

The curious method he used had earlier been discovered by the French naturalist Comte de Buffon, and is known to mathematicians as "Buffon's needle." Rule a series of parallel lines on a sheet of paper, all neighboring lines separated by the same distance k. Obtain a needle shorter than k. Call its length n. If you drop the needle on the grid, so it falls in a random position, the probability it will cross a line is $2n/\pi k$. If the width between lines and the needle's length are the same, the probability reduces to a simple $2/\pi$. Keep dropping the needle, maintaining a record of hits and misses, and you can use the formula for calculating pi. The more often you toss the needle, the closer you get to pi's correct value.

Lazzarini tossed his needle 3,408 times. From this he obtained a value of pi equal to 3.1415929, which is correct through the first six decimal places (the seventh decimal should be 6 not 9). As in Mendel's data, Lazzarini's result is much too good to be true. The odds against getting his value of pi in as few as 3,408 tosses are millions to one. There is no doubt that Lazzarini fudged his results, but no one thought to question them until 1960.[5]

NOTES

1. Fisher's criticism of Mendel is in *Annals of Science,* vol. 1, 1936, p. 115.

2. A good retelling of the story of Beringer is in *Scientists and Scoundrels: A Book of Hoaxes,* by Robert Silverberg (T. Y. Crowell, 1965).

3. Some references on the Levy scandal are: "False Tests Peril Psychic Research," by Boyce Rensberger, *New York Times,* August 20, 1974; "Psychokinetic Fraud," *Scientific American,* September 1974; "Researcher Found Cheating at Psi Lab," *Science News,* August 17, 1974; and "The Psychic Scandal," *Time,* August 26, 1974.

4. On Summerlin's fraud, see *Science,* May 10, 1974, p. 644, and June 14, 1974, p. 1154; *Time,* April 29, 1974; and *Science News,* June 1, 1974. For the possibility that his grafting technique may actually work, see "William Summerlin: Was He Right All Along?" by Lois Wingerson, in *New Scientist,* February 26, 1981.

5. See "Geometric Probability and the Number Pi," by Norman T. Gridgeman, *Scripta Mathematica,* vol. 25, November 1970, pp. 183 ff.

Postscript

My article on science fakes was written at *Esquire's* request. I give it here essentially as I wrote it, not as it was compressed by the editors to make what became short captions for large color illustrations for their October 1977 issue. The last two items were not used by the magazine.

Of course there are many other great fakes I could have included. Here are a few:

—The Piltdown Man. This skull, supposedly discovered in a gravel pit near London in 1911 by Charles Dawson, an antiquarian, was for decades thought to be the most ancient of all fossil remnants of man. In 1953 scientists at the British Museum made chemical tests proving it to be a clever forgery—the skull of a fossil man combined with the jaw of a modern ape. The jaw had been colored to make it look old, and its teeth filed to resemble human teeth. The museum later found that five other antiquities acquired from Dawson and his widow were also forgeries. Moreover, much of Dawson's two-volume history of Hastings turned out to have been copied from an unpublished manuscript in the Hastings Museum.

—Sir Cyril Burt. This distinguished psychologist was editor of the *British Journal of Statistical Psychology* and an ardent promoter of parapsychology. He was best known for his studies of identical twins who had grown up apart. The studies showed that heredity played almost the entire role in determining a person's intelligence. Burt was England's first psychologist to be knighted. In 1976 investigations proved that Burt had shamelessly forged his data. He even invented two mythical women who had supposedly assisted in his research! Most of his test scores were made up. You can read all about it in *Cyril Burt, Psychologist,* by L. S. Hearnshaw (Cornell University Press, 1979).

—Dr. Samuel G. Soal. When he died in 1975 Soal was considered England's top parapsychologist. Rhine called his work "a milestone in ESP research." According to the philosopher C. D. Broad, Soal's work was "outstanding." "The precautions taken to prevent deliberate fraud [were] absolutely watertight," said Broad. Many parapsychologists, and scientists who, like Cyril Burt, accepted the results of parapsychology, considered Soal's work the most reputable and the most convincing in the annals of psi research. Rumors and charges that Soal had falsified data in one of his most famous experiments began to mount shortly before his death. The rumors were intensified when Soal revealed that the original score sheets for the questionable experiment had been lost. Leading parapsychologists defended Soal vehemently until statistician Betty Markwick proved beyond any shadow of a doubt that Soal had deliberately cheated (see Chapter 19).

—The latest scandal involving data fudging was reported in the *New York Times,* June 28, 1980. Dr. John Long, of the pathology department at the Harvard University School of Medicine, had been doing research on antibodies under a $150,000 grant. He resigned after admitting falsifying his results.

Is such cheating more prevalent in parapsychology than in "orthodox" science? I don't know. The pressures to produce are probably greater in certain areas of traditional science than in parapsychology for the simple reason that funding is larger and there is more competition for the loot. On the other hand, the temptation to falsify data may be less in orthodox areas because scientists accept replication failures as evidence against a hypothesis, thus making it easy to discredit falsified research. Parapsychology is almost alone in its remarkable ability to discount a replication failure as being due to some factor, such as a scientist's skepticism, that inhibited the operation of paranormal powers.

Also, if a scientist holds a highly eccentric belief, his compulsion to convince orthodox colleagues, who may regard him with contempt, may provide unusually strong motives for both conscious fudging and self-deception. My overall impression is that cheating and self-deception are greater in parapsychology than in most sciences, especially the physical sciences, but not by much.

11

Fliess, Freud, and Biorhythm

*At Aussee I know a wonderful wood full of ferns and mushrooms,
where you shall reveal to me the secrets of the world of the lower animals and the world of children. I am agape as never before for what
you have to say—and I hope that the world will not hear it before
me, and that instead of a short article you will give us within a year a
small book which will reveal organic secrets in periods of 28 and 23.*

Sigmund Freud, in a letter to Wilhelm Fliess, 1897

One of the most extraordinary and absurd episodes in the history of
numerological pseudoscience concerns the work of a Berlin surgeon
named Wilhelm Fliess. Fliess was obsessed by the numbers 23 and 28. He
convinced himself and others that behind all living phenomena and perhaps inorganic nature as well there are two fundamental cycles: a male
cycle of 23 days and a female cycle of 28 days. By working with multiples
of those two numbers—sometimes adding, sometimes subtracting—he
was able to impose his number patterns on virtually everything. The
work made a considerable stir in Germany during the early years of this
century. Several disciples took up the system, elaborating and modifying
it in books, pamphlets, and articles. In recent years the movement has
taken root in the United States.

Although Fliess's numerology is of interest to recreational mathematicians and students of pathological science, it would probably be

---

Reprinted from *Mathematical Carnival,* by Martin Gardner (Knopf, New York, 1977),
with permission of the publishers.

unremembered today were it not for one almost unbelievable fact: For a decade Fliess was Sigmund Freud's best friend and confidant. Roughly from 1890 to 1900, in the period of Freud's greatest creativity, which culminated with the publication of *The Interpretation of Dreams* in 1900, he and Fliess were linked in a strange neurotic relationship that had—as Freud himself was well aware—strong homosexual undercurrents. The story was known, of course, to the early leaders of psychoanalysis, but few laymen had even heard of it until the publication in 1950 of a selection of 168 letters from Freud to Fliess, out of a total of 284 that Fliess had carefully preserved. (The letters were first published in German. An English translation entitled *The Origins of Psycho-Analysis* was issued by Basic Books in 1954.) Freud was staggered by the news that these letters had been preserved, and he begged the owner (the analyst Marie Bonaparte) not to permit their publication. In reply to her question about Fliess's side of the correspondence Freud said: "Whether I destroyed them [Fliess's letters] or cleverly hid them away I still do not know." It is assumed that he destroyed them. The full story of the Fliess-Freud friendship has been told by Ernest Jones in his biography of Freud.

When the two men first met in Vienna in 1877, Freud was thirty-one, relatively unknown, unhappily married, with a modest practice in psychiatry. Fliess had a much more successful practice as a nose and throat surgeon in Berlin. He was two years younger than Freud, a bachelor (later he married a wealthy Viennese woman), handsome, vain, brilliant, witty, and well informed on medical and scientific topics.

Freud opened their correspondence with a flattering letter. Fliess responded with a gift; then Freud sent a photograph of himself that Fliess had requested. By 1892 they had dropped the formal *Sie* (you) for the intimate *du* (thou). Freud wrote more often than Fliess and was in torment when Fliess was slow in answering. When his wife was expecting their fifth child, Freud declared it would be named Wilhelm. Indeed, he would have named either of his two youngest children Wilhelm but, as Jones put it, "fortunately they were both girls."

The foundations of Fliess's numerology were first revealed to the world in 1897 when he published his monograph *Die Beziehungen zwischen Nase und Weibliche Geschlechtsorganen in ihrer biologischen Bedeutungen dargestellt* (*The Relations between the Nose and the Female Sex Organs from the Biological Aspects*). Every person, Fliess maintained, is really bisexual. The male component is keyed to the rhythmic cycle of 23 days, the female to a cycle of 28 days. (The female cycle must not be confused with the menstrual cycle, although the two are related in evolutionary origin.) In normal males the male cycle is dominant, the female cycle repressed. In normal females it is the other way around.

The two cycles are present in every living cell and consequently play their dialectic roles in all living things. Among animals and humans both cycles start at birth, the sex of the child being determined by the cycle that is transmitted first. The periods continue throughout life, manifesting themselves in the ups and downs of one's physical and mental vitality, and eventually determine the day of one's death. Moreover, both cycles are intimately connected with the mucous lining of the nose. Fliess thought he had found a relation between nasal irritations and all kinds of neurotic symptoms and sexual irregularities. He diagnosed these ills by inspecting the nose and treated them by applying cocaine to "genital spots" on the nose's interior. He reported cases in which miscarriages were produced by anesthetizing the nose, and he said that he could control painful menstruation by treating the nose. On two occasions he operated on Freud's nose. In a later book he argued that all left-handed people are dominated by the cycle of the opposite sex, and when Freud expressed doubts, he accused Freud of being left-handed without knowing it.

Fliess's theory of cycles was at first regarded by Freud as a major breakthrough in biology. He sent Fliess information on 23- and 28-day periods in his own life and the lives of those in his family, and he viewed the ups and downs of his health as fluctuations of the two periods. He believed a distinction he had found between neurasthenia and anxiety neurosis could be explained by the two cycles. In 1898 he severed editorial connections with a journal because it refused to retract a harsh review of one of Fliess's books.

There was a time when Freud suspected that sexual pleasure was a release of 23-cycle energy and sexual unpleasure a release of 28-cycle energy. For years he expected to die at the age of 51 because it was the sum of 23 and 28, and Fliess had told him this would be his most critical year. "Fifty-one is the age which seems to be a particularly dangerous one to men," Freud wrote in his book on dreams. "I have known colleagues who have died suddenly at that age, and amongst them one who, after long delays, had been appointed to a professorship only a few days before his death."

Freud's acceptance of Fliess's cycle theory was not, however, enthusiastic enough for Fliess. Abnormally sensitive to even the lightest criticism, he thought he detected in one of Freud's 1896 letters some faint suspicions about his system. This marked the beginning of the slow emergence of latent hostility on both sides. Freud's earlier attitude toward Fliess had been one of almost adolescent dependence on a mentor and father figure. Now he was developing theories of his own about the origins of neuroses and methods of treating them. Fliess would have little of this. He argued that Freud's imagined cures were no more than the fluctuations of mental illness, in obedience to the male and female rhythms. The two men were on an obvious collision course.

As one could have predicted from the earlier letters, it was Fliess who first began to pull away. The growing rift plunged Freud into a severe neurosis, from which he emerged only after painful years of self-analysis. The two men had been in the habit of meeting frequently in Vienna, Berlin, Rome, and elsewhere, for what Freud playfully called their "congresses." As late as 1900, when the rift was beyond repair, we find Freud writing: "There has never been a six months' period where I have longed more to be united with you and your family. . . . Your suggestion of a meeting at Easter greatly stirred me. . . . It is not merely my almost childlike yearning for the spring and for more beautiful scenery; that I would willingly sacrifice for the satisfaction of having you near me for three days. . . . We should talk reasonably and scientifically, and your beautiful and sure biological discoveries would awaken my deepest— though impersonal—envy."

Freud nevertheless turned down the invitation, and the two men did not meet until later that summer. It was their final meeting. Fliess later wrote that Freud had made a violent and unprovoked verbal attack on him. For the next two years Freud tried to heal the breach. He proposed that they collaborate on a book on bisexuality. He suggested that they meet again in 1902. Fliess turned down both suggestions. In 1904 Fliess published angry accusations that Freud had leaked some of his ideas to Hermann Swoboda, one of Freud's young patients, who in turn had published them as his own.

The final quarrel seems to have taken place in a dining room of the Park Hotel in Munich. On two later occasions, when Freud was in this room in connection with meetings of the analytical movement, he experienced a severe attack of anxiety. Jones recalls an occasion in 1912, when he and a group that included Freud and Jung were lunching in this same room. A break between Freud and Jung was brewing. When the two men got into a mild argument, Freud suddenly fainted. Jung carried him to a sofa. "How sweet it must be to die," Freud said as he was coming to. Later he confided to Jones the reason for his attack.

Fliess wrote many books and articles about his cycle theory, but his magnum opus was a 584-page volume, *Der Ablauf des Lebens: Grundlegung zur Exakten Biologie* (*The Rhythm of Life: Foundations of an Exact Biology*), published in Leipzig in 1906 (second edition, Vienna, 1923). The book is a masterpiece of Teutonic crackpottery. Fliess's basic formula can be written $23x + 28y$, where x and y are positive or negative integers. On almost every page Fliess fits this formula to natural phenomena, ranging from the cell to the solar system. The moon, for example, goes around the earth in about 28 days; a complete sun-spot cycle is almost 23 years.

The book's appendix is filled with such tables as multiples of 365 (days in the year), multiples of 23, multiples of 28, multiples of 23^2, multiples of 28^2, multiples of 644 (which is 23×28). In boldface are certain important constants such as 12,167 [23×23^2], 24,334 [$2 \times 23 \times 23^2$],

36,501 [$3 \times 23 \times 23^2$], 21,952 [28×28^2], 43,904 [$2 \times 28 \times 28^2$], and so on. A table lists the numbers 1 through 28, each expressed as a difference between multiples of 28 and 23 [for example, $13 = (21 \times 28) - (25 \times 23)$]. Another table expresses numbers 1 through 51 [$23 + 28$] as sums and differences of multiples of 23 and 28 [for example, $1 = (\frac{1}{2} \times 28) + (2 \times 28) - (3 \times 23)$].

Freud admitted on many occasions that he was hopelessly deficient in all mathematical abilities. Fliess understood elementary arithmetic, but little more. He did not realize that if any two positive integers that have no common divisor are substituted for 23 and 28 in his basic formula, it is possible to express *any positive integer whatever*. Little wonder that the formula could be so readily fitted to natural phenomena! This is easily seen by working with 23 and 28 as an example. First determine what values of x and y can give the formula a value of 1. They are $x = 11$, $y = -9$:

$$(23 \times 11) + (28 \times -9) = 1.$$

It is now a simple matter to produce any desired positive integer by the following method:

$$[23 \times (11 \times 2)] + [28 \times (-9 \times 2)] = 2$$
$$[23 \times (11 \times 3)] + [28 \times (-9 \times 3)] = 3$$
$$[23 \times (11 \times 4)] + [28 \times (-9 \times 4)] = 4$$
$$\bullet \ \bullet \ \bullet$$

As Roland Sprague recently pointed out in problem 26 of his *Recreation in Mathematics,* 1963, even if negative values of x and y are excluded, it is still possible to express all positive integers greater than a certain integer. In the finite set of positive integers that *cannot* be expressed by this formula, asks Sprague, what is the largest number? In other words, what is the largest number that cannot be expressed by substituting nonnegative integers for x and y in the formula $23x + 28y$? The answer: $(23 \times 28) - 23 - 28 = 593$.

Freud eventually realized that Fliess's superficially surprising results were no more than numerological juggling. After Fliess's death in 1928 (note the obliging 28), a German physician, J. Aelby, published a book that constituted a thorough refutation of Fliess's absurdities. By then, however, the 23-28 cult was firmly established in Germany. Swoboda, who lived until 1963, was the cult's second most important figure. As a psychologist at the University of Vienna he devoted much time to investigating, defending, and writing about Fliess's cycle theory. In his own rival masterwork, the 576-page *Das Siebenjahr (The Year of Seven),* he reported on his studies of hundreds of family trees to prove that such events as heart attacks, deaths, and the onset of major ills tend to fall on certain critical days that can be computed on the basis of one's male and female cycles. He applied the cycle theory to dream analysis, an application that Freud criticizes in a 1911 footnote to his book on dreams. Swoboda also designed the first slide rule for determining critical days. Without the aid of such a device or the assistance of elaborate charts, calculations of critical days are tedious and tricky.

Incredible though it may seem, as late as the 1960s the Fliess system still had a small but devoted band of disciples in Germany and Switzerland. There were doctors in several Swiss hospitals who determined propitious days for surgery on the basis of Fliess's cycles. (This practice goes back to Fliess. In 1925, when Karl Abraham, one of the pioneers of analysis, had a gallbladder operation, he insisted that it take place on the favorable day calculated by Fliess.) To the male and female cycles modern Fliessians have added a third cycle, called the intellectual cycle, which has a length of 33 days.

Two books on the Swiss system have been published here by Crown: *Biorhythm*, 1961, by Hans J. Wernli, and *Is This Your Day?*, 1964, by George Thommen. Thommen is the president of a firm that supplies calculators and charting kits with which to plot one's own cycles.

The three cycles start at birth and continue with absolute regularity throughout life, although their amplitudes decrease with old age. The male cycle governs such masculine traits as physical strength, confidence, aggressiveness, and endurance. The female cycle controls such feminine traits as feelings, intuition, creativity, love, cooperation, cheerfulness. The newly discovered intellectual cycle governs mental powers common to both sexes: intelligence, memory, concentration, quickness of mind.

On days when a cycle is above the horizontal zero line of the chart, the energy controlled by that cycle is being discharged. These are the days of highest vitality and efficiency. On days when the cycle is below the line, energy is being recharged. These are the days of reduced vitality. When your male cycle is high and your other cycles are low, you can perform physical tasks admirably but are low in sensitivity and mental alertness. If your female cycle is high and your male cycle low, it is a fine day, say, to visit an art museum but a day on which you are likely to tire quickly. The reader can easily guess the applications of other cycle patterns to other common events of life. I omit details about methods of predicting the sex of unborn children or computing the rhythmic "compatibility" between two individuals.

The most dangerous days are those on which a cycle, particularly the 23- or 28-day cycle, crosses the horizontal line. Those days when a cycle is making a transition from one phase to another are called "switch-point days." It is a pleasant fact that switch points for the 28-cycle always occur on the same day of the week for any given individual, since this cycle is exactly four weeks long. If your switch point for the 28-cycle is on

Tuesday, for instance, every other Tuesday will be your critical day for female energy throughout your entire life.

As one might expect, if the switch points of two cycles coincide, the day is "doubly critical," and it is "triply critical" if all three coincide. The Thommen and Wernli books contain many rhythmograms showing that the days on which various famous people died were days on which two or more cycles were at switch points. On two days on which Clark Gable had heart attacks, the second fatal, two cycles were at switch points. The Aga Khan died on a triply critical day. Arnold Palmer won the British Open Golf Tournament during a high period in July, 1962, and lost the Professional Golf Association Tourney during a triple low two weeks later. The boxer Benny (Kid) Paret died after a knockout in a match on a triply critical day. Clearly it behooves the Fliessian to prepare a chart of his future cycle patterns so that he can exercise especial care on critical days; since other factors come into play, however, no ironclad predictions can be made.

Because each cycle has an integral length in days, it follows that every person's rhythmogram will repeat its pattern after an interval of $23 \times 28 \times 33 = 21,252$ days, or a little more than 58 years. This interval will be the same for everybody. For example, 21,252 days after every person's birth all three of his cycles will cross the zero line simultaneously on their upswing and his entire pattern will start over again. Two people whose ages are exactly 21,252 days apart will be running on perfectly synchronized cycle patterns. Since Fliess's system did not include the 33-day cycle, his cycle patterns repeat after a lapse of $23 \times 28 = 644$ days. Swiss Fliessians call this the "biorhythmic year." It is important in computing the "biorhythmic compatibility" between two individuals, since any two persons born 644 days apart are synchronized with respect to their two most important cycles.

Postscript

George S. Thommen, president of Biorhythm Computers, Inc., 298 Fifth Avenue, New York, is still going strong, appearing occasionally on radio and television talk-shows to promote his products. James Randi, the magician, was moderator of an all-night radio talk-show in the mid-sixties. Thommen was twice his guest. After one of the shows, Randi tells me, a lady in New Jersey sent him her birth date and asked for a biorhythm chart covering the next two years of her life. After sending her an actual chart, but based on a *different* birth date, Randi received an effusive letter from the lady saying that the chart exactly matched all her critical up and down days. Randi wrote back, apologized for having made a mistake on her birth date, and enclosed a "correct" chart, actually as wrongly dated as the first one. He soon received a letter telling him that the new chart was even *more* accurate than the first one.

Speaking in March, 1966, at the 36th annual convention of the Greater New York Safety Council, Thommen reported that biorhythm research projects were under way at the University of Nebraska and the University of Minnesota, and that Dr. Tatai, medical chief of Tokyo's public health department, had published a book, *Biorhythm and Human Life,* using the Thommen system. When a Boeing 727 jetliner crashed in Tokyo in February, 1966, Dr. Tatai quickly drew up the pilot's chart, Thommen said, and found that the crash occurred on one of the pilot's low days.

Biorhythm seems to have been more favorably received in Japan than in the United States. According to *Time,* January 10, 1972, page 48, the Ohmi Railway Co., in Japan, computed the biorhythms of each of its 500 bus drivers. Whenever a driver was scheduled for a "bad" day, he was given a notice to be extra careful. The Ohmi company reported a fifty percent drop in accidents.

Fate magazine, February, 1975, pages 109–110, reported on a conference on "Biorhythm, Healing and Kirlian Photography," held in Evanston, Ill., October, 1974. Michael Zaeske, who sponsored the conference, revealed that the traditional biorhythm curves are actually "first derivatives" of the true curves, and that all the traditional charts are "in error by several days." Guests at the meeting also heard evidence from California that a fourth cycle exists, and that all four cycles "may be related to Jung's four personality types."

Science News, January 18, 1975, page 45, carried a large ad by Edmund Scientific Company for their newly introduced Biorhythm Kit ($11.50), containing the precision-made Dialgraf Calculator. The ad also offered an "accurate computerized, personalized" biorhythm chart report for 12 months to any reader who sent his birthdate and $15.95. One wonders if Edmund is using the traditional charts (possibly off three days) or Zaeske's refined procedures.

A ridiculous book, *Biorhythm: A Personal Science,* by Bernard Gittelson, was published in 1975 by Arco, and later by Werner books as a paperback. Pocket Books jumped into the action with Arbie Dale's *Biorhythm* (1976). *Reader's Digest* (September 1977) gave the "science" a major boost with Jennifer Bolch's shameless article, "Biorhythm: A Key to Your Ups and Downs."

By 1980 biorhythm had become so popular among the gullible that half a dozen firms were manufacturing mechanical devices, electronic computers, and even clocks that told true believers what to expect each day. See *Fate* for advertisements. *Science 80,* in its January/February 1980 issue, ran an article by Russell Schoch called "The Myth of Sigmund Freud," which included a good photograph of Freud and Fliess together as young men.

For an excellent debunking article see William Sims Bainbridge, "Biorhythm: Evaluating a Pseudoscience," in the *Skeptical Inquirer,* Spring/Summer 1979. A list of thirteen articles reporting tests that failed to confirm any of biorhythm's preposterous claims appeared in the *Zetetic Scholar,* vol. 1, no. 1, 1978. See also Chapter 8, "The Great Fliess Fleece," in James Randi's *Flim-Flam!* (Lippincott & Crowell, 1980).

12

A Skeptic's View of Parapsychology

D. Scott Rogo, struggling to depict parapsychology as a science rapidly gaining respectability, has adopted a familiar tactic. He says nothing at all about the wild, preposterous claims now being made by leading workers in the field. Instead, he describes a few conservative experiments conducted by Charles Honorton and Helmut Schmidt. "I hope I have made it clear," he concludes, "that parapsychologists do not spend their time . . . gawking at Uri Geller."

There is no way a skeptic can comment meaningfully on the Honorton and Schmidt experiments, because there is no way, now that the tests are completed, to know exactly what controls were actually in force. An accurate evaluation, by a skeptical psychologist, would require months of intensive study, analyzing the raw data still available, interviewing all the participants, and so on. No top psychologist has time for such a task unless he is substantially funded; even then, he might not consider it worthwhile. In the few cases where this has been done—C. E. M. Hansel's investigation of the classic Pearce-Pratt tests, for instance—large loopholes in controls have come to light. Instead of acknowledging the loopholes, parapsychologists have reacted with enormous indignation.

A more promising approach is to try to replicate major experiments. Unfortunately, whenever skeptics do this and get negative results, parapsychologists invoke their favorite Catch 22, which is that skepticism inhibits psi. You will look in vain through the writings of psi journalists like Rogo for references to these replication failures. There have been,

Reprinted with permission from the *Humanist*, November/December 1977.

for example, several important failures to repeat the Maimonides dream experiments (see Adrian Parker's book, *States of Mind*). In some cases, negative results were obtained by the original researchers when they collaborated with skeptical outsiders.

Readers of *The Humanist* may be interested in a few facts about Honorton and Schmidt that Rogo does not mention because they conflict with his thesis that modern parapsychology is uncontaminated by Geller-gawking. A few years ago, Honorton announced that he had discovered a young lady, Felicia Parise, who was capable of sliding a plastic pill-bottle across a kitchen counter by psychokinesis. This great event was witnessed only by Honorton and an amateur photographer, Norman Moses, who caught the event on film.

The first thing that occurred to the magician James Randi, when he saw the film, was that a superfine nylon thread was stretched horizontally across the formica counter. A small and empty pill-bottle was standing upright. As Felicia's long-nailed fingers, on either side of the bottle, crept forward, it would have been a simple matter for her to push against the thread. Fine nylon thread is absolutely invisible, even at close range, but it is easily broken and tends to be elastic. When Randi duplicated the test, he found that the nylon thread stretched when it pressed against the bottle, causing the bottle to move forward in a jerky manner, exactly as seen on the film made by Moses. Last year, for a BBC television show in London, Randi replicated the feat, using a stronger thread and a heavy wineglass.

It does not follow, of course, that because Felicia *could* have used nylon thread that she did, in fact, use it. It does follow that, because Honorton was not sufficiently informed about magic to evaluate what happened, his report of the event cannot be taken seriously. Nevertheless, he stands resolutely by his belief that he witnessed a paranormal feat. "I examined the counter," he told a *National Enquirer* reporter (see the December 30, 1975, issue, with its front-page headline: "First American to Move Objects with Her Mind"). "I virtually took it apart to ensure there was no mechanical aid she could have used to move the bottle. But I found none. I even checked the countertop to see if it sloped. I found there was a slight slope—but the bottle had actually moved uphill!"

Honorton was also present during tests of Felicia at Rhine's laboratory in 1972, when she moved a compass needle by PK. (See Honorton's "Apparent Psychokinesis on Static Objects by a 'Gifted' Subject," in *Research in Parapsychology, 1973,* edited by W. G. Roll and others.) It is amusing that Honorton, who is highly skeptical of Uri Geller's PK ability, has spent considerable time Felicia-gawking and defending her powers in lectures and articles. Ms. Parise, by the way, stopped demonstrating her psi abilities several years ago and is no longer available for research. This is another *fait accompli* that no parapsychologist can replicate when a knowledgeable magician is present.

Let's take a quick look at Schmidt. You will never learn from Rogo that Schmidt has long been a Geller-gawker. "Another source of particularly strong PK effects may be available in a few highly selected subjects . . ." Schmidt wrote in Edgar Mitchell's *Psychic Exploration*. "One such person is Uri Geller. . . . Geller can well match the feats of the spectacular performers studied by earlier workers in the field. Critical researchers . . . have seen Geller bend heavy metal objects 'mentally,' just by touching them slightly or even without any touch."

In 1970, Schmidt reported one of the great breakthroughs in modern parapsychology. Cockroaches, he announced, probably have the PK ability to cause a randomizer to turn on a device and give themselves electric shocks more often than chance allows. However, since Schmidt personally dislikes cockroaches, he admits that his experiment was inconclusive. It could have been his *own mind* that influenced the randomizer.

Walter J. Levy, former director of J. B. Rhine's laboratory, found that live fertilized chicken-eggs also could influence a randomizer. In this case the randomizer controlled a heat source. "Psi-hitter" eggs kept the heat on longer than chance dictated, while "psi-loser" eggs influenced the randomizer the other way. When the two kinds of eggs were together in the same box, their PK powers cancelled out, and Levy got chance results. Since Levy resigned his post after being caught faking his records, parapsychologists have stopped citing his egg papers as references. Cockroaches, however, are still suspected of being psychic.

Rogo's claim that parapsychologists do not waste time on Geller is simply not true. Harold Puthoff and Russell Targ spent vast amounts of time Geller-gawking and finally declared Geller to be a genuine clairvoyant. Thelma Moss is another Gellerite who continually praises Uri. Gertrude Schmeidler is on record with similar remarks, not to mention her well-known tests of the PK powers of Uri's chief rival, Ingo Swann.

We all know about Andrija Puharich's Geller-gawking—he wrote an entire book about it. His next book will report on his research with a group of what he calls "Geller children." One of them, Puharich declared last May at a "Towards a Physics of Consciousness" symposium held at the Harvard Science Center, succeeded in materializing a tree. Six of the Geller children, he said, were teleported to his home in Ossining, New York, from as far away as Switzerland. Puharich was followed by E. Harris Walker, a physicist who explains Geller's PK powers by a theory of his own involving quantum mechanics. John Taylor, the British paraphysicist, thinks the "Geller effect" is electromagnetic. Puharich attributes it to tachyons, conjectured particles that go faster than light. But the most popular theory at the moment is Walker's quantum theoretical explanation.

Rhine's associate, William Cox, is one of the loudest of all Geller-gawker squawkers. He has written about the "rigidly controlled" conditions under which Uri caused Cox's watch to start running after Cox had stopped the balance wheel with a tinfoil wedge and about how Uri made his key curl "like a burning match." I exchanged several long letters with Cox about the watch bit. Only Cox was present, and he allowed Uri to handle the watch briefly before it started running. Could Uri, under cover of that misdirection for which he is so famous, have opened the case with a thumbnail and pushed the tinfoil? No way, says Cox. Although I freely confessed that Uri "might" have accomplished this miracle by psi power (though I gave it a low probability), I was never able to extract from Cox the comparable admission that Uri, to *any* degree of probability, could have fooled him with a trick.

Does Rogo, I wonder, consider Edgar Mitchell a "parapsychologist"? Mitchell has often described how Uri materialized part of a long-lost tiepin in a mouthful of ice cream. Indeed, one is hard put to think of many parapsychologists who have *not* been Geller-gawkers. I must add that in recent months their silence about Uri has become deafening as it becomes obvious that Uri has no more psi ability than such performers as Charles Hoy, Kreskin, Dean Kraft, Nina Kulagina, and Matthew Manning.

Rogo ends his article by comparing the reputed iron controls of modern parapsychology with the weaker controls of other sciences. Here he missed an all-essential point. Extraordinary controls are not demanded in experiments that demonstrate ordinary results. If, for example, a psychologist seeks to determine whether gerbils can roll Ping-Pong balls with their noses, unusual controls are not required. Gerbils don't cheat. But if someone says he or she can move a plastic pill-bottle with PK power, that is an extraordinary claim that calls for extraordinary precautions. In view of the long, hilarious history of fraud and gullibility in psi research, it demands at the very least the participation of the only expert on this kind of deception—a knowledgeable magician.

It is Rogo's inability to distinguish between these two kinds of investigations, with their different demands for controls, that has made him such a naive and unreliable reporter of the psi scene. Although he shares Honorton's low opinion of Geller, he has spent many years gawking at haunted houses. In his book *An Experience of Phantoms* (1974), Rogo argues that every person has an etheric double that not only survives death but can project itself as a phantom while the person is still of this world. "Apparitions, as no other phenomenon," he writes, "bring to the fore that man is a spiritual being."

In a 1970 article in *Psychic*, "Photography by the Mind," Rogo praises the laboratory controls of Jule Eisenbud, the psychoanalyst who demonstrated Ted Serios's ability to project thought-pictures onto Polaroid film. "Ted Serios," writes Rogo, "does represent the first case of psychic photography to confront parapsychologists in several years." Rogo regrets that Serios's ability "appears to have deteriorated," but he carefully avoids any mention of the famous *Popular Photography* exposé

(October 1967) of how easily Ted could have cheated with an optical device that he palmed in and out of his "gismo." Almost everybody now (except Rogo and Dr. Eisenbud) believes that Ted stopped doing thought-pictures only because three magicians caught on to the simple method he was using, and poor Ted has been unable to think of any better method.

In a more recent article in *New Realities* (the new name of *Psychic* magazine), Rogo discusses reports of Madonna statues that bleed in their palms or weep human tears. Does Rogo have any doubt about the reality of these paranormal phenomena? No. His only problem is deciding whether the blood and tears (never sweat—apparently statues of Mary don't perspire) are Christian miracles or a form of poltergeist.

And in the latest issue of *Fate*—but I must spare the reader further disclosures about Rogo's adolescent enthusiasms. His phantom-gawking is, I submit, closer to the preoccupations of today's best known parapsychologists than to the few sanitized, unspectacular, and highly dubious experiments that he cites in his article.

Postscript

The foregoing article was written as a reply to D. Scott Rogo's "The Case for Parapsychology" in the same issue of the *Humanist*. Rogo's letter about my article appeared in the January/February 1978 issue:

> I do not wish to reply to Martin Gardner's "A Skeptic's View of Parapsychology," a rejoinder to my article "The Case for Parapsychology," which appeared in the same issue. Most of his arguments need no rebuttal. They are, in fact, only ad hominem arguments. Instead of arguing the issues I raise, Mr. Gardner tries to attack some of the researchers of whom I spoke and does not come to grips with their work. However, I would like to correct some of Mr. Gardner's factual errors—just for the record.
>
> Mr. Gardner states that I did not mention the controversial *Popular Photography* "exposure" of Ted Serios in my 1970 *Psychic* article on psychic photography. As a matter of fact, I discussed this affair quite openly in my article (*Psychic*, April 1970, p. 42). . . . Second, Mr. Gardner also states that only Dr. Jule Eisenbud and I seem to take the Serios work seriously anymore. This is, quite frankly, not true. This research is still highly regarded by many parapsychologists.
>
> Finally, I attempted to make a "case for parapsychology" in my article. Mr. Gardner only tries to make a "case against Rogo" by way of reply. He does this by criticizing some of my previous publications. What this has to do with the scientific value of parapsychology is beyond me. However, I freely admit that I have researched and written on hauntings, out-of-body experiences, religious miracles, and many other even worse things! I have done so because I have come to believe that these are phenomena that we

should seriously study. It strikes me that Mr. Gardner is actually criticizing me only for having the sense to have an open mind on many subjects. I make no apologies for this. I am truly sorry that Mr. Gardner feels that having an open mind is shameful in science.

D. Scott Rogo

My rejoinder was:

Mr. Rogo is right about the first of my two "errors." He did indeed mention *Popular Photography* in his *Psychic* article on thoughtography. He had two sentences about it, which I now quote in full:

"Two photographers, C. Reynolds and D. G. Eisendrath, presented evidence in *Popular Photography* showing that they were able to fake thoughtographs under Eisenbud's supervision. However, they were not able to duplicate the effects obtained by Serios."

This fluffs over a major exposé so rapidly that I had forgotten Rogo mentioned it. Now let's examine three errors that Rogo managed to pack into his two sentences. First, a trivial one: David Eisendrath's middle initial is B. Second, the two photographers did *not* fake any thoughtographs under Eisenbud's supervision, or claim that they did. Third, they *did* later duplicate precisely the effects Serios obtained. The Amazing Randi now does the Polaroid trick even better than Ted used to do it and under more controlled conditions. Note also that Rogo does not mention the issue (October 1967) in which the exposé appeared.

Let me tell a story about this article that illustrates how difficult it is for skeptics to get cooperation from psychic researchers when they investigate someone like Ted. When Reynolds and Eisendrath left New York for their meeting with Ted in Colorado, I made the stupid mistake of mentioning, in a letter to Dr. J. B. Rhine, that three of my magician friends were planning a session with Ted. (The third person was statistician Persi Diaconis. Both Reynolds and Eisendrath are magic buffs.) Rhine instantly shot a copy of my letter by special delivery to Dr. Eisenbud, warning him that the three men en route to see him had a knowledge of magic. Thus when the three arrived, they were confronted with my letter, a situation that almost blew the meeting.

The meeting did take place, but of course Ted was wary. Hundreds of shots were made without a single thoughtograph. Finally, at the close of the last session, Reynolds and Diaconis (Eisendrath had gone back to New York) independently saw Ted appear to palm something and transfer it from hand to hand in a manner familiar to magicians. One shot was made (no results), but Diaconis asked afterwards if he could examine the gismo. (Ted's "gismo" was a cylinder of black paper that he always held in front of the camera lens.) Ted asked why, stepped back, and shoved the gismo into his pocket. He later removed it and gave it, to Dr. Eisenbud.

Dr. Eisenbud had often said that the gismo could be examined at any time. At this point, however, Ted became angry, and Dr. Eisenbud upbraided Diaconis for having upset the "psychic." Dr. Eisenbud could easily have said, "Ted, our guests think you palmed something and now have it in your pocket. May we examine the pocket?"

Dr. Eisenbud said no such thing, and Ted refused to be searched, so the two-day meeting ended without the production of a single thoughtograph. Dr. Eisenbud invoked the old Catch 22—the skeptics had upset Ted; indeed, he later told Reynolds that the visit had set psychic research back fifty years.

My second error, says Rogo, is my statement that "almost everybody" now agrees that Ted has stopped doing thoughtography because his modus operandi was disclosed. When Rogo claims to the contrary that "many parapsychologists" regard the Serios case as still open, I take him to mean that they still believe it is more likely than not that Ted once had the genuine power to stare into a Polaroid camera and implant on its film an exact replica of a photograph he had seen many years before in *National Geographic*.

I would be grateful to Rogo if he would back up this claim by listing the names of prominent parapsychologists who still think this, aside, of course, from himself and Dr. Eisenbud. The only such persons I know are Jan Ehrenwald, Thelma Moss, and Stanley Krippner, but perhaps there are others. Rogo should check with the "others" first, though, before he speaks for them. It would be particularly interesting to know what Rhine now thinks.

As for Rogo's last sentence, it betrays once more how little he comprehends scientific method. Of course a scientist must have an open mind, but there are degrees of probability about the importance of various investigations. For example, one must have an open mind about the reality of fairies. After all, Conan Doyle wrote an entire book about them, complete with excellent photographs. Does the phenomenon of catching elves on film suggest their reality to a degree that warrants government funding for research? It is Rogo's inability to distinguish degrees of credibility, along the spectrum that runs from serious science to Ted Serios, that makes his sensationally written books so hilariously irrelevant.

Martin Gardner

Writing now in 1980 I must concede another point to Rogo. Belief in the authenticity of Ted Serios's thoughtography has persisted among many leading parapsychologists longer than I would have deemed possible. To those mentioned in my letter as not yet willing to declare Ted a charlatan, I must add the names of Gertrude Schmeidler and Charles Tart. When I had occasion in an exchange of letters with Schmeidler to ask her about Ted, she replied: "With Serios, my impression is that he's not smart enough to fool Jule Eisenbud whether Serios is drunk or sober, and I trust and respect Jule. Also Pratt, who's worked with Serios too, and gotten affirmative data." Tart, in conversations with two of my friends, has repeatedly defended the soundness of Eisenbud's controls with Ted.

As for Rogo, he continues to hack out balderdash. One of his latest monstrosities, co-authored with Raymond Bayles, is a book called *Phone*

Calls from the Dead (Prentice-Hall, 1979). It reports on conversations with discarnate souls by telephone and other electronic devices. "We've stumbled on a whole new method of psychic communication!" Rogo exclaims.

Yes, indeedy. And he and Bayles have stumbled on a whole new gimmick for ripping off the yokels. Rogo, in case you didn't know, is a former concert oboist. He should go blow his oboe and stop trying to blow his readers' minds.

I do not know what Rhine's final opinion was about Serios. The letter that I sent to him (see above) was written at a time when I expected to review Eisenbud's book for *Scientific American,* a review I did not write. First, my letter to Rhine, dated May 28, 1967:

> Although we remain on opposite sides in our attitude toward ESP, I hope you don't mind my writing you again about a matter of mutual interest. You will recall our correspondence a few years ago about Pat Marquis, the X-ray eye boy. You helped me run down the *Life* piece on him, and told me about your own tests of him, all of which I made good use of in my *Science* article on DOP. At least we can agree, from our respective biases, that it is good to expose what we both know to be obvious fraud.
>
> I am writing you now because *Scientific American* has asked me to do a long book review of Eisenbud's recent book, *The World of Ted Serios.* I and my magic friends (including the photographer David Eisendrath, whom you know) have thought of many simple ways by which Ted could be getting his results, none of them ruled out by a careful reading of Eisenbud's ridiculous book. We are currently trying to arrange for some tests, at which one or more of us could be present as observers, though I doubt very much if Ted will be willing to perform if he suspects any competent observers are present. It occurs to me that you may have written something about Ted Serios, which you could send me; or, if not, you may have some information about him you would be willing to pass on. I might add that anything you tell me is in strictest confidence unless you specifically give me permission to cite you as a source. So if you are willing to express your personal opinion about Ted, or about Eisenbud, please indicate in your letter if I may or may not quote you in my review.
>
> Finally, if you have no first-hand information about Serios, perhaps you know of someone (on the skeptical side) who has such information, and to whom I could write.
>
> Martin Gardner

Because Rhine gave permission to quote him only if I quoted his entire letter, a photocopy of the letter is shown on the opposite page. Note that Rhine is too cautious to give an opinion about Ted but expresses his high regard for Eisenbud. It is of course impossible to regard Serios as a fraud and simultaneously to praise Eisenbud for his "skill" in psi research. If Ted used a secret optical device—and the evidence is overwhelming that he did—then Eisenbud is as gullible and self-deceiving as Conan Doyle.

FRNM

Foundation for Research on the Nature of Man

THE INSTITUTE FOR PARAPSYCHOLOGY
PARAPSYCHOLOGY PRESS

BOX 6847, COLLEGE STATION
DURHAM, N. C. 27708

May 30,1967

Mr.Martin Gardner
10 Euclid Avenue
Hastings-on-Hudson
New York 10706

Dear Mr.Gardner:

I have written nothing about Ted Serios and have not met him.

Dr. Jule Eisenbud I have known for over twenty years. He and I do not always agree, and I am not hurrying to reach a conclusive position concerning his book about Ted; but I greatly admire Dr. Eisenbud's patience, courage, and skill in handling this investigation of Serios. He has at great cost and effort brought the sciences another challenge - which is what scientists are supposed to welcome. I will watch further developments guardedly, of course, but with keenest interest.

You seem, as always, to be prejudging such new developments. The Scientific American must be afraid of the subject to give the reviewing to a professional denigrator.

Naturally you would not want to quote what I have written, and I would not want you to refer to me in the matter unless you quoted the entire statement.

Sincerely yours,

J.B.Rhine

f/

402 BUCHANAN   FOUNDATION HOUSE   TEL. (919) 688-8241

Rhine's wife, Louisa, sits on the fence when she deals with the Serios controversy in *Mind Over Matter* (1970) and briefly in *Psi: What Is It?* (1975). She clearly wants to believe. Serios's claim was "carefully and extensively investigated" by Eisenbud, she writes, who "reported strong evidence." But she hedges this praise with so many recognitions of the possibility of fraud that her final conclusion is that the question "remains open and no decision is possible."

13

Einstein and ESP

Einstein is frequently mentioned in the literature of parapsychology as a great scientist who, in contrast to so many of his colleagues, believed that psi phenomena had been demonstrated by the work of J. B. Rhine and his successors. In 1930 when Upton Sinclair published his book *Mental Radio,* Einstein contributed a brief preface to the German edition. In the American edition, the preface reads as follows:

> I have read the book of Upton Sinclair with great interest and am convinced that the same deserves the most earnest consideration, not only of the laity, but also of the psychologists by profession. The results of the telepathic experiments carefully and plainly set forth in this book stand surely far beyond those which a nature investigator holds to be thinkable. On the other hand, it is out of the question in the case of so conscientious an observer and writer as Upton Sinclair that he is carrying on a conscious deception of the reading world; his good faith and dependability are not to be doubted. So if somehow the facts here set forth rest not upon telepathy, but upon some unconscious hypnotic influence from person to person, this also would be of high psychological interest. In no case should the psychologically interested circles pass over this book heedlessly.

Parapsychologists, and journalists who write about the paranormal, often refer to this preface, sometimes quoting from it, as evidence of Einstein's belief in ESP. R. A. McConnell, for example, in his influential article "Parapsychology and Physicists" (*Journal of Parapsychology,*

Reprinted with permission from the *Skeptical Inquirer,* Fall-Winter 1977.

vol. 40, September 1976) lists Einstein, along with William Crookes, Oliver Lodge, and other physicists, as one of the "Titans" who were sympathetic toward psi research. A portion of Einstein's preface is quoted.

An even longer quotation appears in Chapter 7 of *Mind-Reach,* the recently published book by Russell Targ and Harold Puthoff. The chapter is about their work with Uri Geller — tests which they are convinced demonstrated beyond any doubt the clairvoyant powers of the Israeli psychic. To put their Geller experiments in perspective, and to argue that Geller's ability is not unique, they bring up Sinclair's book, quote from Einstein's preface, and ask, "Why then has this treasure trove of a book been neglected for the past forty-five years?"

It is not my purpose to explain here why I do not believe Sinclair's book should be taken seriously, because I have summarized my reasons in Chapter 25 of my 1952 book, *Fads and Fallacies in the Name of Science.* If the reader will consult this book's index for page references to Sinclair, they will understand why I regard him as sincere and honest, but incredibly gullible. He had only the dimmest grasp of scientific method and (in my opinion) was an unreliable observer and reporter of the uncontrolled, informal ESP-tests he conducted with his wife.

My purpose now is merely to reproduce, with the permission of the Einstein estate, a letter that Einstein wrote in 1946 to Jan Ehrenwald, and which came into my hands by way of physicists John Stachel and E. T. Newman. Dr. Ehrenwald is a British psychoanalyst now living in New York City, where he is a consulting psychiatrist to Roosevelt Hospital. For thirty years he has been studying psi phenomena and seeking a neurological basis for it. He is the most distinguished of a trio of living psychoanalysts (the other two are Jule Eisenbud and Montague Ullman) who are firm believers in psi. Next year Basic Books will publish Ehrenwald's latest book, *The ESP Experience: A Psychiatric Validation.*

A translation of Einstein's letter (the original is in German) follows:

Dear Dr. Ehrenwald: 13 May 1946

I have read with great interest the introduction to your book,* as well as the story of all the unpleasant experiences you have suffered, as many others among us have. I am happy that you succeeded in emigrating to this country, and I hope that you will find here the possibilities for fruitful work.

Several years ago I read the book by Dr. Rhine. I have been unable to find an explanation for the facts which he enumerated. I regard it as very strange that the spatial distance between (telepathic) subjects has no

*Dr. Ehrenwald had sent Einstein a copy of his book, *Telepathy and Medical Psychology* (London: Allen and Unwin, 1946).

relevance to the success of the statistical experiments. This suggests to me a very strong indication that a nonrecognized source of systematic errors may have been involved.

I prepared the introduction to Upton Sinclair's book because of my personal friendship with the author, and I did it without revealing my lack of conviction, but also without being dishonest. I admit frankly my skepticism in respect to all such beliefs and theories, a skepticism that is not the result of adequate acquaintance with the relevant experimental facts, but rather a lifelong work in physics. Moreover, I should like to admit, that, in my own life, I have not had any experiences which would throw light on the possibility of communication between human beings that was not based on normal mental processes. I should like to add that, since the public tends to give more weight to any statement from me than is justified, because of my ignorance in so many areas of knowledge, I feel the necessity of exercising utmost caution and restraint in the field under discussion. I should, however, be happy to receive a copy of your publication.

> With many regards,
> Albert Einstein

It is worthy of note that Einstein's main reason for skepticism is the fact, so often emphasized by Rhine, that reported psi forces do not decline with distance. All of the four known forces of nature—gravity, electromagnetism, the strong force, and the weak force—diminish in strength as they radiate from a source. Rhine has always considered this proof that psi forces lie entirely outside the bounds of known physical laws. In recent years, attempts to explain psi's independence of distance (as well as time!) have been varied—currently fashionable attempts draw on quantum mechanics—but none has been satisfactory or amenable to confirmation. Einstein found it easier to apply Occam's razor and adopt the simpler explanation: namely that some sort of bias, of which experimenters were unaware, entered into the experimental designs of psi experiments and accounted for the statistical results. If so, the failure of psi to decline with distance and time would be easily accounted for.

Einstein mentioned in his preface one possible source of bias in Upton Sinclair's tests. To spell it out: perhaps Mrs. Sinclair unconsciously suggested to her husband what he should draw, or he unconsciously suggested to her what she should draw. To give another instance, consider the possible role of hand-recorded errors in Rhine's early and poorly controlled PK experiments with dice. If bias were introduced by recording errors on the part of assistants who knew the target number (tests by psychologists have shown how common such recording errors are), it obviously would not matter in the least whether the subject was ten feet from the tumbling cubes, or ten miles, or in a submarine ten fathoms down, or on a spaceship ten thousand miles out. It would not even matter if the dice were shaken and tossed ten hours after the subject had concentrated his PK energy on the target.

This independence of psi from time and space continues to be a dramatic and troubling aspect of psi research. A splendid example is provided by the latest remote-viewing tests of Puthoff and Targ. In their project "Deep Quest," psychic superstars Hella Hammid and Ingo Swann were in a minisub, submerged off the coast of Catalina Island. They managed to describe the target sites, 500 miles away on land, as accurately as they had done in previous tests on land when the targets were nearby. In *Mind-Reach* the authors report that Ms. Hammid also did just as well in her remote viewing when the targets were randomly selected *after* she had made her report.

In his characteristically simple, humble, commonsense fashion, Einstein went directly to the hub of the matter. After a century of reporting of results by parapsychologists, the indifference of psi to all the rules that govern known forces continues to be (along with replication failures by unbelievers) a major reason why the majority of psychologists remain, like Einstein, highly skeptical of the reported extraordinary results.

14

A Second Einstein ESP Letter

The previous chapter was a note on Einstein's attitude toward the brief introduction he wrote for Upton Sinclair's book *Mental Radio*. The note included a letter that Einstein had written to the psychoanalyst and parapsychologist Dr. Jan Ehrenwald.

Dr. Ehrenwald has kindly allowed me to have a copy of a second letter he received from Einstein, which contains further comments on parapsychology. I have obtained permission from the Einstein Estate to publish the following translation:

Dear Mr. Ehrenwald: 8 July 1946

I have read your book with great interest. It doubtlessly represents a good way of placing your topic in a contemporary context, and I have no doubt that it will reach a wide circle of readers. I can judge it merely as a layman, and cannot say that I have arrived at either an affirmative or negative conclusion. It seems to me, at any rate, that we have no right, from a physical standpoint, to deny a priori the possibility of telepathy. For that sort of denial the foundations of our science are too unsure and too incomplete.

My impressions concerning the quantitative approach to experiments with cards, and so on, is the following. On the one hand, I have no objection to the method's reliability. But I find it suspicious that "clairvoyance" [tests] yield the same probabilities as "telepathy," and that the distance of the subject from the cards or from the "sender" has no influence on the

Reprinted with permission from the *Skeptical Inquirer*, Spring-Summer 1978.

result. This is, a priori, improbable to the highest degree, consequently the result is doubtful.

Most interesting, and actually of greater interest to me, are the experiments with the mentally retarded nine-year-old girl and the tests by Gilbert Murray. The drawing results seem to me to have more weight than the large scale statistical experiments where the discovery of a small methodological error may upset everything.

I find important your observations that a patient's productivity in psychoanalytic treatment is clearly influenced by the analyst's "school." This portion of your book alone is worth careful attention. I cannot fail to note that some of the experiences you mention arouse the reader's suspicion that unconscious influences along sensory channels, rather than telepathic influences, may be at work.

At any rate, your book has been very stimulating for me, and it has somewhat "softened" my originally quite negative attitude toward the whole of this complex of questions. One should not walk through the world wearing blinders.

I cannot write an introduction, as I am quite incompetent to do so. It should be provided by an experienced psychologist. You may show this letter privately to others.

Respectfully yours,
A. Einstein

The book, which Dr. Ehrenwald had sent to Einstein in the form of page proofs and for which Einstein declined to write an introduction, was *Telepathy and Medical Psychology.* It was published in England by Allen and Unwin in 1947 and in the United States the next year by W. W. Norton. The introduction was written by Gardner Murphy. (Dr. Ehrenwald's latest book, *The ESP Experience,* was published in 1978 by Basic Books.)

Let me add that I find Einstein's remarks entirely admirable. He is less dogmatic in his negative attitude toward parapsychology than he had been when he wrote his previous letter. He believes one should keep an open mind, but he is still strongly put off by the reported evidence that ESP does not decline with distance. With great tact and politeness he informs Dr. Ehrenwald that unconscious but quite normal sensory channels, rather than ESP, may be causing the effects that Dr. Ehrenwald attributes in his book to telepathic contact between analyst and patient. Finally, with characteristic humility, he points out that the entire field is one in which he has no competence.

I wish to thank Martin Ebon for providing a translation of Einstein's letter. It has been approved and slightly edited by Dr. Ehrenwald.

Postscript

After I published the two Einstein letters in the *Skeptical Inquirer,* Jan Ehrenwald published them again, along with his lengthy and undelivered reply to Einstein. See "Einstein Skeptical of ESP? Postscript to a Correspondence," by Ehrenwald, in the *Journal of Parapsychology,* vol. 42, June 1978, pages 137–142.

15

Geller, Gulls, and Nitinol

Uri is certainly 25 percent fraud and 25 percent showman, but 50 percent is real. For example, the nitinol came back with a different structure. . . .

Arthur Koestler, quoted by Adam Smith in *New York* magazine (Dec. 27, 1976).

It would be hard to invent a more grandiose title for a book about Uri Geller than *The Geller Papers*. The image that floats in the mind is one of a monumental new development in science, so revolutionary that a world congress of experts has convened to discuss the phenomenon. Technical papers are delivered. Here they are, in one impressive volume, published last year by Houghton Mifflin, carefully edited by Charles Panati, former physicist, now a science writer for *Newsweek*.

Panati's subtitle is even more pompous: "Scientific Observations on the Paranormal powers of Uri Geller." The phrase permits not the slightest crack of doubt. The scientific community, it implies, is not concerned with whether Uri Geller, the handsome young Israeli stage performer, has paranormal powers. It has made up its mind. Uri's psychic abilities are not to be questioned. The task now, suggests the subtitle, is to observe those powers in the laboratory, analyze them, and develop viable theories to explain them. One opens Panati's book with trembling fingers.

Among the twenty-two papers assembled in this volume, one stands high above all the others. This is Panati's own opinion. Over and over

Reprinted with permission from the *Humanist*, May/June 1977.

again on radio and TV talk-shows, he has said that the most important chapter in his book is Eldon Byrd's paper, "Uri Geller's Influence on the Metal Alloy Nitinol." This also is the opinion of almost every review of the book I have seen. D. Scott Rogo's review in *Psychic* (September 1976) is typical: "Despite the fact that this book will probably not be too convincing to the confirmed sceptic," Rogo writes, "a few papers are included which offer, to my mind, the best evidence so far published supporting Geller's claims. These contributions do stand in striking contrast to the general run of the accounts. One of these papers is Eldon Byrd's."

Uri himself is exploiting Byrd's paper. A full-page advertisement for Uri in *Variety* (October 27, 1976) has four boxed testimonials beside Uri's picture: one by Werner von Braun, one by Harold Puthoff and Russell Targ (the Stanford Research Institute physicists who claim they have certified Uri's ESP ability but not his PK powers); one by Friedbert Karger, of the Max Planck Institute for Plasma Physics, in Munich; the fourth is a quote from Byrd: "Geller altered the lattice structure of a metal alloy in a way that cannot be duplicated. There is no present scientific explanation as to how he did this."

Byrd, it appears, has written a paper of stupendous scientific value. Its importance is underlined by Panati in his introduction, and in briefer comments before and after Byrd's article. Byrd's paper, writes Panati, "appears here with the official approval of the Naval Surface Weapons Center. . . . The paper represents the first time parapsychological research conducted at a government facility has been released for publication by the Department of Defense."

Parapsychological research at a government laboratory? Officially approved by a department of the U.S. Navy? Such a report is certainly not to be treated lightly. But before we take a close look at exactly what transpired, a brief introduction to Byrd will be helpful.

He was born in 1939 in Winchester, Indiana. After obtaining an engineering degree at Purdue, he went on to get a masters in medical engineering at George Washington University. Since 1968, he has been an operations analyst at the White Oak Laboratory of the Naval Surface Weapons Center (formerly the Naval Ordnance Laboratory) in Silver Spring, Maryland. He is an officer in the Naval Reserve, a member of Mensa (the high-IQ club), and a Mormon. There is a strong connection, he says, between his Mormon views and his belief in paranormal phenomena. He is the author of *How Things Work,* published by Prentice-Hall in 1973.

Byrd has long been interested in all aspects of the paranormal, from the Bermuda Triangle to UFOs. In 1975, he gave a course on psychic phenomena at an occult center in Silver Spring, which we will mention again below. He tells me he believes Geller to be "basically honest." He is open-minded on the question of whether Uri teleported himself from

Manhattan to Ossining, New York, as Uri tells it in his autobiography *My Story.* He is convinced that "hundreds" of children, many of them in Canada, can now bend metals "better than Geller," and he has agreed to serve on a committee to test these children if funding can be obtained.

Byrd's early work on the "Backster effect"—the ability of plants to respond to human thoughts and emotions—is detailed on pages 40–42 of *The Secret Life of Plants,* by Peter Tompkins and Christopher Bird (Harper and Row, 1973). "Byrd was able to demonstrate on television a plant's reaction to various stimuli, including his *intent* to burn it. On camera, Byrd got a plant to respond by shaking a spider in a pill box. . . . He also got a strong reaction when cutting the leaf of another plant."

These reports are now obsolete. Byrd's disenchantment with the Backster effect followed his discovery that he obtained the same reactions when he made similar tests with pieces of Styrofoam. He now believes that the effects measured by Cleve Backster are not produced by the plant's "consciousness," but by electrical fields surrounding persons. Philip J. Klass, a senior editor of *Aviation Week* and author of an excellent book, *UFO's Explained* (Random House, 1975), tells me he once asked Byrd why he so dogmatically ruled out the possibility that Styrofoam may have a low-grade consciousness like a plant. "Because," replied Byrd, "that's ridiculous."

Byrd's first meeting with Uri was on the evening of October 19, 1973, at a laboratory in Silver Spring called the Isis Center. Panati identifies it as the "Isis Center of the Naval Surface Weapons Center." Is not *Isis* a peculiar name for a naval laboratory? The Egyptian mother-goddess was worshipped by members of one of the many mystery cults that flourished in ancient Rome during the decay of traditional religious beliefs and before the Fall. One thinks, too, of Madame Blavatsky's monumental theosophical treatise, *Isis Unveiled.*

I asked Byrd how the Isis Center got such an exotic name. His reply quickly solved the mystery. The Isis Center, which became defunct in 1975, had no connection with the navy. Its full title was "The Isis Center for Research and Study of the Esoteric Arts and Sciences." It had been formed by a group of local occultists, headed by Jean Byrd (no relation to Eldon), who thinks she may be the reincarnation of Isis. The center, on Fenton Street in Silver Spring, had booked Uri for a performance at George Washington University's Lisner Auditorium. Byrd had asked the center to arrange a session with Uri before Uri left for his magic show.

This is what happened. Byrd presented Uri with two pieces of nitinol wire and a small nitinol block. Nitinol is a curious alloy of nickel and titanium, developed many years earlier by a navy metallurgist. It has a "memory." Under intense heat, you can give a piece of nitinol a certain shape. When cold, you can alter its shape; but when it is heated again it goes back to its original form. It has been used for satellite antennas.

Nitinol wire can be coiled into a small area. Once the satellite is in orbit, heat causes the wire to expand to the desired pattern. The alloy has many other uses, both military and commercial.

Byrd was interested in seeing if Uri could alter any of nitinol's properties. He first handed Uri the tiny block. Uri "handled the block for some time," but was unable to change it. He had, he said, no "feel" for the material.

Byrd pocketed the block and gave Uri a straight piece of nitinol wire about 1.5 mm. in diameter. Again Uri "handled it for a while" but without results.

Byrd then gave Uri a five-inch straight piece of nitinol wire 0.5 mm. in diameter. Such wire is black, very thin, and extremely flexible. It is easily bent by the fingers into any desired shape. Uri went into his standard routine for bending keys, nails, spoons, and what have you.[1] After twenty seconds of massage, he produced a small bump in the wire's center.

Stirred by this dramatic paranormal event, Byrd had some boiling water brought to him. Normally a piece of nitinol wire, bent by hand, would spring back to its straight shape when immersed in boiling water, or even hot coffee. Instead, the wire lost its bump and assumed a right-angle shape. "This was an exciting finding," writes Byrd. "I lit a match and held it over the kink, but still the wire did not straighten out." (Byrd is ambiguous in his use of the term *kink* — sometimes he means the bump that first appeared, sometimes the sharp angle that resulted when the wire was heated. I will confine "kink" to mean the sharp angle.) Later Byrd tried to put a similar kink in a piece of wire. He was unable to do so, he writes, "without using Bunsen burner and pliers."

I have no doubt that Byrd's description of what took place at the Isis Center is given to the best of his recollection. Unfortunately, a nonmagician's memory of a magic feat is notoriously unreliable. Even magicians can remember wrongly a trick they have seen for the first time. I consider myself a knowledgeable student of conjuring, yet I am frequently mystified by new tricks. Every time I have a session with Jerry Andrus, a creative magician who lives in Oregon, he completely fools me. After he tells me what he did, I am sometimes amazed at how faulty my memory was of what I had seen. However, quite apart from the untrustworthiness of Byrd's memory, his account contains several misleading assertions.

Item 1: Byrd writes that at the time of his test with Uri, "nitinol was generally not available to the public." His point is that Uri, who might have known in advance (through mutual friends) of Byrd's interest in nitinol, would have had great difficulty obtaining samples. Not true. The Edmund Scientific Company, which advertised then, as now, in popular magazines, listed a "nitinol kit" for five dollars in its catalogs for 1971, 1972, and 1973. For an additional dollar you could get a 96-page *NASA Nitinol Book* that went into details about the alloy's properties. Samples of nitinol wire were distributed free during those years by the Naval Ordnance Laboratory at their lab exhibits to people who toured the lab and in response to written requests.

Not only that, but magicians were familiar with nitinol. In 1972 a New York amateur, Charles Kalish, was mystifying friends with a trick he had invented, which was later sold by a London magic store. A spectator selects a digit from one through nine. A paperclip (actually made of nitinol) is placed in an envelope and the envelope burned. Searching the ashes, one finds that the wire has assumed the shape of the chosen numeral.

Item 2: Byrd's brief account of the Isis tests gives no indication of the confusion that reigned in the Isis "laboratory" when Geller bent the wire. In addition to Byrd and Uri, those present included Jean Byrd, two of her secretaries, Andrija Puharich, and Shipi Shtrang. Puharich is the parapsychologist who discovered Uri in Israel and arranged for him to come to the United States. He is the author of *Uri,* a book that reveals how Uri derives his powers from computers in a flying saucer. Shipi is Uri's frequent companion. According to Shipi's sister Hannah, she and Shipi were confederates who signaled information to Uri during his Israeli stage shows.[2] Uri himself is on record that his powers improve when Shipi is around. Magicians agree. They believe that on many occasions it is Shipi who secretly aids Uri; for example, by "putting in the work" before Uri bends a spoon.

Byrd has freely admitted in letters to me that at the Isis Center conditions were so uncontrolled that "almost anything" could have happened. Distractions were so great, he wrote, that even "wire swapping" could have occurred.

Item 3: In his paper, Byrd says that "several metallurgists" at the naval center tried to remove the right-angle kink in the Gellerized wire by putting the wire "under tension in a vacuum chamber" and heating it until it glowed. When the wire cooled, the kink returned. "They had no explanation for this behavior."

According to the Naval Surface Weapons Center, this test was never made. Indeed, the center was so annoyed by Panati's false assertion that the Isis Center was a government laboratory that they asked their public-affairs office to prepare a four-page memorandum (dated July 19, 1976) to send in response to serious inquiries. The memorandum stated that the experiments with Uri "were undertaken by Mr. Byrd as a personal interest, on his own time, and at no cost to the government."

On page 4 of his introduction, Panati writes that Geller demonstrated his psi powers "for physical scientist Eldon Byrd at the Naval Surface Weapons Center," and on page 5, "Geller arrived at the Naval Surface Weapons Center in October of 1973." "This is an error," the memorandum states. "Geller has never been on the premises of the Naval Surface

Weapons Center." Byrd's chapter had been reviewed by the Naval Ordnance Laboratory only from the standpoint of accuracy about the properties
of nitinol and for compliance with military security. The laboratory, says the
memo, "assumed no responsibility for the outcome or implications of
Mr. Byrd's experiments." Approving the release of Byrd's paper "neither
confirmed nor denied the parapsychological aspects" of the paper.

As for Byrd's claim that "several metallurgists" tried to remove the
kink in a vacuum chamber, the memo states: "The occurrence of that test
could not be confirmed by laboratory records or by metallurgists at the
laboratory." Dr. Frederick E. Wang, the navy's top nitinol expert, was
the man who Byrd thought had made such a test. Wang cannot remember
making it.

Since no magicians were present during the chaotic session at which
Uri bumped and kinked one piece of wire, it is impossible to do more
than speculate on possible nonparanormal explanations. One scenario is
that Uri came prepared with samples of wire to which he had previously
given permanent kinks (in a way to be explained below) and then straightened.[3] He passed up the wire of larger diameter because he had not
brought wire of that size. The smaller-diameter wire was then switched,
by Uri or Shipi, for Byrd's sample. Uri bumped the wire while stroking
it—easily done by a push with the thumbnail. Then when Byrd heated the
wire it naturally lost the bump and assumed its permanent kink.

However, it is not necessary to suppose that Uri came with prepared
wire. Nitinol wire is now harder to get than in 1973, but I finally obtained
a sample about a foot long of the 0.5 mm. wire. I cut off a small piece,
and I swear by all that is holy that my very first experiment was a whopping success. Using two small pairs of pliers I bent the wire at a sharp
angle. I straightened the wire, then by holding the wire between thumb
and first two fingers, and pressing with my thumbnail, I created a bump
at the wire's center. I put the wire in a bowl and poured boiling water
over it. The bump vanished and the wire assumed the shape of an angle,
almost 90°, with a sharp vertex. The angle was unaltered by applying a
match flame.

Excited by this unexpected success, I tried producing a sharper angle
(about 3°), but when I straightened the wire it snapped in half. I then
repeated the experiment with a third piece, this time using nothing more
than two pennies to grip the wire, and a third penny to force an acute
angle. I straightened the wire, letting the angle remain as one side of the
bump. When boiling water was poured over this wire the bump dis-
appeared and the wire assumed an angle with a sharp vertex of about
75°. I have it before me as I type. It is indistinguishable from the wire in
plate 4 of Panati's book.

There are no signs of scratches on the wire. A cloth over the wire will,
of course, eliminate all possibility of scratching. A close look at the more

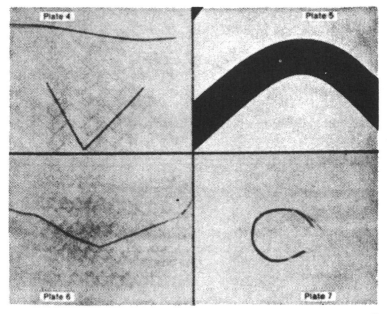

Panati's Plate 4. Two pieces of nitinol wire. Upper: Shape before Geller rubbed it. Lower: Shape after he rubbed it and after it had been heated to restore straight configuration. The wire is now permanently deformed. Plate 5. Shadowgraph of one piece of nitinol bent by Geller. Radius of curvature was found to be less than 1 mm. Plate 6. This piece of nitinol rubbed by Geller developed multiple *2-dimensional* permanent bends. Plate 7. Geller's influence on this piece of nitinol induced a *3-dimensional*, permanent bend. After Geller rubbed it, it took the shape of an ellipse. The only known way to get this result is to twist the wire into an ellipse, constrain it, and then heat it to about 500°C.

obtuse kink in the first wire I bent shows the kink to be indistinguishable from the shadowgraph of Panati's plate 5. Again, a match had no effect on the 75° kink, although the wire glowed red in the flame. In less than ten minutes of experimentation, without a Bunsen burner, I had permanently altered the memory of two nitinol wires!

But, you may say, how could Geller have prepared a wire in this way under Byrd's eagle eyes? There are several scenarios. While Uri tries unsuccessfully to alter one piece of wire, Shipi surreptitiously picks up the other piece, excuses himself to go to the bathroom, prepares the wire, straightens it, then returns and leaves the wire where he found it. Uri discards the wire for which he has no "feel," picks up the other one, and creates the bump while stroking it. When Byrd puts the wire in boiling water, it kinks.

Another possibility: Byrd had obtained the five-inch piece, his paper says, by snipping a longer piece of wire into three parts. Assume that while he is working with Uri the other two pieces are at spot *X*. Shipi picks up one, takes it to the washroom, where he puts in the work, returns, and at a suitable moment switches it for the wire near Uri. He then replaces the wire he now has at spot *X*.

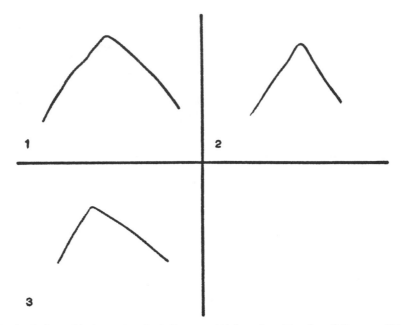

Gardner's three nitinol experiments. 1. Permanent kink produced by pliers. 2. Permanent kink produced with pennies. 3. Permanent kink produced by biting.

A third scenario: A sharp-angle kink cannot be put into nitinol wire with the fingers because the wire forms a rounded hump that must be squeezed together by two hard surfaces. This is easily done by the teeth, but fingers are not sufficiently firm. It would be possible, however, to make a tiny device, readily palmed, do the job. All you need is a hinged piece of hard plastic, its surface grooved to keep the wire from turning. Twist the wire's center to form a small loop, give the rounded bend a pinch, and the deed is done.

Such a kink is permanent in the sense that it becomes the wire's new memory. However, it is always possible to put such a kinked wire into a vacuum or gas chamber and under extreme heat reanneal the wire so that its memory is straight again. This is the test that Byrd mistakenly thought Dr. Wang had performed. Had this been the case, Uri would have altered the wire's memory in a way that is, in Byrd's phrase, "beyond technology"— in other words, a feat unexplainable by science even on the assumption that Uri may have cheated. Dr. Wang was unwilling to make such a test without funding, so Byrd sent the wire to his friend Ronald S. Hawke at the Lawrence Livermore Laboratory, in Livermore, California.

The great test was finally made on January 31, 1977. Hawke removed the kink. Thus the most sensational claim in Byrd's paper—indeed the most sensational "fact" in Panati's book—proved to be an error. Geller had done nothing "beyond technology." The kink he put in the Isis wire was no different from the kind of kink a child can put in nitinol wire by biting it.

Did Uri or Shipi bring prepared wires to the Isis Center, or did Uri or Shipi kink a wire while Byrd's attention was on something else? Who knows? Since Byrd himself regards the Isis test as uncontrolled, we need waste no more speculations on it.

Byrd's second experiment with Uri occurred a month later. This time Uri produced permanent memory changes in two pieces of nitinol wire. Plates 6 and 7 in Panati's book show how the wires looked after being "rubbed gently by Uri Geller" as the caption has it. Any reader of the book would assume, from Byrd's text and Panati's photo captions, that Byrd himself had witnessed this paranormal feat.

Being curious to know who else was present during this second miracle, I wrote Byrd for details. I could hardly believe my eyes when I read his answer. Byrd didn't know! He had given samples of wire to Uri when he was at the Isis Center. Uri took them home, then later brought two back in the distorted forms shown in the plates. How did Byrd and Panati know that the wires had altered when Uri gently rubbed them? Because that's what Uri said had happened!

Not to tell the reader that Uri had taken these wires home with him is the kind of omission that, in any report claiming to be scientific, stamps the author as naive, disingenuous, or both. As sociologist Marcello Truzzi likes to say, extraordinary laboratory results, violating all known laws of science, demand more than extraordinary controls; they demand extraordinary care in reporting. Since this second test obviously had *no* controls, we may dismiss it at once.

Let us move on to test three, the climax of our comedy. It is the only test, insists Byrd, that had controls of utmost rigidity. This great experiment took place in October 1974, one year after the Isis test. In what laboratory? Well, not exactly a laboratory. It took place at the Connecticut home of the writer John G. Fuller. As all Geller-watchers know, Fuller is the leading author of books on the occult. His *Arigo: Surgeon of the Rusty Knife* is about a Brazilian psychic surgeon who operated according to instructions whispered in his left ear by a dead German doctor. Fuller's latest book, *The Ghost of Flight 401*, tells how the spirit of an Eastern Airlines officer, killed in a plane crash, kept appearing on Eastern Airlines flights. Fuller is the man who "edited" Uri's wild autobiography. He ghosts Uri's newspaper column. He is a confirmed Gellerite who buys almost every aspect of the current psi scene.

Who was present on that memorable day at Fuller's house? Uri, Byrd and his wife, Fuller, Ronald Hawke, and two of Uri's ladyfriends, Solvej Clark and Melanie Toyofuko. Hawke is the paraphysicist at the Lawrence Livermore Laboratory who recently did the reannealing test on the

Uri-kinked wire. He, too, contributed to Panati's book: a short paper about a test in his laboratory in 1974, when Geller erased a pattern on a magnetic program card. At Fuller's house, only Byrd and Hawke witnessed the nitinol tests with Uri.

Byrd had brought with him three pieces of 0.5 mm. nitinol wire, each about four inches long. In his paper he writes that he cut these wires before he left for Connecticut. This conflicts with what Byrd told Klass in a phone conversation on October 11, 1976. Byrd then said several times that the wire was cut into four pieces at Fuller's house. When Klass asked if Uri had been in the room when the wire was cut, Byrd said he could not recall. In a second phone conversation with Klass, November 29, 1976, after being reminded of what he had said in his paper, Byrd went back to the story that the wire had been cut before going to Connecticut.

As Byrd recalled it in November, the original wire was about twenty inches long. He cut off one four-inch piece, which he left at his laboratory as a "control." In New York, where (as we shall see) he spent the night before going to Connecticut, he cut the remaining piece of wire into four pieces. One of these was then put aside as a *second* control piece. It remained in an envelope in his suitcase.

Byrd's paper makes no mention of a second control piece. "Prior to leaving for Connecticut," he writes, "I had cut the wire into four pieces. . . . One piece was used as a control and was not taken to Connecticut." As he now recalls it, there were five pieces of wire, two of which were controls. One piece was left at Silver Spring. One was taken to Connecticut, but not touched by Uri. These are not trivial details, because they show how carelessly Byrd wrote up the "experiment," and how confused his memory of the details became.

At Fuller's house, Uri stroked the three wires in his usual manner and produced sharp kinks in each. Byrd reports no bumps this time—just kinks. But in a letter to me, Byrd explicitly states that a bump formed each time in the wire, exactly as at the Isis Center. The bump turned into a kink when Byrd applied a match flame. In his paper, Byrd says he held the first wire at both ends, the second wire at one end, and the third "was given to Geller to do with as he pleased. He rolled it between his thumb and forefinger and it kinked sharply (see plate 4)." The three kinks, he told Klass, were about 60°, 90°, and 110°. We are not told how much time elapsed during this test.

"How did Geller achieve such results?" Byrd asks in his paper. "At the present I have no scientific explanations for what happened. . . . I can say that the possibility of fraud on Geller's part can be virtually ruled out."

Byrd's account of this test is notable mainly for his failure to supply any details about controls. His descriptions of tests 1 and 2, which gave no clue to the chaotic conditions prevailing at the Isis Center and no indication that he had not observed what Uri did in test 2, hardly inspire confidence that he is giving an accurate account of what went on at Fuller's house. He simply assures us he took "extra precautions." The entire session was *audio* taped. This isn't much help to magicians trying to reconstruct nonparanormal explanations. Hawke refuses to discuss the controls with anyone, having agreed to let Byrd be the sole spokesman. Once more, we are forced to rely on the easily distorted memory of one physical scientist who firmly believed before the test began that Uri had paranormal powers and who was as anxious to vindicate that belief then as he is now to defend his controls.

In my correspondence with Byrd, and in going over careful notes taken by Klass of his phone conversations with Byrd, many facts have come to light that cast grave doubts on the adequacy of Byrd's controls. These are facts that should have been in Byrd's paper but that he did not think important enough to mention.

The night before the test, Byrd and his wife and Ms. Toyofuko slept in a Manhattan apartment belonging to a friend of Melanie's who was out of town. The test wire (or wires) was in a suitcase. It would have been a simple matter for Melanie to switch the wires for previously prepared wires of the same length, diameter, and die marks.[4] It would also have been easy for Melanie to borrow the wires and put in the work herself. When Byrd and his wife left the apartment with Melanie, to have dinner out, the suitcase remained unguarded at the apartment. What was to prevent Shipi from entering the apartment and putting in the work? Byrd is, of course, persuaded that Uri's friends would never take part in such skulduggery. But according to Shipi's sister, Shipi is quite capable of such things; besides, the mere possibility proves how weak Byrd's "extra precautions" were.

Byrd tells me in a letter that his procedure at Fuller's house called for a close watch on Uri to make sure he did not get to the test wires in Byrd's briefcase. Did the procedure also call for carefully monitoring the two young ladies? Did Hawke and Byrd watch like hawks and birds the paths followed by Ms. Toyofuko and Ms. Clark whenever one of them went to the bathroom? I asked Byrd if his protocols included this. He never answered.

Byrd has not a line in his paper about testing the straight memory of each wire before allowing Uri to handle it, but in letters to me he strongly maintains that he did test each wire with a match flame. How trustworthy is Byrd's memory on this score? I tried to obtain confirmation of the pre-kink testing from Hawke, but he did not reply to my letter.

Let us give Byrd the benefit of the doubt and assume that he did indeed test each of the three wires with a match flame before allowing Uri to touch it. Such testing would have been advisable, not only to make sure the wires had not been switched, but also to put the wires (which get slightly bent when carried about) in a perfectly straight condition. Uri would surely have anticipated this. His plan would be to switch each wire *after* it had been straightened by a match flame. What are the most likely scenarios?

One obvious scenario is that the night before, while Byrd and his wife were having dinner out with Melanie, someone went into the apartment, opened Byrd's suitcase, and replaced his wire samples with duplicates — all with straight memories. Byrd's wires were carefully given a permanent kink by mechanical means, then either straightened, or straightened with a small bump in the center.

When Ms. Clark drove the Byrds and Melanie to Fuller's house the next morning, she could have carried with her these wires that contained "the work" and secretly passed them to Uri. Uri had been swimming in the river in back of Fuller's house when the Byrds arrived and (according to what Byrd told Klass) came into the house wearing only bathing trunks. Uri now has Byrd's original wires with him, concealed under the belt of his trunks or perhaps in his hair.[5]

I have no doubt that Byrd and Hawke will vigorously deny that Uri could have switched the wires that were flame tested by Byrd for the wires with the altered memory. They would be less sure of their ability to detect such a switch if they would spend several hours watching a good close-up magician do some tricks that involved switching. Switching has been developed into a subtle art. Nonmagicians simply do not know what to look for or when to look for it; and as any magician will tell you, scientists are easier to fool than children.

Imagine that Byrd has taken the first test wire from its envelope. It is not his original wire, but there is no way he can tell without an electron microscope; and even a similarity of die marks would not guarantee it to be the original. He tests the wire with a match flame, allows it to cool, then without letting go of both ends he permits Uri to stroke the wire. But nothing happens. Uri pretends to be tremendously disappointed. He doesn't yet "feel" for it. Could he have a glass of water? Byrd momentarily allows Uri to hold the wire while he reaches for the glass at the end of the coffee table. At that moment of distraction — a moment of what card-hustlers call "shade" — Uri makes the switch. There is no reason on earth why Byrd would remember he had momentarily released the wire. At the time it would have seemed totally irrelevant, and he is now completely honest when he says that, to his best recollection, he "never let go of the wire." Hundreds of people whose keys Randi has bent will tell you the same thing: that they never let go of the key, when actually they did. Seemingly irrelevant details fade quickly from the memory. Uri is a master at precisely this kind of "one ahead" timing, followed by remarks carefully designed to leave a false memory of what happened.

Magicians have devised dozens of ways for switching small objects. I do not care to get into trouble with my magician friends by saying too much here, so let me give only one simple way Uri could have handled it. He had the duplicate wire palmed in the same hand he had used for stroking the wire. When Byrd did whatever Uri requested to get his moment of shade, Uri's hand momentarily lowered and the unprepared wire was simply dropped on the floor. In my experiments with nitinol, I accidently dropped a four-inch piece on a rug. It fell noiselessly, bounced, and it took me five minutes to find it again. Nitinol wire of 0.5 mm. is finer than a hairpin. A vacuum cleaner picks it up like a piece of thread.

Byrd immediately seized the wire again, and today has no memory of having released it. He was now holding one of his original wires, but one with a kinked memory. It may also have had a bump in the center. Without letting go of the wire, Byrd applied a match flame and the wire went into its kink.

Since we are doing no more than speculating on possible scenarios, let's try a different tack for the second wire. Flushed with success, Uri bumped the second wire (which Byrd held at only one end) almost immediately. Byrd was eager to apply a match flame. Now surely it takes two hands to open a match folder, take out a match, close the folder, and strike the match.[6] Where was the wire? Was it on a table? In Uri's hand? Did Uri hand the wire to Hawke? In each case the opportunity for a switch is apparent. Byrd applied the flame, the wire kinked. It was put into another envelope for lab testing.

As for the third wire, there is no problem about when Uri made the switch, because Byrd himself wrote in his paper that this wire "was given to Geller to do with as he pleased." Our scenario contains nothing beyond the ability of a clever magician. It accords completely with the facts as Byrd recounts them. Is this how Geller handled it? That is not the point. The point is that such scenarios make nonsense of Byrd's claim that the tests at Fuller's house were carefully controlled.

The possibility of distractions that would permit switching rises when we learn that much more went on during that fall afternoon at Fuller's than we are told about in Byrd's paper. In talking to Klass, Byrd has mentioned four other tests that took place, and perhaps there were still others he has not yet seen fit to mention. The first experiment of the day, he told Klass, involved a germanium crystal about the size and shape of a Hershey chocolate kiss. A piece of it broke off in Uri's hands, but this happened when neither Byrd nor Hawke was looking. This tendency of things to Gellerize when no one is looking is so common that John Taylor, the eminent British Gellerite, calls it the "shyness effect." "Oh, look!" Uri exclaimed. "It's broken!" Later lab tests showed no change in the crystal. The test was declared a failure.

Another test was with a very thin silicon wafer. It had previously been smashed into thousands of pieces, which had been put in a "poly" bag. One piece had been kept out as a control. "What do you want me to do?"

Uri asked. "Well," said Byrd, "if you can put these pieces together again, that would be pretty cool." Uri, said Byrd, smiled faintly. Nothing happened to the pieces. Later lab tests showed them unchanged. This test was failure number two.

A third test, which Byrd reported to both Klass and me, used Byrd's large brass office key. Hawke held one end. After Uri rubbed the key, its shank was seen to have been slightly curved. The key was placed on a sheet of white paper on a piano bench, where Byrd and Hawke watched it for half a minute. Byrd tells me that the key continued to bend "visibly" while they watched.

This phenomenon has often been reported. After bending a key, Uri usually puts it aside, points to it, and shouts: "Look! It's still bending!" This is exclaimed so convincingly that people actually imagine they see the key continuing to bend. Randi has had exactly the same results when he bends a key for anyone who is strongly suggestible and who believes Randi has bent it by some mysterious force that continues to act on the key. If the reader has any doubts about the effect of belief on what even a trained scientist "sees," I suggest he look into one of astronomer Percival Lowell's books about the canals on Mars that he "saw" so clearly that he was able to sketch detailed maps of them.

Success with the key inspired Uri to declare that he was getting "hot" and was now eager to try the nitinol. We do not know how much time elapsed before the wire tests were made or how much time elapsed during the wire tests. Byrd told Klass he was with Uri for about five hours, of which about three were recorded on cassette tape. We do not know if the room in which the nitinol tests were made was locked. We do not know how often Byrd, Hawke, and Geller were interrupted by others looking in or wandering in and out of the room.

We do know, from Byrd's conversations with Klass, that sometime during those five hours Uri executed another little miracle. He bent the tweezers of Hawke's Swiss army knife. This time Uri used his familiar "under water" bit. When people are watching too closely, Uri often says that objects sometimes bend better under water. In moving to the nearest sink he gets his needed shade. As Byrd tells it, Uri held the tweezers under a faucet, and the blades could be seen to "curl up" under the running water.

Until now, I have given only scenarios that require switching, but perhaps we are underestimating Uri's dexterity. Almost a year had passed since Byrd had given Uri samples of nitinol to take home. Geller had plenty of time to construct a small device, such as I described earlier, but more ingeniously made. We must now consider the possibility that Uri, in the very act of stroking a wire, used such a gimmick to put in the work.

"Gimmick" is a magician's term for any small device that is kept concealed from the audience but is essential to a trick's working. That Uri sometimes uses gimmicks is beyond doubt. Bob McAllister, a New York magician, spotted a palmed magnet in Uri's hand on one occasion when Uri altered the time on a digital watch. When Uri made a compass needle jump, on a television show, it was obvious from his head movements that he had a magnet either in his mouth or on his clothing near his chin. When he produced a Geiger-counter burst at Birkbeck College, in London, his gimmick was probably a concealed source of beta radiation. We cannot, therefore, rule out the possibility that Uri used a palmed gimmick to kink the wire while massaging it.

Magicians know many ways to design a gimmick so that it can be finger-palmed. It would, of course, be painted flesh color to make it difficult to see even if one caught a glimpse of the inside of the hand. In the act of rubbing the wire, the gimmick would pinch a small bend, then Uri's thumb would convert the kink to a bump, using the kink as one side of the bump. The other hand could take over the rubbing, allowing the hand with the gimmick to dispose of it in ways familiar to magicians. Both hands would thus be "clean" at the time the bump is first revealed.

If Uri had such a gimmick, he would probably not have used it the first time he stroked a wire. He would have seen to it that Byrd and Hawke got a clear view of the palm side of both hands. If they made no effort to inspect his hands, Uri could have introduced the gimmick a moment later.

That such a gimmick can be made would be inconceivable to Byrd and Hawke. Why should they even think to inspect Uri's hands? Suppose, however, that Byrd had designed his experiment more carefully and that his protocols called for a close inspection of Uri's hands each time he began to rub a wire. What would Uri have done? The answer is simple. Nothing. Unlike magicians, whose tricks must always work, psychics are under no such handicap. Paranormal power comes and goes in mysterious ways, its wonders to perform. Uri would simply not have had the "feel" that day for nitinol, just as he had no feel for the germanium crystal and the smashed silicon wafer. Uri is, above all, an opportunist who takes things as they come and plays them in whatever way is best under the circumstances. If nothing extraordinary happens, Gellerites even take that as evidence that Uri is a genuine psychic and not a magician!

Hundreds of clever gimmicks have been designed for close-up magic and, in the hands of a good performer, are never detected. My personal opinion, however, is that Uri did *not* use a gimmick. He has enormous skill at psychological misdirection. If his spectators are believers, he can get away with things no magician would dare attempt. There is no evidence that Uri is particularly skillful with his hands, and my guess is that the use of a gimmick would be beyond his manipulative abilities. The previous scenario, in which Byrd's original wires are borrowed and later given back to him, would be easier for Uri to carry out, and more in his style.

There is, of course, no way to know exactly how Uri handled it. On this point, let me quote from a marvelous essay by Lucian, the second-century Greek satirist. His "Alexander the Oracle Monger" is a detailed exposé of methods used by one of the Uri Geller's of his time. Writing about Alexander's performance before a group of Pamphilogian "fatheads," Lucian adds: "It was an occasion for a Democritus . . . a man whose intelligence was steeled against such assaults by skepticism and insight, and who, *if he could not detect the precise imposture,* would at any rate have been perfectly certain that, though this escaped him, the whole thing was a lie and an impossibility."

Let me sum up, Byrd describes, elliptically and inadequately, three sloppily designed, informal tests of Geller's ability to influence nitinol. The first test had almost no controls. The second had no controls of any sort. The third, which Byrd naively insists had rigid controls, turns out to be as crudely controlled as the first. Almost everything Panati says about Byrd's paper is wrong, nevertheless he is right about one thing. Byrd's paper is the most impressive in the book.

NOTES

1. The best accounts of Uri's methods of bending things are in *The Magic of Uri Geller,* by James Randi (Ballantine paperback, 1975); *Mediums, Mystics, and the Occult,* by Milbourne Christopher (Crowell, 1975); and *Confessions of a Psychic,* by Uriah Fuller, 1975 (obtainable from the publisher, Karl Fulves, Box 433, Teaneck, N.J. 07666).

2. Hannah gave this information to an Israeli journalist who reported it in an Israeli paper. A translation of his article can be found in Randi's book, cited in the previous footnote. Uri has since said that Hannah made up her story to spite him because he had ceased to consider her his number-one girlfriend.

3. On March 25, 1974, Philip Klass asked Byrd if he had told Uri in advance that he was bringing nitinol wire to the Isis Center. Byrd replied: "I don't think I mentioned it was nitinol. I just said it was a metal with a memory.

4. Die marks are striations produced on nitinol wire when it is drawn through a die. They are similar to ballistic marks on a fired bullet and can be seen only under high magnification, preferably by an electron microscope.

One of the most confusing aspects of my correspondence with Byrd is that, like nitinol wire stroked by Uri, his memory keeps altering. When I first suggested (May 1976) that Uri might have come to the Isis Center with a prepared wire, Byrd replied that this was ruled out because, had Uri switched wires, the kinked wire would not have shown the same die marks as the wire from which it had been cut. In a later letter (December 1976), I asked Byrd if it was he who had checked the correspondence of die marks on the Isis wire and the control wire. No, he answered, it was Hawke.

I also asked if I could purchase photomicrographs showing the correspondence of die marks on the three Fuller wires and the control wire. He had earlier written (June 1976) that although Uri could have brought prepared wires to Fuller's house, no switching was possible because "the die marks on the Uri-bent piece were the same as the control piece. The probability of his being able to obtain wire from the same die that I got my piece from is *almost nil. . . .*" (my italics).

To my vast surprise, Byrd responded to my request for photomicrographs by saying that no check had been made yet of die marks on the Fuller wires! Since in our best scenario Byrd gets his original wires back, one would expect the marks to match. If they did not match, it would be positive proof of switching, and one of the other scenarios would become more viable. I wondered why a control wire had been kept at all if no check of die marks had been made.

I then asked Byrd if he would mind checking the die marks on the Fuller wires and let me know the results. His next letter was an even greater surprise. He had been mistaken in his previous letter, he said. One Fuller wire *had* been checked by electron microscope and "appeared to be from the same die."

And the Isis wire? Byrd now recalled it had not been checked at all! Since this wire was sent to Hawke for reannealing, it will not ever be possible to check its former die marks.

An exchange of letters with the Public Affairs Office of the Naval Surface Weapons Center disclosed that any correspondence of die marks is virtually meaningless. Byrd obtained all his wire from the Public Affairs Office, where samples have always been available upon request. The office maintains one reel from which it dispenses samples, and each reel lasts many months. Had Uri (or a friend) obtained samples at about the same time Byrd did, the probability that all the samples would have the same die marks is extremely high— not "almost nil."

In January 1977, I asked Byrd if he could check the die marks on all the Fuller wires with an electron microscope. He replied: "This can be done but what would it prove? Most nitinol from this lab comes from the same die, therefore, I do not place a lot of emphasis on the fact." So much for die marks, and Byrd's "rigid controls"!

5. To this day, concealing small objects in the hair is a common practice of East Indian psychics who specialize in "materializations." The hidden object is palmed under cover of a casual brushing of the hair with the hand, then the object is produced as if it came from another world. Uri, too, specializes in the materialization of small objects. Astronaut Edgar Mitchell is firmly convinced that one day in a Stanford cafeteria the head of a tiepin he had lost four years earlier suddenly materialized in a spoonful of ice cream that Uri was about to swallow.

6. On December 31, 1976, Byrd played for Klass, on the telephone, the portion of the tape dealing with the nitinol tests. It was, Klass tells me, unintelligible. Machines have a way of malfunctioning when Uri is around, Byrd explained, and for some reason the voices on the tapes are so garbled that it would be impossible to obtain a typescript. Apparently the recorder (which had been brought by Hawke) was continually binding. There is no way to tell if the recording is continuous, or if the machine was stopped at certain points. Nothing can be heard about when or how the wires were pretested with a match flame, or when and how the flame was applied to kink them. The portion of the tape heard by Klass ends when Fuller enters the room.

Klass asked Byrd where each of the first two wires were at the moment he lit the match that produced their kink. Byrd said he recalls holding both wire and match folder in one hand while he struck a match with the other. Accurate memory or wishful memory? As I have said elsewhere, rats and electrons don't cheat. Superpsychics do. It is precisely such details as this that are crucial in evaluating a test with a man who, as even Koestler now recognizes, is "certainly 25 percent fraud." And they are details that only a knowledgeable magician is capable of recognizing as relevant.

Postscript

Eldon Byrd's letter, commenting on my article, was published in the *Humanist,* September/October 1977:

I almost ignored the invitation to respond to Martin Gardner's article "Geller, Gulls, and Nitinol," but decided that it needed to be on record that I did not 100 percent agree with it. First of all I am not sure if the title had to do with a previous article Mr. Gardner had written about Puthoff and Targ or was aimed at my "Mormon" point of view.

I believe that there are certain physical principles that all earthly events must adhere to, but that is not to say that we know what all those physical principles are. The very foundations of science include relativity, quantum mechanics, and the uncertainty principle. These seem to be the mechanistic equivalent of existentialism, irrationalism, and irresponsibility.

The scientific method is a narrow technique for arriving at truth. It is without doubt the most broadly useful tool man has devised for solutions to his problems; however, it is not without its flaws. The very basis upon which we approach certain problems with the method preclude arriving at truth. For example, the method dictates that hypotheses be tested *after* an observed event has occurred or a theory is established. This implies that scientists who observe an event make educated guesses at what the possible mechanisms are which produced the event. If the event is "paranormal" and the scientist "believes" in such stuff, his experiments are liable to be slanted toward supporting his hypotheses. If he does not believe, his experiments will be slanted toward trying to support his hypotheses, also. Therefore, it is reasonable to expect the Martin Gardners of the world to generate hypotheses and "experiments" that appear to be different from those of others who have a more open mind about the possible existence of events that we as yet do not understand.

History has demonstrated that things have always been around that man has misunderstood. As knowledge increases we find that the "mystical" becomes understood. I believe the same will ultimately happen with current "paranormal" events. But not if people shout long enough and hard enough from the rooftops that there is no need to examine things thoroughly that we do not understand. The attitude that "if I can generate a logical alternative scenario, then that proves the event is understood" will get us nowhere. These kinds of people with their narrow thinking have always been around, but thankfully they have not kept progress at a standstill, just slowed it down.

Just because Martin Gardner says Uri is a fake does not prove him to be. Just because he "proves" it is *possible* for him to be a fake does not prove him to be. Just because I feel there is a possibility unexplained events occur doesn't mean they do. I just hope that the investigation needed to find out is not turned off because of people like Martin Gardner.

There were many errors in Mr. Gardner's article. A few were easy to make, such as referring to me as an operations analyst instead of a physical scientist and spelling *Ordnance,* "Ordinance." However, others were made on purpose to mislead the reader. Most of your readers are probably unaware of Martin Gardner's journalistic techniques because they may share his viewpoints. I feel that Martin Gardner believes that if he can discredit the work of anyone who supports the possibility that Uri Geller or anyone like him can produce "paranormal" events then the end justifies the means. (Be it journalistic "license" or enlisting the aid of a third party like Mr. Klass to act as an intelligence gatherer.)

I had an opportunity to ask Hannah (Shipi Shtrang's sister) if she ever said that she assisted Uri dupe an audience. I doubt that Martin Gardner asked her directly about it, relying on third party information. It is my understanding that when Hannah was approached by the reporter from Israel she wouldn't even talk to him. Therefore an alternative possibility to footnote #2 in Mr. Gardner's article is that the reporter made up the story because Hannah would not talk to him.

Another error was made by referring to Dr. Hawke as a paraphysicist. Still another was that I told Mr. Klass that I may have told Geller that I had metal with a memory, but did not recall if I specifically said it was Nitinol. I never had any contact with Uri Geller prior to the October 1973 meeting at the Isis Center.

However, there exists a third type of error in the article by Mr. Gardner that is inexcusable. Mr. Gardner *knew* that Charles Panati's editorial comments in *The Geller Papers* were in error, yet he *deliberately* made it appear to the reader that I condoned them. There are other errors I could cite but they would only serve to add more of the same.

I am not trying to prove to the world that Uri Geller is "real," nor am I trying to say that my scientific techniques are flawless. It is apparent that there are those, however, who are vigorously pursuing a course of action similar to the Salem witch-hunts to try to convince people that the Uri Gellers of the world and their friends should be drowned.

Eldon A. Byrd

My reply to Byrd was in the same issue:

Mr. Byrd speaks of "many errors" in my article. He lists what I presume are the six he considers the most horrendous. I will comment briefly on each:

1. Mr. Byrd is frequently referred to in the parascience literature as an "operations analyst." I cite one instance from Peter Tompkins's great scientific work, *The Secret Life of Plants* (page 40): "Eldon Byrd, an operations analyst with the Advanced Planning and Analysis Staff of the Naval Ordnance Laboratory at Silver Spring."

2. Yes, *Ordnance* is spelled "Ordnance."

3. Uri himself has explained Hannah's interview as the result of her being mad at him at the time. Readers should check the interview in James Randi's book *The Magic of Uri Geller* and should decide for themselves if the reporter is lying. Apparently Mr. Byrd believes anything that Uri, Shipi, or Hannah tells him.

4. A "paraphysicist" is a physicist who investigates the paranormal. Dr. Hawke is a physicist at Lawrence Livermore Laboratory. He contributed to Panati's book a paper on his investigation of Geller's paranormal ability to erase magnetic patterns. If he isn't a *paraphysicist,* what does the term mean?

5. When Mr. Byrd says he "never had any contact with Uri Geller," he must mean that he had not previously met him in the flesh. Does he wish to

deny that he communicated with Geller, through a third party, prior to the October 1973 meeting? If so, I have information to the contrary.

6. I did indeed know that Panati's editorial comments were in error, but not until Mr. Byrd told me. Nowhere did I suggest that Byrd condoned those errors.

Byrd writes: "I am not trying to prove to the world that Uri Geller is 'real,' nor am I trying to say that my scientific techniques are flawless." This is a doubly false statement. His paper is the strongest argument in Panati's book for the genuineness of Geller's powers. In letters to me, he repeatedly referred to his tests with Uri at John Fuller's house as "rigidly controlled." To this day he has not admitted the slightest "flaw" in his experimental design.

The sad truth is that Mr. Byrd is another of Uri's casualties. Geller has used him the way he has used many other sincere but highly gullible scientists, and it is a tragedy that Mr. Byrd does not yet have the courage to admit it.

Martin Gardner

16

The Extraordinary Mental Bending
of Professor Taylor

No one can say that John G. Taylor, professor of mathematics at Kings College, University of London, is not a brilliant and colorful personality. He was born in 1931 at Hayes, Kent, the son of an organic chemist. After getting his doctorate at Cambridge University, he taught mathematics and physics at a number of colleges in England and the United States, including a stint as professor of physics at Rutgers University. His technical papers (more than a hundred) display a wide range of interests that include pure mathematics, particle physics, cosmology, and brain research.

There is another side to Professor Taylor that I can best characterize as that of a ham actor who thrives on crowd adulation and personal publicity. When in the United States, he studied acting at the Berghof Herbert Studio, in Manhattan, and for a while was "sex counselor" for *Forum* magazine. In England, his constant appearances on radio and television shows made him such a celebrity that in 1975, when the respected British magazine *New Scientist* conducted a poll of readers to determine the world's top twenty scientists, Taylor made the list. The magazine's cover ran his picture alongside Archimedes, Darwin, Einstein, Galileo, Newton, and Pasteur!

Taylor also enjoys writing popular books about science, of which his best known was the international best-seller *Black Holes* (1973). It is not a bad introduction to black-hole theory, but toward the end of the book Taylor indulges in lots of freaky conjectures. He thinks it quite possible, for example, that Earth was visited in the distant past by extraterrestrials,

Reprinted from the *Skeptical Inquirer*, Winter 1979–80.

who may have come in spaceships driven by "black-hole power generators." Saturn, he tells us, is the most likely planet that "high-gravity aliens" could have used as a way-station in their explorations of our solar system.

In his last chapter, Taylor considers the possibility that we have souls that are structured forms of energy capable of moving from one body to another. The universe, he reminds us, has two possible destinies. It may expand forever to die the familiar thermodynamic "heat death," or it may go into a contracting phase and eventually be crushed out of existence by a black hole. In either case, no matter will be left "which could realistically be said to be worth having a soul." However, the universe may bounce back from the big crunch. "The only chance of immortality then is in an oscillating universe. Even in that, everlasting life will not be of the usual form but one in which there may be no relation at all between one cycle and the next due to the enormous re-scrambling of matter in the collapsed phase. It could well be that souls will have to cast lots as to which of the variety of bodies they will inhabit in subsequent lives. That is, of course, unless the hand of God intervenes, his wonders to perform."

There is one other possibility of immortality. If one fell into a black hole, says Taylor, he might emerge in a parallel universe. This, however, has a big shortcoming. If two "close friends" fell into different holes, they could find themselves in separate universes with no possibility of reunion. "So there is always the chance that the immortality gained by falling through a rotating black hole may be a very lonely one."

In view of such quirky speculation, it was not surprising that in 1973, when Taylor appeared on a BBC television show with Uri Geller, he was so stunned by Geller's magic that he became an instant convert to the reality of ESP and PK. Geller did his familiar trick of duplicating a drawing in a sealed envelope. "No methods known to science can explain his revelation of that drawing," wrote Taylor with his usual dogmatism. The professor's jaw dropped even lower when Geller broke a fork by stroking it. "This bending of metal is demonstrably reproducible," Taylor later declared, "happening almost wherever Geller wills. Furthermore, it can apparently be transmitted to other places—even hundreds of miles away."

"I felt," said Taylor in his most often quoted statement, "as if the whole framework with which I viewed the world had suddenly been destroyed. I seemed very naked and vulnerable, surrounded by a hostile and incomprehensible universe. It was many days before I was able to come to terms with this sensation."

Although Taylor was supremely ignorant of conjuring methods, and made not the slightest effort to enlighten himself, he at once set to work testing young children who had developed a talent for metal bending after seeing Geller on television. Taylor's controls were unbelievably inadequate. Children, for example, would put paper clips in their pockets

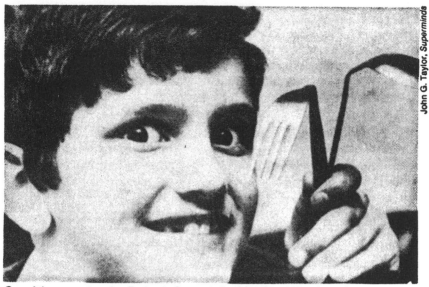

One of the youngsters that Taylor, in *Superminds*, claimed could bend metal.

and later take one out twisted. Nevertheless Taylor was persuaded that hundreds of youngsters in England had the mind power to deform metal objects. Curiously, Taylor never actually *saw* anything bend. One minute a spoon would be straight, later it would be found twisted. Taylor named this the "shyness effect." Metal rods were put inside sealed plastic tubes and children were allowed to take them home. They came back with the tubes still sealed and the rods bent. One boy startled Taylor by material-izing an English five-pound note inside a tube.

So certain was Taylor that his high IQ, combined with his knowledge of physics, gave him the ability to detect any kind of fraud that he rushed into print a big book called *Superminds* (published here by Viking in 1975).[1] It will surely go down in the literature of pseudoscience as one of the funniest, most gullible books ever to be written by a reputable scien-tist. It is even funnier than Professor Johann Zöllner's *Transcendental Physics,* inspired by the psychic conjuring of the American medium Henry Slade. Taylor's book is crammed with photographs of grinning children holding up cutlery they have supposedly bent by PK, tables and persons floating in the air during old Spiritualist seances, glowing ectoplasmic ghosts, psychic surgeons operating in the Philippines, Rosemary Brown displaying a musical composition dictated to her by the spirit of Frederic Chopin, and numerous other wonders.

Not the least peculiar aspect of Taylor's volume was his argument that all paranormal feats, including religious miracles, are explainable by electromagnetism. "The Geller effect is a case in point. Will it ever turn out that the miracles of Jesus Christ also dissolve in scientific speculation. . . . This book has presented the case that for one modern 'miracle,' the Geller effect, there *is* a rational, scientific explanation. This explanation is also claimed to allow us to understand other apparently miraculous phenomena—ghosts, poltergeists, mediumship, and psychic healing. What, then, of other miracles? Can they too be explained by these newly discovered powers of the human body and mind, and the properties of matter broadly described in the book?"

After writing *Superminds,* of which let us hope he is now super-ashamed, Taylor slowly began to learn a few kindergarten principles of deception. When the Amazing Randi visited England in 1975, Taylor refused to see him, but Randi managed to call on him anyway, disguised as a photographer-reporter. You'll find a hilarious account of this in Chapter 10 of Randi's Ballantine paperback, *The Magic of Uri Geller.* Taylor proved to be easier to flimflam than a small child, and his "sealed" tubes turned out to be so crudely sealed that Randi had no trouble uncorking one and corking it again while Taylor wasn't looking. Randi even managed to bend an aluminum bar when Taylor's attention was distracted, scratch on it "Bent by Randi," and replace it among Taylor's psychic artifacts without Taylor noticing.

Another crushing blow to Taylor's naive faith in Geller was a test of the "shyness effect" by two scientists at Bath University. They allowed six metal-bending children to do their thing in a room with an observer who was told to relax vigilance after a short time. All sorts of bending at once took place. None was observed by the observer, but the action was secretly being videotaped through a one-way mirror. The film showed, as the disappointed researchers wrote it up for *Nature* (vol. 257, Sept. 4, 1975, p. 8): "*A* put the rod under her foot to bend it; *B, E* and *F* used two hands to bend the spoon . . . while *D* tried to hide his hands under a table to bend a spoon."

Slowly, as more evidence piled up that Geller was a charlatan and that the "Geller effect" never occurs under controlled conditions, Taylor began to have nagging doubts. After several years of silence, he suddenly announced his backsliding. Of course he didn't call it that. Instead, he and a colleague at Kings College wrote a technical article for *Nature,* "Can Electromagnetism Account for Extrasensory Phenomena?" (vol. 276, Nov. 2, 1978, pp. 64–67; also *Skeptical Inquirer,* Spring 1979, p. 3.)

In *Superminds,* after considering all possible ways to explain psi phenomena by known laws, Taylor concluded that only electromagnetism offered a viable possibility. The *Nature* paper reinforces this view. Electromagnetism, the authors decide, "is the only known force that could conceivably be involved." They then report on a series of carefully controlled tests of ESP and PK using talented subjects. No psi phenomena occurred. When controls were eased, the phenomena did take place but the experimenters could not detect a whiff of electromagnetic radiation. Their conclusion is that all the phenomena they investigated, metal bending in particular, have normal explanations.

More was to come. In *Nature* (vol. 279, June 14, 1979) the same authors published a sequel to their first paper. In this sequel, titled "Is There Any Scientific Explanation of the Paranormal?" they again stress the fact that "on theoretical ground the only scientific explanation [for psi forces] could be electromagnetism." Their conclusion is that neither electromagnetism "nor any other scientific theory," including quantum mechanics, can explain dowsing, clairvoyance, or telepathy. "In particular there is no reason to support the common claim that there still may be some scientific explanation which has as yet been undiscovered. The successful reductionist approach of science rules out such a possibility except by utilization of energies impossible to be available to the human body by a factor of billions. We can only conclude that the existence of any of the psychic phenomena we have considered is very doubtful."

Now it is pleasant for skeptics like me, who also regard psi phenomena as possible but "very doubtful," to welcome Taylor back to our ranks. But surely his reasons are as shaky as those that converted him to the paranormal six years ago. The history of science swarms with observed phenomena that were genuine but had to wait for centuries until a good theory explained them. A lodestone's magnetism was sheer magic until the modern theory of magnetism was formulated, and even today no physicist knows why the acceleration of electrical charges inside atoms causes magnetic effects. It is not even known why electricity comes in units of positive and negative charge, or whether magnetic monopoles exist as theory seems to demand.

Kepler correctly decided, on the basis of confirmable correlations, that the moon causes tides; but in the absence of a theory, even the great Galileo refused to believe it. One could add hundreds of other instances in which a phenomenon was authenticated long before a theory "explained" it. On this I find myself in full agreement with J. B. Rhine and other parapsychologists who regard the lack of a physical theory as no obstacle whatever to the acceptance of psi.

Science cannot absolutely rule out the possibility of anything, but it can assign low degrees of probability to unusual claims. In my view, which is the view of most psychologists, the classic psi experiments are more simply and plausibly explained in terms of unconscious experimenter bias, unconscious sensory cuing, fraud on the part of subjects eager to prove their psychic powers, and, on rare occasions (such as those recently disclosed about S. G. Soal), deliberate fraud on the part of respected investigators.

The central point is this. When science assigns a low degree of credibility to an extraordinary claim, it does so by evaluating the empirical evidence. Geller and the spoon-bending children are indeed frauds, but the reasons for thinking this have nothing to do with the fact that the supposed "Geller effect" is unsupported by an adequate physical theory. It is because the conjuring techniques for fraudulently bending metal are now well known, and because the metal invariably refuses to twist whenever the controls are commensurate with the wildness of the claim.

NOTE

1. For my review of *Superminds,* along with Randi's *The Magic of Uri Geller* mentioned later, see Chapter 27.

17

Quantum Theory and Quack Theory

Earlier this year, at the annual meeting of the American Association for the Advancement of Science, Dr. John Archibald Wheeler startled his audience by asking the AAAS to reconsider its decision (made ten years ago at the insistence of Margaret. Mead) to dignify parapsychology by giving its researchers an affiliate status in the association. Here is the background to Wheeler's explosive remarks.

John Wheeler, director of the Center for Theoretical Physics at the University of Texas, is one of the world's top theoretical physicists. In 1939 he and Niels Bohr published a paper on "The Mechanism of Nuclear Fission" that laid the groundwork for atomic and hydrogen bombs. Wheeler later played major roles in their development. He named the black hole. In 1968 he received the Enrico Fermi award for "pioneering contributions" to nuclear science. When Richard Feynman accepted a Nobel Prize for his "space-time view" of quantum mechanics (QM), he revealed that he had gotten his basic idea from a phone conversation with Wheeler when he was a graduate student of Wheeler's at Princeton.

No one knows more about modern physics than Wheeler, and few physicists have proposed more challenging speculative ideas. In recent years he has been increasingly concerned with the curious world of QM and its many paradoxes which suggest that, on the microlevel, reality seems more like magic than like nature on the macrolevel. No one wants

This article, which appeared in the *New York Review of Books,* May 17, 1979, and the letters quoted in the Postscript are reprinted with permission from the New York Review of Books. © 1979 NYREV, Inc.

to revive a solipsism that says a tree doesn't exist unless a person (or a cow?) is looking at it, but a tree is made of particles such as electrons, and when a physicist looks at an electron something extremely mystifying happens. The act of observation alters the particle's state.

In QM a particle is a vague, ghostly, formless thing that cannot even be said to have certain properties until measuring it causes a "collapse of its wave packet." ("Wave packet" refers to the total set of waves, defined in an abstract multidimensional space, that constitutes all that is known about a particle.) At that moment nature makes a purely random, un- caused decision to give the property (say the electron's position or its momentum) a definite value predicted by the probabilities specified in the particle's wave function. As Wheeler is fond of saying, we no longer can think of a universe sitting "out there" as if separated from us by a thick plate of glass. To measure a particle we must shatter the glass and alter what we measure. The physicist is no mere observer. He is an active participator. "In some strange way," Wheeler has said, "the universe is a participatory universe."*

---

*The problem of saying what a quantum particle looks like when no one is looking at it is something like the problem of what a mirror looks like when no one is looking at it. Here is how the late J. A. Lindon, of Addlestone, England, expressed his befuddlement in a poem he sent me which I here print for the first time:

Look and See

I thought I knew my physics, though my knowledge isn't deep,
But now a problem haunts me that is ruining my sleep;
All night I toss and turn about, I stare into the gloom:
What does a mirror look like *when there's no one in the room?*

It can't reflect its owner, if the owner isn't there,
But surely it must still reflect his table and his chair,
His picture on the wall behind, the window and the door,
The potted fern, the chandelier, the carpet on the floor?

For these are there, in lots of light, before the mirror's gleam;
They're bound to be reflected, if it's real and not a dream;
But everything reflected forms an image in the glass,
That seems to come and move along and vanish as you pass.

So how a mirror looks *to you* depends on your position,
And here the books are guilty of a serious omission;
Away with "how it looks to you," put "seeming" on the shelf:
What does a mirror look like *when reflecting by itself?*

There stands a mirror by the wall, a candle-flame before it
Reflected with such clarity a *worm* could not ignore it;
I take my toothbrush and my towel, yet wonder more and more
How it will look *to no one there* when I have shut the door.

This is not a new suggestion, because Niels Bohr constantly emphasized the need to redefine reality on the microlevel, always hastening to add that on the macrolevel of the laboratory classical physics still holds. It is easy to understand, however, how QM would appeal to physicists who are into Eastern religions and/or parapsychology. Consider a spoon. Because its molecules are made of particles it can be regarded as a quantum system. If particles are influenced by observation, may we not suppose that a super-psychic, observing a spoon, could in some mysterious way alter the system and cause the spoon to bend?

In the past, parapsychologists have had an extraordinary lack of success in trying to explain "psi"—i.e., parapsychological—phenomena by familiar forces such as electromagnetism and gravity. One difficulty—it was the main reason for Einstein's skepticism about psi—is that all known forces weaken with distance, whereas, if the results of parapsychology are valid, there is no decline of ESP with distance. Is it possible that QM can provide a workable theory of psi?

Parapsychologists who are not physicists (J. B. Rhine for instance) take a dim view of explaining psi by any aspect of physics, but there is a growing number of paraphysicists—physicists who believe in and are investigating paranormal phenomena—for whom QM opens exciting possibilities. This approach was given a boost a few years ago by experiments involving a famous paradox of QM known as the EPR paradox, after the initials of Einstein and his friends Boris Podolsky and Nathan Rosen. In 1935 they published a thought experiment designed to prove that QM is not a complete description of nature on the microlevel but needs to be incorporated in a deeper theory in a manner similar to the way that Newtonian physics became incorporated in relativity theory.

The EPR paradox involves pairs of "correlated" particles. For example, when an electron and positron meet and annihilate one another, two photons, A and B, go off in opposite directions. No matter how far apart they get they remain correlated in the sense that certain properties must have opposite values. If A is measured for property x, its wave packet collapses and x acquires the value of, say, $+1$. The corresponding value for B is at once known to be -1 even though B is not measured. Measuring A seems somehow to collapse the wave packet of B even though A and B are not in any way causally related!

Einstein hoped that his paradox could be resolved by a hidden variable theory—a theory that assumes a mechanism within both particles that keeps them correlated like two Frisbees simultaneously tossed left and right with both hands so that they spin in opposite ways. A person catching one Frisbee and noticing that it rotated clockwise would instantly know that the other Frisbee spun the other way even though nobody caught it. Alas, the formalism of QM rules out this possibility. If, for example, two correlated particles have opposite spin, you cannot say particle *A* has either kind of spin until it is measured. Not until the instant of measurement does nature "decide" what spin to give it.

In 1965 J. S. Bell hit on an ingenious proof, now known as "Bell's theorem," that no local hidden variables (local means in or near each particle) could explain the EPR correlations. It leaves open the possibility that the particles remain connected, even though light-years apart, by a nonlocal subquantum level that no one understands. Moreover, Bell's theorem provided for the first time a way of testing EPR correlations in a laboratory. Many such tests have been made and almost all confirm the EPR paradox. Most physicists have little interest in trying to explain the paradox—they simply accept QM as a tool that works—but physicists concerned with theoretical interpretations of QM are very much in a quandary over what to make of the new results.

For many paraphysicists the EPR paradox suggests that quantum information can be transferred instantaneously (or almost so) from any part of the universe to any other, otherwise how does one particle "know" what happens when its twin is measured? (Relativity theory is not violated because no energy is transferred, only information.) This is the view of paraphysicist Jack Sarfatti, who heads a small San Francisco organization called The Physics/Consciousness Research Group, initially financed by Werner Erhard of est. (Sarfatti and Erhard have since had a violent falling out, and Sarfatti is devoting much of his time to attacking Erhard as a native "fascist.") For Sarfatti's far-out views see his article "The Physical Roots of Consciousness" in Jeffrey Mishlove's wild book, *The Roots of Consciousness* (published by Random House in a fit of absence of mind), and an interview with Sarfatti in *Oui*, March 1979. Last year Sarfatti applied for a patent (disclosure number 071165) on a device he hopes can send faster-than-light messages to any part of the universe.

Five years ago interest in QM as a basis for psi was so widespread that, at the suggestion of Arthur Koestler, an international conference on QM and parapsychology was held at Geneva in the fall of 1974. The *Proceedings* were published the following year by the Parapsychology Foundation, New York City. This quaint volume opens with a long paper by Evan Harris Walker, an American physicist who has made the most elaborate attempt to develop a QM theory of consciousness and psi. Gerald Feinberg of Columbia University spoke on precognition. Harold Puthoff and Russell Targ, the two Stanford Research Institute physicists who "verified" the clairvoyant powers of the Israeli magician Uri Geller, also gave papers. Both are sold on QM as the most likely explanation of psi. Other speakers included Ted Bastin, Helmut Schmidt, and O. Costa de Beauregard.

Costa de Beauregard, a French physicist, has the most eccentric of all explanations for the EPR paradox. He believes that information from the measurement of particle *A* travels backward in time to the origin of the particle-pair, then forward in time to particle *B*, arriving there at the exact instant it left *A*. Among leading physicists who did not attend the Geneva meeting but who believe QM is behind psi, there are England's Nobel-Prize-winner Brian Josephson and Richard Mattuck of the University of Copenhagen.

What does all this have to do with Wheeler? The answer is important and amusing. For many years Wheeler's views on QM have been widely cited by parapsychologists as strengthening their own. If you check Sarfatti's paper mentioned earlier you'll find Wheeler's name constantly invoked. Wheeler has found this increasingly irritating. Asked to speak in Houston at last January's annual meeting of the American Association for the Advancement of Science, he chose the topic "Not Consciousness But the Distinction Between the Probe and the Probed as Central to the Elemental Quantum Act of Observation." Wheeler hoped he could make clear his agreement with Niels Bohr that acts of QM measurement are made by devices which can be monitored by computers, and thus disassociate himself from those who argue that human consciousness is essential to QM observation. To his amazement he found himself sharing a panel with Puthoff and Targ, and parapsychologist Charles Honorton of Maimonides Medical Center in Brooklyn.

In his paper Wheeler went into considerable detail about the EPR paradox and its perplexing implications. It is a marvelous, subtly argued essay woven around the central theme: "no elementary phenomenon is a phenomenon until it is an observed phenomenon." Wheeler closed his lecture with these strong words: "And let no one use the Einstein-Podolsky-Rosen experiment to claim that information can be transmitted faster than light, or to postulate any 'quantum interconnectedness' between separate consciousnesses. Both are baseless. Both are mysticism. Both are moonshine."

Two appendices that Wheeler added to his paper have shaken the world of parapsychology more than any remarks made by a distinguished scientist in the past half-century. Here are the appendices, accompanied by Wheeler's letter to the president of the AAAS:

DRIVE THE PSEUDOS OUT OF THE WORKSHOP OF SCIENCE

J. A. Wheeler

The author would be less than frank if he did not confess he wanted to withdraw from this symposium when—too late—he learned that so-called extrasensory perception (SCESP) would be taken up in one of the papers. How can anyone be happy at an accompaniment of pretentious pseudoscience who wants to discuss real issues about real observations in real science? How can pseudoscience fail to profit in prestige and acceptability by being on the same platform as science? And how can science fail to lose? That is why the author, then on the AAAS Board of Directors, voted against the majority of the much larger Council at that time and against the admission of "parapsychology" as a new division of the American Association for the Advancement of Science at its meeting in Boston in 1969. That is why, with the decade of permissiveness now well past, he suggests that the Council and the Board of Directors will serve science well to vote "parapsychology" out of the AAAS.

It is not the slightest part of this proposal to prevent anyone from working on "parapsychology" who wants to. Neither does the author yield to anyone in his respect for the idealism and good intentions of some he has known in that field. Nor is there in this proposal any intention to deny investigators full freedom of speech and a forum for their fribbles. There is forum enough already in a country that can afford 20,000 astrologers and only 2,000 astronomers. There is forum enough in a Parapsychological Association, a Boston Society for Psychical Research, an American Society for Psychical Research, an International Society for Psychotronic Research, and a Parapsychology Foundation. No one would think of interfering with the freedom that anyone has to publish in the *International Journal of Parapsychology,* the *Journal of the American Society for Psychical Research,* or the *Journal of Parapsychology.* Neither is it part of this proposal to interfere with the fund-raising that keeps parapsychology going in the United States to the tune of from $1 million to $20 million a year.[1] Faith healers can be prosecuted, confidence men can be sent to jail, but no one would propose that parapsychologists be prevented from soliciting—even soliciting for government support. But why should the name "AAAS-Affiliate" be allowed to give those solicitations an air of legitimacy?

Surely when so much is written about spoon bending, parapsychology, telepathy, the Bermuda Triangle, dowsing, and when others write on "quantified etherics," bioactochronics, levitation, and occult chemistry there must be *some* reality behind those words? Surely where there's smoke there's fire? No, where there's so much smoke there's smoke.

Every science that is a science has hundreds of hard results; but search fails to turn up a single one in "parapsychology." Would it not be fair, and for the credit of science, for "parapsychology" to be required to supply one or two or three battle-tested findings as a condition for membership in the AAAS?

Self-delusion or conscious fraud was Houdini's diagnosis of psychic phenomena. "He threw down a challenge . . . offering any medium five thousand dollars if he could not duplicate any phenomenon of alleged spirits himself. . . . Early in 1926 Houdini made a pilgrimage to Washington to enlist the aid of President Coolidge in his campaign 'to abolish the criminal practice of spirit mediums and other charlatans who rob and cheat griefstricken people with alleged messages.'"[2]

Hudson Hoagland, in an editorial in *Science* magazine,[3] tells us:

> A famous case was that of a Boston medium in the 1920s, who had a wide following. She was the wife of an eminent surgeon and claimed communication

with her dead brother. The old *Scientific American* magazine had offered a prize of $5,000 to anyone who could demonstrate supernormal physical phenomena to a committee of its choosing. At her request, she was investigated in 1924 by this committee, composed of several Harvard and M.I.T. professors along with Harry Houdini, the magician. The committee reported that evidence for her supernormal powers was inconclusive, although Houdini denounced her as fraudulent.

Following wide press publicity, a group at Harvard, of which I was one, later investigated her in a series of seances in the psychological laboratories and found not only that the phenomena were due to trickery, but also how the tricks were done. Our findings, published in an article by me in the *Atlantic Monthly* of November 1925, resulted in violent recriminations and denunciations of us in published pamphlets and press statements by her followers. Our exposure enhanced her publicity and she gained more adherents. She was skillful in modifying her mode of operation, depending upon the gullibility of her audience and other circumstances. On several subsequent occasions she was also exposed by other scientists, but at no time until her death did she lose a diminishing circle of devoted believers.

The basic difficulty inherent in any investigation of phenomena such as those of psychic research or of UFO's is that it is impossible for science ever to prove a universal negative. There will be cases which remain unexplained because of lack of data, lack of repeatability, false reporting, wishful thinking, deluded observers, rumors, lies, and fraud. A residue of unexplained cases is not a justification for continuing an investigation after overwhelming evidence has disposed of hypotheses of supernormality, such as beings from outer space or communications from the dead. Unexplained cases are simply unexplained. They can never constitute evidence for any hypothesis.

Let parapsychology pass, or try to pass, the *Scientific American*-Houdini test with one or two or three of its findings. Is there any more searching way to make a first trial whether there is anything in parapsychology worth further scrutiny?

For every phenomenon that is proven to be the result of self-delusion or fraud or misunderstanding of perfectly natural everyday physics and biology, three new phenomena of "pathological science" spring up in its place. The confidence man is able to trick person after person because so often the victim is too ashamed of his gullibility or too mouselike in his "stop, thief" to warn others. Happily a journal now exists called the *Skeptical Inquirer*[4] which provides a list of some of the items of pathological science currently in vogue. Some other references which the reader may want to consult are Gardner's *Fads and Fallacies*[5] ("the curious theories of modern pseudoscientists and the strange, amusing, and alarming cults that surround them; a study in human gullibility with topics including flying saucers, Atlantis, Bridey Murphy, Alfred Korzybski, eccentric sexual theories, Dr. W. H. Bates, Wilhelm Reich, L. Ron Hubbard, psionics machines"), Condon's *Scientific Study of Unidentified Flying Objects*,[6] and Jastrow's *Error and Eccentricity in Human Belief*[7] ("The author chronicles one episode after another from the record of human credulity . . . to support his central contention, that man tends to fashion his beliefs out of his desires, not out of rational thought").

Robert Buckhout's article[8] on "Eyewitness Testimony" remarks "although such testimony is frequently challenged, it is still widely assumed to be more reliable than other kinds of evidence. Numerous experiments show, however, that it is remarkably subject to error." Irving Langmuir's colloquium talk at the General Electric Company's Knolls Research Laboratory[9] on December 18, 1953, tells of his own experience investigating delusions, conscious and unconscious. Langmuir analyzes the Davis-Barnes effect, N-rays (for which also see especially the famous encounters between R. W. Wood[10] and R. Blondlot), mitogenetic rays, characteristic symptoms of pathological science, Allison effect (see also a recent review),[11] extrasensory perception and flying saucers. Langmuir's table of symptoms of pathological science are as appropriate today as they were when he gave his lecture in 1953:

1. The maximum effect that is observed is produced by a causative agent of barely detectable intensity, and the magnitude of the effect is substantially independent of the intensity of the cause.

2. The effect is of a magnitude that remains close to the limit of detectability; or, many measurements are necessary because of the very low statistical significance of the results.

3. [There are] claims of great accuracy.

4. Fantastic theories contrary to experience.

5. Criticisms are met by *ad hoc* excuses thought up on the spur of the moment.

6. Ratio of supporters to critics rises up to somewhere near 50 percent and then falls gradually to oblivion.

There's nothing that one can't research the hell out of. Research guided by bad judgment is a black hole for good money. No one can forbear speaking up who has seen $10,000 cozened out of a good friend, $100,000 milked out of a distinguished not-for-profit research organization, and $1,000,000 syphoned away from American taxpayers—all in the cause of "research" in pathological science.

Where there is meat there are flies. No subject more attracts the devotees of the "paranormal" than the quantum theory of measurement. To sort out what it takes to define an observation, to classify what it means to say "no elementary phenomenon is a phenomenon until it is an observed phenomenon" is difficult enough without being surrounded by the buzz of "telekinesis," "signals propagated faster than light," and "parapsychology."

Now is the time for everyone who believes in the rule of reason to speak up against pathological science and its purveyors.

Notes

1. Order of magnitude of ~200 actively working in the field. Costs per full-time Ph.D. investigator per year in industry, ~$100,000 per year; perhaps half of this in academic work when ancillary costs are included; figures for less than full-time workers tapering down to a few $1,000 per year; rough average adopted here,

~$20,000 per year; this times ~200 gives ~$4 million per year or, with uncertainties, a number in the range of $1 million to $20 million a year.

2. B. R. Sugar, "Houdini," Braniff Airlines *Flying Colors 5*, No. 2, pp. 31-39 and 58 (1975); the quotation comes from p. 39. The papers of Houdini are on deposit in the library of the University of Texas at Austin.

3. Hudson Hoagland, "Beings from outer space — corporeal and spiritual," *Science 163*, p. 625 (February 14, 1969).

4. *The Skeptical Inquirer* (published by the Committee for the Scientific Investigation of Claims of the Paranormal), Box 29, Kensington Station, Buffalo, NY 14215.

5. M. Gardner, *Fads and Fallacies in the Name of Science* (Dover, 1957; first published in 1952 as *In the Name of Science*).

6. E. U. Condon, *Scientific Study of Unidentified Flying Objects*, edited by D. S. Gillmor (Bantam, 1969).

7. J. Jastrow, *Error and Eccentricity in Human Belief* (Dover, 1962).

8. R. Buckhout, "Eyewitness Testimony," *Scientific American 231*, pp. 23-30 (December 1974).

9. I. Langmuir, "Pathological Science," R. N. Hall, ed., Colloquium at the Knolls Research Laboratory, December 18, 1953, 13 pp. (On deposit with the Manuscript Division of the Library of Congress as a microgroove disk recording.)

10. R. W. Wood, "The N Rays" (Letter exposing delusion), *Nature 70*, p. 530 (1940). W. Seabrook, *Doctor Wood* (Harcourt, Brace and Co., 1941).

11. H. Mildrum and B. Schmidt, "The Allison method of chemical analysis," United States Air Force Aeropropulsion Laboratory Technical Report AFA PLTR-66-52, 1966. Contains extensive bibliography.

'A DECADE OF PERMISSIVENESS'

Dr. William D. Carey
American Association for the
  Advancement of Science
1776 Massachusetts Avenue, NW
Washington, D.C. 20036

Dear Bill:

Quite innocently I found myself drawn into a controversy at the session on Science and Consciousness at the meeting of the American Association for the Advancement of Science in Houston Monday morning, January 8. I had been asked to talk on the relation between quantum mechanics and consciousness. I discovered to my dismay after the program had been cast in concrete that Eugene Wigner and I, two people from the world of physics, were being put together on a panel with several parapsychologists. What is more, one of them and many of the audience were ready to call on the most extreme ideas out of physics. I am writing as a concerned member of the AAAS and as a former member of the board of directors and as a former president of the American Physical Society to ask that a five man committee of review be appointed by the board of directors and the council jointly to review the work of the section of parapsychology of the AAAS to determine:

(a) Whether this field of investigation by now has produced *any* "battle tested result";

(b) To report on the advantage gained in fund raising by workers in the field of parapsychology by their association with the AAAS;

(c) To report on the effect of this association on the public image of the AAAS;

(d) To advise whether this section should be left "as is," suspended until the field has produced some "battle tested" results or deleted outright from the AAAS.

I know that the views of our late and beloved Margaret Mead were strong in getting parapsychology admitted to the AAAS. I was present at the meeting where it happened. The opinion that I had and many others had was overridden by the permissiveness of the time. The words might not have been used, but the idea was there of that old phrase, "Marry him to reform him." Now the decade of permissiveness has passed.

Moreover, in the quantum theory of observation, my own present field of endeavor, I find honest work almost overwhelmed by the buzz of absolutely crazy ideas put forth with the aim of establishing a link between quantum mechanics and parapsychology—as if there were any such thing as "parapsychology." A young person who wants to work in this field does so at his risk. He runs the danger of earning, not reputation, but snickers. In this sense the association of "parapsychology" with the AAAS puts a strain on the progress of an important field of investigation. That is the origin of my concern and the reason I appeal to you for your good offices in setting up the "Committee for the Review of Parapsychology in the AAAS."

More background for this letter will be found in Appendices A and B of the attached paper, "Not consciousness, but the distinction between the probe and the probed, as central to the elemental quantum act of observation."

We have enough charlatanism in this country today without needing a scientific organization to prostitute itself to it. The AAAS has to make up its mind whether it is seeking popularity or whether it is strictly a scientific organization. Admiral Hyman G. Rickover has just this minute telephoned to back my position on making a clean break between the AAAS and parapsychology and authorizes me to quote him so.

Many thanks for your consideration.

John Archibald Wheeler

Director
Center for Theoretical Physics
The University of Texas at Austin
Austin, Texas

Postscript

Four professional physicists, all firm believers in the reality of ESP, including precognition, and in the reality of PK, signed the following letter, which was published in the *NYR* (June 26, 1980):

To the Editors:

In a recent article,[1] J. A. Wheeler has violently attacked parapsychology, calling it a "pathological science" and a "pretentious pseudoscience" and he suggests that it "will serve science well to vote 'parapsychology' out of the American Association for the Advancement of Science." In addition, he criticizes physicists who are investigating a possible connection between quantum theory and parapsychology,[2] stating that "in the quantum theory of observation, my own present field of endeavor, I find honest work almost overwhelmed by the buzz of absolutely crazy ideas put forth with the aim of establishing a link between quantum mechanics and parapsychology— as if there were any such thing as 'parapsychology,'" and "Where there is meat there are flies. No subject more attracts the devotees of the 'paranormal' than the quantum theory of measurement." Wheeler's attack has been reproduced in an article by Martin Gardner, entitled "Quantum Theory and Quack Theory," published in *The New York Review of Books.*

The authors of the present note are all physicists who have for some years been engaged in research on a possible connection between quantum mechanics and parapsychology. We are very much shocked by Wheeler's remarks, which we feel show no trace of the open-minded, imaginative, rational approach to science for which Wheeler is otherwise so famous. We will now answer Wheeler's objections in turn.

1. Wheeler calls parapsychology a "pseudo" or "pathological" science on the grounds that "every science that is a science has hundreds of hard results, but search fails to turn up a single one in parapsychology."

In our opinion, no new science can be expected to present "hundreds of hard results" in its infancy. There are even older, accepted sciences which cannot meet this criterion, such as, e.g., general relativity, where there are only three or four "hard" confirmations of the theory. What entitles a field of research to be called "science" is not "hard results" but rather the intention and care with which its investigations are carried out, and the competence of its investigators. We feel that there are several pieces of research in parapsychology in which these criteria have been met. For example, there is Dr. C. Crussard and Dr. J. Bouvaist's investigation of the French medium, Jean Pierre Girard.[3] Girard produced large changes in the physical properties of metal bars, without the use of physical agents, under what appear to be rigorously controlled conditions. For instance he increased the hardness of an aluminum bar by ca. 10 percent, without using any known physical means. The experiment was repeated four times in three different laboratories, two in France, one in England.

A second example is the investigation of remote bending produced by English schoolchildren,[4] carried out by Professor J. B. Hasted, chairman of the physics department at Birkbeck College, University of London. Under controlled conditions, the children produced large bending and stretching signals in metal objects equipped with strain gauges, without being in contact with the objects. The signals were of a character such that they could not have been produced by any known physical forces under the given experimental conditions. A third example is Dr. H. Schmidt's investigation of the influence of selected subjects on the output of a random number generator based on radioactive decay.[5] For example in rigorously controlled experiments, Schmidt found two subjects who could, by an effort of the will, cause the generator output to be non-random. The probability that the result was due to pure chance was less than one chance in ten million. A fourth example is Dr. H. Puthoff and R. Targ's investigation of remote viewing.[6] In their experiments, several subjects were able to acquire statistically significant amounts of information about randomly chosen targets blocked from ordinary perception by distance or shielding.

If Wheeler has any concrete criticisms of the above experiments, we would like to hear them. Moreover, we challenge any magician to duplicate these results under the given controlled conditions.

2. Wheeler talks of "crazy ideas put forth with the aim of establishing a link between quantum mechanics and parapsychology—as if there were any such thing as 'parapsychology.'" We feel that the above experiments are of sufficiently high quality to warrant the assumption that there is indeed such a thing as parapsychology. However, assuming the existence of paranormal phenomena, we seem to lack a way of putting these phenomena into our present picture of the physical universe. In fact, this lack is probably one of the main reasons for the irrational attacks on parapsychology. Therefore, we feel that it is imperative to try to extend the framework of modern physics—in particular, quantum mechanics—in order to include the new phenomena in a rational and coherent fashion. We feel that this requires a new approach in physics in which consciousness plays an important role, and we are trying to find such an approach.[7] The theories we are working on are completely rational, and lead to results which can be tested in the laboratory, although so far there have been only preliminary attempts in this direction.

3. Wheeler states his belief that "not consciousness but the distinction between the probe and the probed [is] central to the elemental quantum act of observation." That is, in contrast to us, consciousness is *not* a part of Wheeler's model. In fact he states "I would have felt very uncomfortable if Bohr had used the term 'consciousness' in defining the elemental act of observation. I would not have known what he meant."[8] Therefore, we find it indeed regrettable that, as Gardner puts it, "Wheeler's views on quantum mechanics have been widely cited by parapsychologists as strengthening their own." This serves only to confuse the issue, and we sympathize fully with Wheeler's irritation on this point. The issue as we see it is this: Assuming that the phenomena of parapsychology are real, then which model— Wheeler's, or ours, or some other model—gives the best description of these phenomena? We believe that this question can only be answered by further experiment, not by attempting to legislate parapsychology out of existence as a respectable field of research by removing it from the AAAS.

4. Wheeler states that Langmuir's "table of symptoms of pathological science" is appropriate to parapsychology. We do not believe this. For example, one "symptom" is that "the effect is of a magnitude that remains close to the limit of detectability." As pointed out in 1. above, Girard produced an easily detectable change in hardness of a metal bar. Hasted's

bending signals were also well above the noise level. Another symptom is "fantastic theories contrary to experience." Do we have to remind Wheeler that many new theories looked "fantastic" when they were first proposed — for example, relativity and quantum theory? The criterion for accepting or rejecting a theory is not how "common sense" or "fantastic" it appears, but rather, how well it describes the observed data and gives them coherence and meaning.

It would be a good idea for Wheeler to reread p. 38 of his own book *Gravitation.*[9] On that page is a quote from the great physicist Galileo Galilei ridiculing Kepler's belief that the moon is the cause of the tides:

> Everything that has been said before and imagined by other people [about the tides] is in my opinion completely invalid. But among the great men who have philosophised about this marvellous effect of nature the one who surprised me the most is Kepler. More than other people he was a person of independent genius, sharp, and had in his hands the motion of the earth. He later pricked up his ears and became interested in the action of the moon on the water, and in other occult phenomena, and similar childishness.
>
> Galileo Galilei (1632)

So Wheeler is taking quite a risk in ridiculing parapsychology!

5. Wheeler writes that parapsychology "siphons" between 1 and 20 million dollars per year away from the American taxpayer. We would like to point out that this sum is negligible compared with the amount of money going into other areas of science. Assuming 50,000 scientists in all other fields of science, with an average cost of 100,000 dollars per year per scientist yields $5 billion. Thus, less than half of one percent of research money is going into parapsychology in the United States.

In conclusion, we find that Wheeler's claim that parapsychology is a "pseudo" or "pathological" science is unsupported. Unless he is able to prove that the experiments described in part 1. of this rebuttal were carried out incompetently, we feel that his argument has no basis in fact. With his immoderate attack on an embryo science, we believe that Wheeler is in grave danger of repeating the mistake of the great French chemist Lavoisier, who declared, after examining a meteorite which others had seen fall on a meadow September 13, 1768: "We must conclude therefore that the stone did not fall from the sky. The opinion which seems to us the most probable and agrees best with the principles accepted in physics is that this stone was struck by lightning."

Finally, Wheeler concludes with "now is the time for everyone who believes in the rule of reason to speak up against pathological science and its purveyors." On the contrary, we feel that all those who believe in the "rule of reason" should examine the research on paranormal phenomena in an open-minded fashion, and start thinking of how one might extend the borders of our present theories so as to include these phenomena within them.

Olivier Costa de Beauregard
Institut Henri Poincaré
University of Paris, Paris, France

Richard D. Mattuck

Physics Laboratory I
University of Copenhagen
Copenhagen, Denmark

Brian D. Josephson

Cavendish Laboratory
Cambridge University, Cambridge, England

Evan Harris Walker

Department of Mechanics and Materials
Sciences, Johns Hopkins University
Baltimore, Maryland, and Ballistics Research
Laboratory, Aberdeen, Maryland

Notes

1. J. A. Wheeler, appendix to lecture delivered at the January 1979 meeting of the AAAS, reproduced in "Quantum Theory and Quack Theory" by Martin Gardner, *NYR,* May 17, 1979.

2. J. A. Wheeler, in letter to Wm. D. Carey, ibid.

3. C. Crussard, and J. Bouvaist, *"Etude de quelques deformations et transformations apparemment anormales de metaux,"* *Mémoires Scientifiques Revue Metallurgie,* February 1978, p. 117.

4. J. B. Hasted, "Physical aspects of paranormal metal bending," *J. Society for Psychical Research, 49,* 583 (1977); "Paranormal metal-bending" in *The Iceland Papers* (see under ref. 7).

5. H. Schmidt, "Instrumentation in the parapsychology laboratory," p. 13 in *New Directions in Parapsychology,* ed. J. Beloff, Scarecrow Press, Metuchen (1975).

6. H. E. Puthoff and R. Targ, "A Perceptual Channel for Information Transfer over Kilometer Distances: Historical Perspective and Recent Research," *Proc. IEEE, 64,* p. 329 (1976); "Direct Perception of Remote Geographical Locations," in *The Iceland Papers* (see under ref. 7).

7. See, for example, E. H. Walker, "Foundations of Paraphysical and Parapsychological Phenomena," P. 1 in *Quantum Physics and Parapsychology,* Ed. L. Oteri, Parapsychology Foundation, 29 W. 57th St., NY (1975); O. C. de Beauregard, "Time Symmetry and the Interpretation of Quantum Mechanics," *Found. Phys., 6,* 539 (1976), "S-matrix, Feynman Zigzag and Einstein Correlation," *Phys. Lett., 67 A* 171 (1978). See also R. D. Mattuck and E. H. Walker, "The Action of Consciousness on Matter: A Quantum Mechanical Theory of Psychokinesis," in *The Iceland Papers: Experimental and Theoretical Research on the Physics of Consciousness,* Essentia Research, Amherst, Wisc. Ed. A. Puharich (1979), O. C. de Beauregard, "The Expanding Paradigm of the Einstein Paradox," ibid., B. D. Josephson, "Conscious Experience and its place in Physics," paper presented at *"Colloque International Science et Conscience,"* Cordoba, 1–5. October 1979.

8. J. A. Wheeler, "Frontiers of Time," in *Problems in the Foundations of Physics,* Ed. N. Toraldo di Franca and Bas van Fraassen, North Holland, Amsterdam, 1979 (International School of Physics "Enrico Fermi," Varenna, LXXII Course 1977).

9. C. W. Misner, K. S. Thorne, and J. A. Wheeler, *Gravitation* (W. H. Freeman, San Francisco, 1975).

To this letter I replied:

One may have the highest respect for the signers of the above letter—one of them, Brian Josephson, is a Nobel Prize winner—at the same time recognizing that knowledge of physics no more qualifies a scientist to evaluate psychic claims than does knowledge of chess or medieval Latin.

The comparison of parapsychology with general relativity is singularly inapt. Special relativity was initially confirmed by hundreds of tests. General relativity, which extended the theory to accelerated motion, had an enormous elegance and unifying power (the equivalence of gravity and inertia alone made it persuasive); soon it, too, was being confirmed by all tests capable of refuting it. More to the point, it was confirmed by skeptics. In contrast, after a century of research parapsychology has only vague suggestions for theories, and has yet to produce a single experiment that can be reliably replicated by unbelievers.

The letter signers cite four investigations they consider outstanding. It is a curious list. First we have the testing of Jean-Pierre Girard by Charles Crussard, a French metallurgist. Like Uri Geller, Girard began his career as a conjuror. Marcel Blanc's article, "Fading Spoon Bender" (*New Scientist,* February 16, 1978) reproduces a photo of Girard from a 1975/76 *Magicians' Annual* which shows him doing the now-standard bent-key trick. In the accompanying autobiographical remarks Girard says his specialty is "devising tricks based on optical illusions." Gérard Majax, a French magician, reveals in his recent book on cheating in parapsychology that Girard once told him he planned a gigantic joke to show how easily leading scientists could be fooled.

The American magician James Randi had no difficulty detecting Girard's simple methods when he saw Crussard's films, and in 1977, in a series of tests based on controls proposed by Randi, Girard failed to bend a single piece of metal. (See Blanc's article and Randi's book, *Flim-Flam!*) Crussard remains convinced of Girard's power. He has stated that Randi also has it, and secretly used it to inhibit Girard during the 1977 tests! Like Geller, Girard performs a variety of standard magic feats, such as driving a car while "securely blindfolded." That four distinguished physicists could consider him a "French medium" is almost beyond belief.

It is worth noting that had their letter been written a few years ago Geller would have been heralded as the star demonstrator of the "Geller effect" (psychic metal bending). In *Quantum Physics and Parapsychology* (Parapsychological Foundation, 1975), the proceedings of a 1974 Swiss conference, Geller's name is never mentioned without respect. On page 274 Walker, a signer of the letter, praises Uri's PK ability, and on page 279 he tells of once seeing Geller fail to produce PK effects because the "powerful wills" of unbelievers in the audience were "directed in the opposite direction."

All four writers contributed articles (two are cited in their notes 4 and 7) to *The Iceland Papers,* an anthology edited by Andrija Puharich. This is the Puharich whose notorious book, *Uri,* claims that Uri gets his powers from extraterrestrial spacecraft, and who believes that Uri once teleported himself from Manhattan to the back porch of Puharich's house in Ossining. Why is Geller, who started the metal-bending flap, so thunderously missing from the letter? Can it be because Geller is now discredited whereas Girard is still almost unknown outside of France?

Next we are told about England's spoon-bending children as reported in Puharich's book by John Hasted. I suggest that interested readers look up this hilarious paper to judge for themselves whether Hasted is a competent psychic investigator. Physicist John Taylor, Hasted's London colleague, was so bamboozled by Uri and by spoon-bending youngsters that he wrote an entire book about it, *Superminds.* As a result of learning some kindergarten magic, and making a few better-controlled tests, Taylor is now persuaded that the Geller effect does not exist, and that there is no evidence whatever for ESP and PK. See his just published Dutton book, *Science and the Supernatural,* in which he details his disenchantment. Hasted's work is demolished by pointing out that Hasted failed to take into account amplification by his sensitive strain gauges of slight static charges produced by body movements.

Next we have Helmut Schmidt's testing of psychics who seem to influence his random number generators. This work is considered "rigidly controlled" only by himself and by true believers. Schmidt seldom works with another investigator; skeptics have not had access to his raw data, nor have they been able to replicate his experiments. There also have been failed replications by sympathetic parapsychologists. For a probing of the weakness of Schmidt's experimental designs see C. E. M. Hansel's *ESP and Parapsychology: A Scientific Reevaluation,* recently published by Prometheus Books, pages 220–233. Schmidt is best known in psi circles for his research on the PK powers of cats and cockroaches. He, too, was once a Gellerite. In his paper in Edgar Mitchell's anthology, *Psychic Explorations,* he speaks of Uri as a "particularly strong" source of PK, whose ability to bend "heavy metal objects 'mentally,' just by touching them slightly or even without any touch" has been observed by "critical researchers."

Finally we have the remote-viewing (clairvoyance) experiments by Harold Puthoff and Russell Targ. No hint is given of the fast-growing literature on the carelessness of this work, especially as detailed in *The Psychology of the Psychic,* by psychologists Dick Kammann and David Marks. The latest failure to replicate was an extremely rigorous experiment, following all the original protocols, by four researchers at Metropolitan State College, in Denver. They reported their negative results at the 1980 annual convention of the American Association for the Advancement of Science, in San Francisco last January.

Reminders of Galileo's ridicule of Kepler, and of scientists unable to believe stones fell from the sky, were tired clichés even in 1952 when I mentioned them in my book *Fads and Fallacies.* They only prove what everybody knows, that great scientists can be mistaken. But as hard evidence accumulated for the lunar theory of tides and for elliptical planetary orbits (which Galileo also refused to accept), and for the fall of meteorites, no one suggested that beliefs were necessary for confirmations. This Catch 22 is peculiar to parapsychology, making it difficult in principle for skeptics to disconfirm any claim.

Instead of thinking of themselves as having the great insights of a Kepler, the writers should ponder their close resemblances to those eminent physicists who not so long ago were convinced that mediums could photograph the faces of departed spirits and exude luminous ectoplasm from their noses. If the four investigations listed in their letter are the best evidence they can muster for the reality of psi, their letter is a sad reinforcement of what John Wheeler had to say.

Martin Gardner

The remarks I made above about Charles Crussard produced the following letter from Crussard and his associate. It appeared in the *NYR*, December 18, 1980:

After being challenged by Mr. Martin Gardner in an article which recently appeared in your journal under the title of "Parapsychology: An Exchange," we should like to take advantage of our right of reply to complete the information of your readers about our experiments, since the remarks made about them by Mr. Gardner are astonishingly incomplete and only present partial views on incidental aspects of the problem.

The experiments carried out with J. P. Girard were the subject of a scientific report accepted by the reading committee of a specialist review* ending with a statement by an academician and former president of the Académie des Sciences in which he says, in particular, "in the absence of any proof to the contrary, it is not possible to provide a rational explanation for all these experiments, most of which were video recorded with a considerable wealth of controls. . . . Having had an opportunity to follow these experiments fairly closely, I agreed to add these few lines simply to vouch for the scientific rigor with which they were carried out by the authors."

Various factors preclude any possibility of faking:

—sometimes the experiment was filmed from end to end with a video camera: this was the case of a bending produced by J. P. Girard in a stoppered tube: it was the person making the experiment who took back the stoppered tube from J. P. Girard's hands at the end of the test, removed the stopper and took out the test specimen himself, *noting that it had been bent.*

—sometimes the tested metal bar was of such dimensions that, even a very strong man (weighing 140 kgs!) could not bend it with his two hands. But J. P. Girard bent strong light alloy bars of a diameter of 17 mm on four occasions.

—in yet other cases, the very nature of the phenomenon precludes faking. This is true of several test specimens in which J. P. Girard produced structural transformations without deformation, martensitic transformation or hardening, by introducing numerous dislocation loops into the metal. It would take too long to describe here the experiments and counter tests subsequently carried out, but it is important to note that it was these tests and controls which most convinced the metallurgists.

Moreover, precautions were taken to ensure that there was no possibility of the substitution which illusionists claim to see everywhere, and

*C. Crussard and J. Bouvaist, "Mémoires Scientifiques," *Revue de Metallurgie,* February 1978, p. 117.

these are described in a passage of our report, as follows, "marks had been engraved in the body of the bar, and the positions of small characteristic defects had been observed. It had been conveyed to the experimental station in a different car from the one which had brought J. P. Girard . . . (after the experiment) we first checked in the laboratory that all the marks, scratches and defects initially in the bar were present on the bar returned after the experiment, so that it can be stated unequivocally that there had been no substitution."

The eight experiments described in this report were selected from about twenty highly significant tests which were in turn selected from about 150 tests carried out with J. P. Girard. Naturally, among all these tests, some failed, were less sure, or suspect. Contrary to what some people appear to think, we knew from the very beginning all about J. P. Girard's talents as an illusionist and, in all fairness, he had previously warned us. We consulted other illusionists before and after our tests and we know the tricks which they use to imitate psychokinesis.

Other demonstrations carried out with J. P. Girard were even followed by no less than seven reputed illusionists who saw the deformations of the metal but were unable to find any sign of faking, and they have witnessed to this.

The most important fact is that in the two years and more since our article appeared, nobody has suggested any trick or normal explanation for the phenomena—not even Mr. Gardner!

C. Crussard and J. Bouvaist

To the above I replied as follows:

Since no one I consider knowledgeable about the methods of psychic charlatans was present during the tests described by Charles Crussard, I can make only general comments.

When I reported that Jean-Pierre Girard, Crussard's superpsychic, was a former magician I did so only because the letter on which I was commenting (signed by four paraphysicists) referred to Girard only as a "French medium." When a magician turns "psychic" it not only is important to let the public know of his conjuring skills, it is more important to conduct no tests with him that are not carefully designed by a well-informed magician, and with that magician there as an observer. Whenever this procedure has been followed, Girard has failed to produce results. Several such controlled tests are reported by Marcel Blanc in his article on Girard, "Fading Spoon Bender," in *New Scientist,* February 16, 1978.

Crussard is firmly convinced that both Girard and his counterpart, Uri Geller, have genuine psi powers, but since both are former magicians, both sometimes cheat so as not to disappoint an audience. In NBC's outrageous pseudo-documentary, "Exploring the Unknown" (featuring as narrator that eminent scientist, Burt Lancaster), Girard is shown making an aluminum bar bend slowly. It is apparent to any magician that the tube was first held so its bend could not be seen, then while one hand caressed the air above it, the other hand slowly rotated the tube through a right angle to bring the bend into view. Crussard's position on psychic cheating comes down to this. When a psychic is caught using fraud, that is when he uses it. When he is not caught, that is when he uses genuine psi powers.

Crussard speaks of seven magicians who "saw the deformations . . . but were unable to find any sign of faking." Note how carefully this is worded. We are not told who the magicians are, whether they helped design the tests, or whether they were there during the actual bending. Watching a videotape of a miracle is no substitute for being present when it occurs.

Crussard typifies a small, sad class of scientists who are experts in their field, passionate believers in psychic forces, supremely ignorant of methods of deception, yet convinced of their ability to detect fraud. They will watch a conjuror vanish an elephant on a brightly lit stage, and readily admit they cannot explain how he did it. Next day they will watch an ex-magician move an empty pill bottle three inches and instantly declare that no conjuring techniques could possibly have been used!

But I waste time. So persuaded is Crussard of Girard's ability to produce the "Geller effect" that all efforts to disenchant him are like trying to write on water. I suspect, however, that the feeling is rapidly growing among better-informed parapsychologists, hornswoggled for years by fake metal benders, that the Gellers and Girards of the world are doing more damage to their cause than anything a skeptic can say.

Martin Gardner

The letter from Crussard and his assistant was accompanied by the following letter from paraphysicist John Hasted:

I write to defend myself against your correspondent, author Martin Gardner, who calls my competence into question in his rather outdated reply to the letter of four physicists, who were kind enough to single out for mention my no-touch dynamic strain gauge experiments on paranormal metal-bending children.

Apart from their alleged hilarity, the only feature of these experiments specifically mentioned is that mathematician John Taylor, also of London, demolishes them by pointing out that "Hasted failed to take into account amplification by his sensitive strain gauges [*sic*] of slight static charges produced by body movements."

To show that no-touch strain gauge signals were of electrostatic origin would require a demonstration that triboelectric charges were generated and also that they were coupled capacitatively into the sensitive part of the circuitry. The second factor is crucial.

Naturally we started with careful screening and earthing, and conducted our own tests for artifacts both with tribo- and with current electricity. With both the children and myself about ten feet away from the metal, normal tribo-effects could not be detected. Any experimental physicist would do the same, and would not even bother to mention it in his articles, since many referees are hard on the inclusion of details which are standard practice.

However, with subsequent children producing effects mostly at shorter distances, about a foot from the metal, some precautions were deemed necessary. We therefore included a dummy strain gauge and amplifier, responsive to electrical artifacts but not to strain. The very few strain gauge signals synchronous with dummy channel signals were always rejected. At a later date a common mode channel was included, more as a protection against touch than against electrical artifacts.

Experience with the common mode channel has indeed revealed no-touch electrical artifacts, but because of the small area of the miniaturized strain gauges, these seldom appear synchronously in the strain gauge channels. Moreover these artifacts do not synchronize with body movements, and occur only in the presence of the child subjects; they are occasionally accompanied by a pricking or tingling sensation in the subject's hands, and are of entirely different time duration to the effects of (normal) emission of ions by the human skin, which we are also studying. They occur even in an electrically screened room with metal floor and furniture, and it might be argued that they were a paranormal phenomenon in their own right. Their time structures differ from those of most strain gauge signals. John Taylor has not discussed any of these developments, since he has not kept in touch with my experiments, and has operated independently, despite, or perhaps because of, the defection of several families from him to me, the parents being critical of his rather casual methods.

Experimental physicists and engineers, both in England and in other countries, have replicated the no-touch strain gauge detection method, with varying degrees of success. Several have been present at my own sessions.

Much of Gardner's reply shows that he is still at the stage of reporting, at second or third hand, statements by unqualified investigators, conjurors and media specialists who would be utterly lost in the instrumenting of micro-effects. But science, as usual, has moved on, and it would be wise to recognize this.

J. B. Hasted

My response to Hasted's letter was:

Hasted's letter is intended to snow laymen with technical details impossible to check without being there. Recently he sent James Randi the circuit diagrams of his set-up, plus additional details about his latest protocols. Randi had this material, along with Hasted's published papers, evaluated by Dr. Paul Horowitz, a Harvard physicist with special expertise in strain gauge technology. Horowitz's opinion was that Hasted had only a dim understanding of how to use these sensitive devices. For details, see Randi's book *Flim-Flam!*

In Hasted's paper on "Paranormal Metal-Bending" in *The Iceland Papers* (edited and published by Andrija Puharich), the funniest picture is a photo of a glass globe containing dozens of paper clips that have been paranormally "scrunched" into a wild tangle of twisted wires by "Andrew G," one of Hasted's superkids. Why is there a hole in the globe? All Hasted reveals is: "We have found it necessary that a small orifice be left in the glass globes in which wires are bent."

Has anyone actually *seen* paper clips in the act of bending, or recorded it on videotape? No, a youngster just takes a globe home, or goes into another room, and comes back with the scrunch. Mysteriously, clips never scrunch in globes without holes or when someone other than the child is watching. Other experimenters have had no difficulty twisting paper clips and pushing them into such globes where they intertwine to form tight scrunches, and to do it in just a few minutes.

Teleportation sometimes accompanies metal bending. Hasted reports that "under good witnessing" a dozen crystals were "observed" to teleport in and out of small capsules. Well, not actually *seen* going in and out. In two excerpts from Hasted's unpublished "Geller Notebooks" in *The Geller Papers,* you can read about how half of a tiny vanadium carbide foil vanished from a capsule during Hasted's celebrated tests of the first metal-bender, Uri Geller. How trivial this now seems in the light of Uri's ability to teleport a dog through a wall of Puharich's house, as Puharich himself "observed," not to mention Uri's teleportation of himself from Manhattan to Puharich's home in Ossining.

For years Hasted's boundless gullibility and bumbling experiments have been almost as embarrassing to parapsychologists as to his Birkbeck colleagues. Until his strain gauge tests are reliably replicated by competent and skeptical physicists, not just by a handful of true believers, who except Crussard and a few other naïve paraphysicists can take them seriously?

Martin Gardner

There is a ridiculously easy way to test the hypothesis, put forth by us hard-nosed skeptics, that children shake paper clips out of Hasted's globes, twist them, then shove them back in. Just film the phenomenon secretly through a one-way mirror. Hasted gives no indication that he ever tried this. If he did, he has released no information about the outcome.

For readers unfamiliar with QM and who wish to learn more about its paradoxes and philosophical implications, I know of no better or more up-to-date nontechnical book than *Other Worlds,* by Paul Davies (Simon and Schuster, 1981). On the EPR paradox and Bell's theorem, see Bernard d'Espagnat's splendid article, "Quantum Theory and Reality," in *Scientific American* (November 1979), and the more technical paper, "Bell's Theorem: Experimental Tests and Implications," by J. F. Clauser and Abner Shimony, in *Reports on the Progress of Physics* (vol. 41, 1978, pp. 1881–1927).

A discussion of the two kinds of psi — the parapsychologist's psi and the psi-function of QM — will be found in my article on "Parapsychology and Quantum Mechanics" in *Science and the Paranormal,* edited by George O. Abell and Barry Singer (Scribner's, 1981). Consult also Chapter 36 of the book you now hold.

Strange and awesome things indeed take place on the microlevel of the universe. We do not yet know whether the paradoxes of QM will someday be resolved in a way that better conforms to our intuitions about space, time, and causality, or whether the universe, on the particle level, is behaving in a way that will never be free of seeming irrationality. All of which has nothing whatever to do with ex-magicians and crafty children who bend spoons and scrunch paper clips by methods so crude that self-respecting prestidigitators would be ashamed to use them.

18

Tart's Failed Replication

A recent paper by Dr. Charles T. Tart, a parapsychologist at the University of California at Davis, casts some revealing light both on Tart and on a sensational earlier experiment by him that was the topic of a spirited debate in 1977 in two issues of *The New York Review* (see Chapter 31).

The debate began with my note, "ESP at Random" (*NYR,* July 14), in which I criticized Tart's book, *Learning to Use Extrasensory Perception* (University of Chicago Press, 1976). In this book Tart reported success in ESP scoring that far exceeded anything obtained before in the history of parapsychology.

My note reproduced a letter from three of Tart's colleagues at Davis, mathematicians Aaron Goldman, Sherman Stein, and Howard Weiner. Impressed by the results in Tart's book, they had asked to see the raw data. Going over it they found that the alleged random-number device did not produce random numbers for the target sequence. "Until the experiment is done again," they wrote in their letter, "we are in the position of a chemist who at the end of an experiment discovers that his test tube was dirty. . . . The experiment has to be executed with a clean test tube."

My note also pointed out a glaring flaw in Tart's experimental design. His "Ten-Choice Trainer" (TCT) was constructed so that a "sender" and "receiver" who wished to cheat could easily do so by signaling what magicians know as a time-delay code.

This article, which appeared in the *New York Review of Books,* May 15, 1980, and the letters quoted in the Postscript are reprinted with permission from the New York Review of Books. © 1980 NYREV, Inc.

To see how such a code could have been used it is necessary to describe again the TCT's basic working. A sender in one room is in front of a console that displays a circle of playing cards from ace through ten. Next to each card is a pushbutton and a pilot light. When the electronic randomizer selects a card value, the sender pushes the button that turns on the light by the corresponding card. This actuates a "ready light" on a duplicate console in another room where the receiver is stationed. As soon as the receiver sees the ready light he begins an ESP search for the target card, usually by moving a hand around the dial. The sender observes this sweep on a TV monitor above his console. This arrangement is intended to help him telepathically "urge" the receiver to stop at the target. When the receiver has made a choice he pushes a button by the card. The target card's pilot light immediately goes on and a chime sounds if the guess is correct. This immediate feedback is supposed to keep the receiver's interest high and to stimulate the training of his psi powers.

In my 1977 note I explained how a sender could transmit the value of each card by varying the time between the receiver's last choice and the activation of the ready light. It would, however, be foolish to send individual numbers because what is wanted is not a perfect score but only a significant score. This permits such simple coding that wrist watches are not even necessary, just a little practice in counting seconds mentally. For example, a delay of under ten seconds could mean an even card, a delay of more than ten seconds an odd card. Transmission of just this one bit of information raises an expected score of 50 hits to 100. If the cards are divided into three groups the code is almost as simple. This would increase the expectation from 50 to 150 hits. To raise a score higher than that would be too suspicious.

Tart himself recognized the possibility of time-delay coding in a footnote that begins: "After the completion of the Training Study, I realized that this procedure allowed a possibility of sensory cueing. If a particular experimenter showed a differential time delay between reading the output of the random number generator and switching on various newly selected targets, a subject might become sensitive to this and artifactually increase his score. . . . This possibility should be eliminated in future work."

Tart's superstar was a girl identified only as S3. Her overall TCT score far exceeded that of any other subject: 124 hits when only 50 were expected by chance. Her experimenter-sender was Gaines Thomas, one of Tart's students. John Sladek, a London correspondent, sent me a plausible conjecture that derives from Thomas's vivid account of how he fixed the targets in his mind. First he entered the number on a score sheet, then he "silently repeated the number" to himself. Finally he "positioned" the card in his mind by keeping it just "posterior to the upper part of my ears." Success, he adds, "very often corresponded with a numbing feeling in that location. Once I felt I had the number positioned, I would turn on the proper target switch. . . ."

Is it possible, Sladek asks, that Thomas took less time to fix behind his ears an image of a card with a low value? It certainly is easier to visualize one of the five or six low cards than to visualize a seven, eight, nine, or ten. If this were the case, a sensitive receiver might unconsciously learn, as testing continued, a correlation between certain cards and a short or long delay. Thomas's five subjects, each of whose scores were better than those of any student who worked with other senders, had an overall expectation of 250 hits. A binary code would raise this to 500. The actual hits were 466. S3 would be the student with the greatest skill in picking up, subliminally or otherwise, Thomas's time-delay pattern.

After Tart became aware of time-delay coding he examined the scores of Thomas's five subjects without finding any consistent pattern "as to which targets they scored best or worst on." But of course such a check would not catch a binary code unless one had a record of the delay for each target, which Tart apparently does not have. However, Sladek points out, the following examination of the raw data should be made. Check the misses to see if these cards tend to be related to the target by a two-part grouping of the ten numbers.

The most likely division that Thomas would unconsciously make would be into high versus low cards. If a subject responded to such a grouping, misses would tend to be displaced a short distance clockwise or counterclockwise around the dial. This was actually the case. Tart reports significant "displacement" scores on adjacent cards. For Tart this was the result of poor ESP "focusing," but such displacements are just as easily explained by a binary code. Sladek ran a simulation test of 500 trials, using a binary code. After sixty-seven trials his subject had correctly divided the numbers into two groups. Results for hits, as well as for positive and negative displacement to an adjacent number, were remarkably close to the overall scores of Thomas's five subjects.

Thomas also tells us that during the subject's sweep, as he watched it on the monitor, he would sometimes "orally coax the image on the screen, or swear at the near misses." The receiver's cubicle was just across a four-foot-wide corridor. (The doors of the two cubicles were ten feet apart.) One assumes that the sender's cubicle, which Tart calls a "sound-attenuated" Faraday cage, was sufficiently sound attenuated to prevent a subject with sharp ears from hearing Thomas's swearing, though I know of no checks by outsiders that were made on this. I suppose it depends on how loud Thomas swore. Of course any kind of sensory feedback from Thomas would explain high scores on both direct hits and displaced targets as readily as the time-delay hypothesis. In either case, it is significant that Thomas's five students scored 466 hits when only 250 were expected, whereas the other five students in the TCT test, who had other senders, scored 256 hits when 250 were expected.

Tart replied to my *NYR* note in a letter (*NYR,* October 13, 1977) to which I responded in the same issue. I asked if videotapes had been made of S3's performance. If so, a study of them could confirm or refute the time-delay theory. If videotapes were not made, this would be another design defect. I urged that S3 be tested under better controlled conditions in another laboratory. I assume no tapes were made and that the girl has not been tested again even though her scores, if based on genuine ESP, would make her one of the most talented psychics in the history of parapsychology.

When Tart wrote about his TCT experiment in *Psychic* magazine ("ESP Training," March-April, 1976) he began as follows:

> Strong criticism has been leveled at ESP research over the years because the phenomena could not be repeated regularly. Since they could not, skeptics gloated that they did not exist.
>
> Now, a research breakthrough soon may shelve such criticism. A study carried out under my direction at the University of California at Davis Psychology Department has taken a big step toward repeatability of ESP by helping people understand how ESP works and how it can be controlled.

Tart did replicate his "breakthrough," and the report, "Effects of Immediate Feedback on ESP Performance: A Second Study," by Tart and two associates, John Palmer and Dana J. Redington, appeared in *The Journal of the American Society for Psychical Research* (vol. 73, April 1979, pp. 151–65). The study was funded by est and by the Parapsychology Foundation of New York. Sixteen students participated. They were selected by a screening of high scorers from 2,424 tested persons. Overall scores for the screening process were at chance levels. As the authors put it: "There is no evidence that more percipients scored significantly above chance than would be expected if no ESP were operating."

The high scorers were used in replicating the earlier experiment even though the screening provided no basis for assuming they were better at ESP than any of the others. The replication used three machines: a four-choice trainer called Aquarius, the TCT in its improved form, and ADEPT (Advanced Decimal Extrasensory Perception Trainer) which Tart describes as a ten-choice machine similar to the TCT but with more sophisticated circuitry.

Overall results for both ten-choice machines did not differ significantly from chance. There was some evidence of ESP improvement on the part of some subjects who used the Aquarius, but since only three students completed this part of the experiment it contributed nothing of statistical value.

In view of the astonishing contrast between the replication's chance results and the near miraculous scores of the original experiment, one would expect Tart to end his report by withdrawing his former results. A chemist or physicist would feel ethically obliged to do so, especially if flaws had been pointed out in the original experiment. On the contrary, Tart expresses his belief that the second experiment failed mainly because "too few talented percipients were selected by the screening process."

Here are Tart's three reasons for the failure:

> With respect to psychological interpretations, several people who have had close contact with students at the University of California, Davis, over the past three or four years have told us of a dramatic change in the attitudes of students during that period. In the last year or two, students have become more serious, competitive, and achievement-oriented than they were at the time of the first experiment. Such "uptight" attitudes are less compatible with strong interest and motivation to explore or develop a "useless" talent such as ESP. Indeed, we noticed that quite a few of our percipients in the present experiment did not seem to really "get into" the experiment and were anxious to "get it over with."
>
> The situation also was different for the student experimenters in the two experiments. Experimenters in the first experiment could legitimately feel that they were embarking on a new adventure. Despite our best efforts to create the same enthusiasm in the second group of experimenters, there was no way to deny the fact that we were asking them to simply repeat an experiment designed and executed by others before they ever arrived on the scene. It is understandable that they did not feel as intensely involved in the experiment as did the first group of experimenters, and this factor could have been responsible for the relatively poor performance of their percipients. Indeed, several of the more seriously involved experimenters later told C.T.T. [Charles T. Tart] that they were quite disturbed by the attitude of some other experimenters who "just wanted to get it done with."
>
> Finally, we were constantly plagued by machine malfunctions . . . and this was a source of continual annoyance and inconvenience to all concerned.

Let me summarize. Tart reported in a book, written with unbounded confidence, results so extraordinary that they far exceeded those obtained in similar testing by any other researcher. His TCT machine was found to have a flawed randomizer, and a design that permitted time-delay coding. A replication of the experiment, with both flaws eliminated, showed no significant departures from chance. Tart attributes this primarily to his inability to find sufficiently psychic students. As for the original experiment: "Because the level of scoring in the first experiment was so high, it would be absurd to argue that the results of the second experiment mean that the results of the first experiment were a mere statistical fluke." Nowhere does he even mention the possibility that the first experiment was invalid because of a defective randomizer, or fraud, or unconscious time-delay or sensory cuing.

Tart's last statement leaves me so staggered that I can respond only with a parable. A parapsychologist finds a psychic who can levitate a table forty feet. He investigates this under poorly controlled conditions, but is so convinced the phenomenon is genuine that he writes a book about it. The book is published by a gullible university press. After skeptical magicians—those terrible spoilsports!—patiently explain how the levitation could have been accomplished by trickery, the parapsychologist agrees to test the psychic again, this time with adequate controls. The table does not rise at all. The parapsychologist then writes a formal report that concludes: "In view of how high the table rose during the first experiment, it would be absurd to contend that the failure of the second experiment in any way casts doubt on my previous observations."

During the academic year 1978-1979, Tart was on leave from Davis to work with Harold Puthoff and Russell Targ, at Stanford Research Institute, on remote viewing (clairvoyance) experiments presumably funded by the military. SRI is a private research organization not affiliated with Stanford University. As a matter of policy it will not reveal the sponsors of research projects. Experiments in parapsychology, which has yet to establish itself as a science, should be open to the scrutiny of the scientific community. When open research in parapsychology is conducted in shoddy fashion, we have a right to be concerned about the quality of clandestine ESP research. From what we know of their previous work, the grant in support of the new work by Targ and Puthoff will likely be a total waste. Unfortunately, the money comes from taxpayers such as you and me.

Postscript

Tart's reply to my criticism of his failed replication appeared in the *NYR* (February 19, 1981):

> I see that Martin Gardner is again using this popular literary journal as a vehicle to attack my scientific research that was reported in my *Learning to Use Extrasensory Perception* (University of Chicago Press, 1976) [*NYR,* May 15]. As a working scientist, I am committed to reporting and dealing with all of the facts in my studies, whether they agree with my cherished beliefs or not. Data is primary. Gardner, by contrast, apparently knows what's true and false in some absolute way, so when inconvenient facts run counter to his beliefs he suppresses them or rationalizes them away. He knows that ESP is impossible, so when he is presented with evidence for it, he imagines some way in which the experimenters are fools, frauds, or both. Mr. Gardner doesn't need actual evidence for this, his suspicions are sufficient. Most people would consider his casual and unsupported accusation of fraud against one of my more successful experimenters, Gaines Thomas (now a professional psychologist), as malicious libel, but I suppose Mr. Gardner believes he's just protecting us gullible people from ourselves.

Gardner demonstrates how his absolute convictions allow him to take liberties to protect us from ourselves in presenting his apparently ingenious theory of a deliberate timing code used by a fraudulent experimenter being responsible for the high level of ESP shown in my study. He cites a publication of mine and my colleagues in the *Journal of the American Society for Psychical Research* (1979, *73*, 151–165), indicating his familiarity with that *Journal,* but he does not mention several earlier communications of mine in that same *Journal* (see 1977, *71*, 81–102; 1978, *72*, 81–87; 1979, *73*, 44–60) reporting precognitive ESP effects in the experiment he attacks, which could not be accounted for in any way by his time code model. Again I stress the obligation genuine scientists (and genuine critics) have of dealing with all the facts in a case, not just those they find convenient. Gardner has presented a clearly inadequate theory to a literary audience as if it were valid. The interested reader is invited to look at the above communications to ascertain the facts for himself. There are other distortions in Gardner's article that I shall not bother to waste our time correcting here: they are, unfortunately, typical of Gardner's writings on parapsychology.

When real scientists have criticisms of each other's work, the standard procedure is to submit the criticisms to the appropriate technical journal. The submission is reviewed by other scientists for basic competency and relevance, and then published. I doubt that Mr. Gardner's article would have stood up to this refereeing process in a legitimate scientific journal. A thoughtful reader might begin to wonder, then, why Mr. Gardner presents such a distorted and selectively incomplete picture of serious scientific research to the general audience represented by readers of this *Review.*

The implications of ESP for understanding human nature are enormous, and call for extensive, high quality scientific research. A recent survey of mine showed hardly a dozen scientists working at it full time, on a most inadequate budget of only a little over half a million dollars a year for the entire United States. The subject is too important and too under-researched to waste further time with pseudo-critics like Mr. Gardner who are covertly trying to manipulate public opinion, rather than contributing anything to scientific progress.

Charles T. Tart

Tart's letter was followed by my rejoinder:

The funny thing about Tart's letter is that he devotes most of it to attacking my article as misleading, malicious, and distorted, but nowhere replies to its central point; namely, that his first experiment had major flaws, he corrected some of them, repeated the test himself, got negative results, but has refused to retract his former sensational claims.

Precognition is the paranormal perception of future events. Psi-missing is making such a low score on an ESP test that it indicates paranormal inhibition. When Tart went over the data for his original flawed test of clairvoyance he found that subjects who did extremely well on "hitting" target cards scored significantly low on the next card to be selected.

There is a simple explanation. One of the grave defects of Tart's first experiment was that his machine did not automatically record the numbers selected by his randomizer. When mathematicians found a strange absence of doublets (such as 2,2 or 7,7) in his target sequences, Tart explained this by saying that when assistants pushed the button to obtain a new number, and noticed that the displayed number did not change, they sometimes thought they hadn't pushed hard enough, so they would push again before hand-recording the number! Clearly this freedom to keep pushing permits a sender to push again if the displayed number matches the subject's last guess. Tart himself tells us that subjects almost never guessed the same number twice in a row. Knowing this, senders would have a strong unconscious urge to alter a random number if it matched the last guess. If done every time it would produce zero matching of guesses with +1 targets. Done occasionally it would significantly lower precognitive hits on +1 cards as well as raise hits on "real-time" targets. Tart has the chutzpah to claim that, because he found some precognitive psi-missing in both experiments, this transforms the obvious failure of his replication into a whopping success! One's mind reels at his capacity for self-deception.

There are two false statements in Tart's letter. First, he accuses me of knowing ESP is impossible. I know no such thing. I firmly believe it is possible. I do not believe it has been demonstrated by evidence commensurate with the extraordinary nature of its claims.

Second, Tart says I accused his former assistant, Gaines Thomas, of deliberate fraud. I did nothing of the kind. I did show that Tart's first experiment failed to guard against simple time-delay codes, and that in binary form such codes could operate without sender or subject being aware of their use. If there were collusion between sender and subject, which I doubt, the freedom to keep pushing the randomizer button also provides endless simple ways of beefing up a score.

When reputable scientists correct flaws in an experiment that produced fantastic results, then fail to get those results when they repeat the test with flaws corrected, they withdraw their original claims. They do not defend them by arguing irrelevantly that the failed replication was successful in some other way, or by making intemperate attacks on whoever dares to criticize their competence.

Martin Gardner

Part Two

19

"ESP: A Scientific Evaluation"

Ever since Joseph Banks Rhine, a botanist-turned-parapsychologist, began his systematic study of psi (his term for "psychic") phenomena, he has enjoyed an unusually favorable popular press and an unusually unfavorable academic one. Long, laudatory articles about him have been appearing for decades in mass circulation magazines (e.g., "A Case for ESP" by Aldous Huxley, *Life*, January 11, 1954). Arthur Koestler has compared Rhine's discoveries to the Copernican Revolution. Today's skeptical scientists, Koestler says in *The Sleepwalkers*, resemble those Italian philosophers who refused to look through Galileo's telescope at Jupiter's moons because they knew in advance that such moons did not exist. Many otherwise sophisticated people, I would guess, take it for granted that ESP and other psi powers have been conclusively demonstrated by workers in the field, and that only a few pigheaded professors refuse to look through Rhine's telescope at the towering mountain of scientific evidence.

For thirty years professional psychologists, using sophisticated modern techniques, have been trying to duplicate the experiments of the parapsychologists, and they remain unconvinced. Unfortunately, their monotonously negative results are too dull to interest *Time* or *Newsweek;* to learn about them one must subscribe to the academic journals. Last year, for instance, the *Journal of Psychology* (vol. 60, pp. 313–18),

This review, which appeared in the *New York Review of Books,* May 26, 1966, and the letters quoted in the Postscript are reprinted with permission from the New York Review of Books. © 1966 NYREV, Inc.

reported on a carefully designed series of ESP tests by Richard C. Sprint-hall and Barry S. Lubetkin. Fifty subjects were divided into two equal groups and each group was given a standard ESP test. One group took the test without "motivation"; the other was told that anyone who guessed twenty out of twenty-five ESP test cards correctly would immediately be given a hundred dollars. No one won any money. There was no significant difference in the results obtained from the two groups, and neither showed any evidence of ESP.

This test had been prompted by Rhine's repeated assertions that financial reward provides strong motivation for ESP, and that "subject motivation to score high has long stood out as the mental variable that seems most closely related to the amount of psi effect shown in test results." Indeed, the most sensational result ever obtained by Rhine occurred during the Depression when he kept offering Hubert Pearce, one of his star subjects, a hundred dollars for each top card he could call correctly in a pack of ESP cards. They halted the test by mutual consent after Pearce had correctly named twenty-five cards in a row. No one else was present on this occasion, and Rhine's published accounts of exactly what happened are vague. (I once tried to get a few easily remembered details out of Pearce by correspondence—he is now a Methodist minister in Arkansas—but he flatly refused to discuss the incident.) Nevertheless, Rhine always cites this in his lectures as the most remarkable demonstration of clairvoyance he has ever witnessed, giving the odds of 298,023,223,876,953,125 to 1 that it could have happened by chance. (I always felt sorry for Pearce with respect to this event. He was poor at the time and needed the money, but when the test was over, and he was owed $2,500, Rhine explained that he had just been joking.)

No one can deny that some of the most remarkable results in the history of psi research were obtained when subjects were strongly motivated. Can it be that strongly motivated subjects are often strongly motivated to cheat? This is the opinion of C. E. M. Hansel, Professor of Psychology at the University of Manchester, and author of *ESP: A Scientific Evaluation* (Scribner's, 1966). After a careful study of the most important ESP experiments by Rhine and his British counterpart, S. G. Soal, Hansel has become convinced that more hanky-panky has been going on than even the skeptics have suspected.

Consider the classic series of tests that Soal made in the early forties with a photographer named Basil Shackleton. When Hansel subjected the records to a statistical analysis, a curious anomaly turned up. Soal's score sheets had been ruled with a double line after every five blanks, and Shackleton's "hits" were concentrated (with odds greater than 100 to 1 against such a concentration) on the third and fourth lines of each group of five. It is hard to think of any reason why ESP would conform to the pattern of ruled lines on score sheets, but easy to understand if someone had gone over the scores to beef them up a bit and had been too stupid to make the beefing random. Hansel's discovery set off a noisy dispute among British psychic researchers when he published it in *Nature,* in 1960, although not a line about this appeared in the U.S. press. When Hansel tried to get a look at Soal's original score sheets, on which chemical tests would, of course, reveal any tampering, he was told by Soal that they had all been lost on a train in 1946. (Soal had written in 1954 that the original records had been preserved and "could be rechecked by anyone at any future time.") Moreover, one of Soal's assistants in the Shackleton tests told Hansel that she had glanced through a hole in a screen and had seen Soal altering figures.

When Rhine heard of these disclosures, he invited Hansel to visit his laboratory, then affiliated with Duke University, to look over *his* records. This proved to be bad judgment. Hansel did visit Rhine's laboratory, and stayed until he was asked to leave. Every major series of tests that he investigated in depth turned out to have gigantic loopholes, hitherto unnoticed (or unmentioned) by Rhine, that permitted skulduggery of the most elementary sort.

Rhine's most respected series of tests, to which he refers constantly in his later writings, was a series of long-distance tests with Pearce, conducted in 1933–34 by Rhine's long-time assistant, Joseph Gaither Pratt. Pearce and Pratt met in Pratt's office, synchronized their watches, fixed a time for the test to start, then Pearce walked across the Duke quadrangle to the library, where he sat in a cubicle in the stacks. Pratt went through a supply of fifty ESP cards, taking them one at a time and placing each card face down in front of him for one full minute. Then he turned over all the cards, made two records of their order, and sealed one in an envelope and delivered it to Rhine. Pearce also made duplicate records of his guesses and delivered one to Rhine. There were thirty-seven such sittings. Pearce's scores, throughout, were much too high to be explained by chance.

Anyone who reads the informal accounts of the Pearce-Pratt tests, in books by Rhine and Pratt, cannot but be impressed by the lengths to which Rhine went to rule out collusion between Pratt and Pearce. But the one thing neither Pratt nor Rhine ruled out was the possibility that Pearce did not stay in his cubicle. He could have sneaked back across the campus, entered a vacant room across the corridor from Pratt's office, stood on a chair, and peeked through the transom and a clear-glass hallway window, just behind Pratt's shoulder, to get a good view of the cards while Pratt recorded them. While Hansel was at Duke, he asked one of Rhine's researchers to run through a pack of ESP cards while he (Hansel) locked himself in an office down the hall. Hansel tip-toed back, stood on a chair, and peeked through a crack over the door. He scored twenty-two hits out of twenty-five cards, to the complete mystification of the researcher. Hansel does not say that Pearce cheated in a similar manner, or in any of several other easy ways permitted by the experiment's amateurish design. He does say that, because this obvious bias factor was not guarded against, the entire Pearce-Pratt series is now highly suspect. Hansel puts it this way:

> One would expect that anyone in Pratt's position would have examined the room carefully and have taken elaborate precautions so that no one could see into it. At least he might have covered the windows leading to the corridor. Also, the cards should have been shuffled after they were recorded, and the door of the room might well have been locked during and after the tests. These experiments were not a first-year exercise. They were intended to provide conclusive proof of ESP and to shake the very foundations of science. If Pratt had some misgivings, there is no evidence that he ever expressed them. . . . Again, Rhine might well have been wary of trickery, for neither he nor Pratt were novices in psychical research.

Hansel's book reaches its climax in the chapter on Soal's last great series of tests, his experiments with two Welsh schoolboy cousins, Glyn and Ieuan Jones. Soal wrote up his results in a book called *The Mind Readers,* which was favorably reviewed by every leading newspaper in England except the *Manchester Guardian.* Even the distinguished Sir Cyril Burt, editor of the *Journal of Statistical Psychology,* praised Soal for the care in which he conducted his tests and called them "unrivaled in the whole corpus of psychical research."

They were unrivaled, it is true, but not in "care." The Jones boys were repeatedly caught signaling to each other, using various visual and auditory codes. As soon as the boys improved their signaling methods, Soal promptly concluded that he had succeeded in persuading them to stop cheating. Hansel's analysis of Soal's book demolishes everything except the sad, comic, unintentional revelation of one parapsychologist's extraordinary naiveté.

Rhine's published work on PK (psychokinesis), his term for the ability of the mind to move such objects as falling dice or to levitate a table, is far more revolutionary than his work on ESP. Not even Soal has been able to find evidence for PK, and it has long been a joke among psychologists that PK somehow fails to operate in British laboratories. Skeptics are always asking Rhine: If PK is strong enough to control a rolling die, how come it can't move an eyelash on a smooth surface in a vacuum, or rotate a tiny needle, suspended magnetically so that there is virtually no friction? Hansel writes that when Rhine was asked this in 1950, after a lecture at Manchester, he replied by saying that such a test was a splendid idea and one he might get around to trying sometime. Hansel later learned that Rhine had been making such tests for years, back at Duke, but when Rhine gets negative results he likes to keep them under wraps. To this day, the failure of PK to display itself in such a simple, direct fashion is one of those subtle psi mysteries that true believers find the hardest to explain.

In the cold light of Hansel's analysis, what should be the layman's attitude toward the claims of parapsychology? First, he should realize that the claims are claims of fact, not theory. There *is* no theory of psi phenomena. If the facts are true, they are independent of all known laws of science, and no new laws have been formulated to explain them. Second, he should realize that these facts are historical questions. Psi experiments are not repeatable in the way that other experiments in physics or psychology are. Star subjects such as Pearce and Singleton invariably and inexplicably lose their former powers. One does not ask: "Is this Methodist minister, Hubert Pearce, clairvoyant?" but "Did Pearce, at a certain time in his youth, actually call twenty-five ESP cards correctly by clairvoyance?" Such a question is of the same type as: "Did the medium D. D. Home, in 1868, actually float horizontally through an open third-floor window in London, turn around in the air, and float back in again, feet first?" Such claims must be approached in the spirit recommended by David Hume in his famous essay on miracles: One must ask if present evidence for the alleged event is so strong that any other explanation of the evidence would be even more miraculous.

It is also important to realize that one's attitude toward parapsychology should be completely independent of one's metaphysics. Sigmund Freud, who believed in ESP, was an atheist. Most people I know who admire Rhine are atheists. On the other side, I once heard a devout theist argue that one of God's great gifts to humanity is the insulated brain in which thoughts can be kept inviolate. The questions raised by parapsychology must be answered, in the only way they can be answered, by considering all the evidence bearing on them and making an estimate of the probability of ESP that is free of emotional bias. Because the claims of parapsychology run so strongly against the entire corpus of known physical laws, the burden of proof is surely on the claimants. They may win over most laymen by means of uncritical reports in the popular press, but they are unlikely to impress other psychologists until they produce evidence strong enough to justify what Koestler is absolutely right in calling (if the facts are true) a "Copernican Revolution."

So far, the strongest evidence has come from the work of Rhine and Soal, but I do not think that anyone can read Hansel's book with an open mind and believe that evidence compelling. There is one hopeful new development. The U.S. Air Force Research Laboratories has devised a type of ESP experiment in which a computer called VERITAC is substituted for the experimenter and his assistants in such a way as to rule out such common sources of bias as fraud and recording errors. "If 12 months of research on VERITAC can establish the existence of ESP," Hansel concludes, "the past research will not have been in vain. If ESP is not established, much further effort could be spared and the energies of many young scientists could be directed to more worthwhile research."

Postscript

Two letters attacking my review, one from Bob Brier and one from J. G. Pratt, appeared in the *New York Review of Books,* July 28, 1966:

Mr. Gardner says that "professional psychologists, using sophisticated modern techniques, have been trying to duplicate the experiments of the parapsychologists, and they remain unconvinced." The reviewer implies that there is a dichotomy between parapsychologist and professional psychologist. Mr. Gardner is either unaware of or ignores the fact that such scholars as Dr. Gardner Murphy, Dr. R. H. Thouless, Dr. Gertrude Schmeidler, Dr. Hans Bender, Dr. J. G. Pratt, Dr. Maurice Marsh, etc., are only a few of the "professional psychologists" who have made important contributions to the field of parapsychology. Thus there are many professional psychologists who are not "unconvinced." The gap between psychologist and parapsychologist does not exist as Martin Gardner would have his readers believe. It is unfortunate that Mr. Gardner assumed the responsibility to speak for professional psychologists.

It is also regrettable that Dr. J. B. Rhine is referred to in a demeaning way as "a botanist-turned-parapsychologist." Dr. Rhine is well-trained in psychology, and Mr. Gardner ignores the fact that for twenty years Rhine taught psychology, not botany, at Duke University.

The above distortions are superficial when compared with the methodological error created by Hansel and perpetuated by Gardner. Mr. Gardner mentions that Hansel discovered that the scores of Dr. Soal's experiment appeared in a curious pattern and that this would be "easy to understand if someone had gone over the scores to beef them up a bit and had been too stupid to make the beefing random." There is no reason to delve so deeply for such an explanation. If one takes any data and examines them for all possible anomalies, one is almost certain to discern something which seems unusual. This is merely due to chance. For this very reason, parapsychological experiments, to be considered valid, predict precisely what kind of anomaly shall appear. Both Hansel and Gardner commit an error in basic experimental procedure. Hansel looked at the data and turned up one anomaly. By chance one must expect anomalies; this does not indicate fraud. It is surprising that the author of a column dealing with mathematics should make an error in statistics.

There are other errors in the review, but it is not the intention of this letter to correct all mistakes made. Rather, I should merely like to point out that parapsychology is a rigid discipline utilizing scientific method. It is not guilty of committing the errors which those unskilled in the field are so ready to criticize.

Bob Brier

Among the critics of parapsychology, a small number have always approached the subject as if ESP just has to mean "error some place." The review by Martin Gardner of the book by C. E. M. Hansel seems to indicate that the reviewer belongs along with the author in this minority group. They both write as if they are rendering to mankind the great service of helping to rid science of the fallacy of ESP and that any means is worthy in the pursuit of this end. Certainly your readers could easily see that many of the criticisms in Mr. Gardner's review go beyond proper scientific and ethical bounds. In other instances, however, the facts will be known only to an insider in parapsychology. May I give a few illustrations?

Mr. Gardner says: "I once tried to get a few easily remembered details out of Pearce by correspondence . . ." *Easily remembered*—after more than 20 years? Psychologists have long known that details regarding even the most vivid experiences are not accurately recalled over a period of time. What could Pearce have said beyond expressing confidence in the experimenter and his report, and what valid purpose could it have served if he had tried to do more?

The reviewer continues: ". . . but he flatly refused to discuss the incident." I must thank Mr. Gardner for informing me about his 1951-56 correspondence with Pearce. Hubert answered one letter saying that he did not wish to write about his ESP work at Duke. When the second letter, the one referred to in the review, was received Pearce apparently did not reply. But surely there are many *innocent* reasons why he may not have done so, including the fact that his earlier letter had already given his answer. Under the circumstances his failure to write again was hardly a flat refusal to discuss the matter. Yet in the review the omission is treated—could anyone fail to get the implication?—as a sign of a guilty conscience.

The reviewer accepts without question Hansel's statement that Hubert could have peeked to see the cards by hiding in a room across the corridor from the experimenter's room in the Pearce-Pratt series. Hansel shows a floor plan of the rooms and corridor that he admits is "not to scale." But how is the reader to know that in his plan the positions of the illumination windows have been changed and that a true floor plan would show that the supposed peeking was impossible?

I strongly protest against the tone of the whole review, which is one of scorn and ridicule for an on-going field of research that the reviewer has not accepted. Mr. Gardner has every right, of course, to his opinions. But am I alone in feeling that the type of critical attack exemplified by his review is foreign to the spirit of scientific inquiry? The methods he (and Professor Hansel) use for guarding the present frontiers of science against ESP could keep out any new discoveries that would require revolutionary changes in our views of the nature of the universe.

J. G. Pratt

I replied to both letters as follows:

Mr. Brier's first point is a linguistic one. I meant to call attention to the well-known fact that the overwhelming majority of professional psychologists are not only skeptical of ESP, but have performed scores of ESP tests, regularly reported in the journals, that have failed to support the findings of parapsychology. Those psychologists who are also parapsychologists have, naturally, made tests that support parapsychology, but their results are published almost entirely in parapsychology journals. The gap between this small group, and the large group of psychologists who are not parapsychologists, not only exists, but is so large that it is the object of constant complaints by the parapsychologists themselves.

Mr. Brier's second point is conceded. There is no reason why a person who received his Ph.D. in botany, as Rhine did, could not become a competent experimental psychologist.

And he is correct in saying that if one goes over a table of, say, random numbers, and looks deep enough for statistical anomalies, he is likely to find them. This is a point that I stress in the chapter on Rhine, in my book *Fads and Fallacies,* where I point out (pages 304–305) that many of the positive results reported by Rhine may actually be no more than statistical anomalies of just this sort. The question of how strong such a pattern must be in order to be taken seriously as indicating a bias in an experimental situation is a question that touches on deep aspects of inductive logic. There are no simple answers. Suppose that someone has recorded a thousand throws of a die and found that it came up an ace more often than it should. The hypothesis that the die is loaded has a probability, or what Carnap calls a "degree of confirmation," that depends on how many times the die showed an ace. If it rolled 1,000 aces, the degree would be close to certainty that the die is not a fair one. Suppose, however, that the bias is small. Then you discover, after checking the records more carefully, that the bias is confined entirely to every tenth roll of the die, and that records are on a sheet that divides the scores into groups of ten. This anomaly would be harder to explain than the overall bias. Since there is no connection between the symmetries in the experimental situation and the number 10, you would be led toward hypotheses that concerned the score sheets rather than the die.

This is similar to the situation presented by Soal's score sheets. Soal himself, in published replies to Hansel's criticism, has not regarded the anomaly as merely accidental. He considers it a "segmented salience" effect of unknown origin. Hansel reports in his book on a test he made with a group of students, asking them to place dots at random on the spaces of a form similar to Soal's score sheets. The patterns they produced showed salience effects similar to those on Soal's records of tests with Shackleton. Of course this *proves* nothing. But it does suggest the hypothesis that the records were doctored by someone who did not make the doctoring sufficiently random; a suggestion which gains credence by Soal's statement that he lost the original records, and by the charges of Mrs. Gretl Albert, who was one of three assistants in these tests. (The other two were Soal's wife and his barber.)

I was incorrect in saying that Mrs. Albert told Hansel personally that she had seen Soal doctoring the records. She reported this to Mrs. K. M. Goldney, Soal's collaborator and a council member of the Society for Psychical Research. The society did not publish Mrs. Albert's charge until about eighteen years later, when it appeared in their journal (vol. 40, p. 378, 1960) along with Soal's reply. (It was said at the time that Soal was being harassed by two enemies, Hansel and Gretl.) Nor was I quite correct in saying that Hansel had been requested by Rhine to terminate his visit at Duke University. Hansel has since informed me that he left a week before he was supposed to because Rhine refused to continue to let him have original data sheets unless he signed a form saying that he would publish nothing about them without Rhine's permission. This Hansel refused to do. Since there was no longer any point in staying, he left.

Professor Pratt excuses Pearce from replying to my letter of September 7, 1953, on the grounds that it would not be easy for Pearce to recall details of that historic occasion on which he, alone with Rhine, correctly named 25 ESP cards without seeing their faces. Since this was by all odds the most sensational ESP demonstration ever witnessed by Rhine, so unequivocally unexplainable by known natural laws, then surely it is a scandal that neither Rhine nor Pearce immediately recorded every detail so that the event could later be described with great accuracy. In his infuriatingly vague description, Rhine says that "each card was returned to the pack and a cut made." I asked Pearce if this meant that, each time he guessed a card correctly, Rhine was holding that card and looking at its face. It is hard to believe that Pearce would not remember this aspect of the procedure. Pratt writes that Pearce "apparently" did not answer my letter. From later correspondence, which I photocopied and sent to Pratt, it is quite clear that Pearce did not answer that letter.

In his first letter, Pearce had said that he would be delighted to do whatever he could to assist me, but he added that his replies would have to be confidential, because the leaders of his church (Pearce is a Methodist minister) did not look with favor on his connection with ESP work. I would not have written him a second time unless I thought that he was willing to answer a few questions off the record. I am responsible for Pratt's misunderstanding on this point, since I did not include my first exchange of letters with Pearce in the group of subsequent letters that I photocopied.

As for the floor plans of Pratt's old office, Hansel says in his book that when at Duke he had asked for the details of the structural alterations. "These details were to be forwarded to me," writes Hansel, "but I never received them. I wrote again requesting them, but had no reply." If Pratt possesses the "true floor plan," it would have been helpful if he had made it available to Hansel. Indeed, ruling out all possibilities of peeking into the room was so absolutely essential to Pratt's test with Pearce that his failure to print the room's floor plan, when he first published his results, remains a serious blot on that report. However, the exact floor plan is now seen to be amusingly irrelevant. Even if it had not been possible for Pearce to stand across the corridor, in another office, and peek through a transom into Pratt's room, there was nothing in Pratt's clumsy experimental design to prevent Pearce or a collaborator, had he wished to do so, from standing on a chair in the corridor and peeking directly through Pratt's own transom.

My reply provoked another exchange between Pratt and me. It ran in the September 22, 1966, issue of the *NYR*:

I was glad to see that my letter on the Hansel-Gardner attack on ESP was published in spite of delays caused by my parapsychological journey around the world. I was glad to see also from Mr. Gardner's reply that the space limitations under which Mr. Brier and I labored have been removed, because there are a few things more that must be said in reply to Mr. Gardner's letter.

Incidentally, he admirably used more than half of his reply admitting a number of errors in his review. Why did you not suggest to him the happy alternative of achieving brevity by simply withdrawing the entire review? If he had done so, this would have saved him the embarrassment of further errors that he has now made in his reply to my letter.

Mr. Gardner says: "Nor was I quite correct in saying that Hansel had been requested by Rhine to terminate his visit to Duke University. Hansel has since informed me that he left a week before he was supposed to because Rhine refused to continue to let him have original data sheets unless he signed a form saying he would publish nothing about them without Rhine's permission. This Hansel refused to do. Since there was no longer any point in staying, he left."

I spoke to Dr. Rhine in Durham, N.C. this forenoon and read him the above quotation. He gave me permission to quote him as denying categorically Professor Hansel's revised statement.

Mr. Gardner says that the photocopies he sent me of his later correspondence with Reverend Pearce make it "quite clear that Pearce did not answer that [previous] letter." It is true that Mr. Gardner's letter to him three years later contains the statement: "You did not reply to my letter of September 7, 1953." But Reverend Pearce's answer to this letter makes no reference to his not having replied to the former one. As a scientist, I am justified by these circumstances only to say that Pearce "apparently did not reply." How do I know that Pearce did not write a letter that was lost in the mail? Or how do I know that Gardner did not receive a letter which he has since lost and forgotten about? This is a small point and not important except as an illustration of the difference between the scientist who must carefully choose his words and the popular science writer who can take greater liberties in the use of language.

Mr. Gardner is incorrect in implying that I said I had a "true floor plan." I checked the corridor itself when Professor Hansel first offered his "not-to-scale" plan for publication in the *Journal of Parapsychology* (June, 1961), and I did not need a floor plan to see that those two windows did not line up as Professor Hansel claimed. It is true that this question is not relevant for the evaluation of the experiment; but it is extremely relevant for the evaluation of the objectivity and care shown by Professor Hansel in making his "scientific" evaluation of ESP.

Now Mr. Gardner prefers to place Pearce (or a collaborator) on a chair in a busy corridor peeking at me through my own transom as I recorded the cards. Is Mr. Gardner quite sure that the door to my room had a transom? If so, how did he learn it? And if he is not sure, would he not feel safer to have Pearce or his collaborator move the chair a few feet along the corridor and peek through the illumination window? But even Professor Hansel could not bring himself to propose such a daily public display of cheating, and that it is why he relocated the windows to give the impression that someone could hide across the hall and peek without having to look around corners.

It is no good for Mr. Gardner next to imagine a crack over my door. Professor Hansel has already looked and found none; and for *his* demonstration of *his own* capacity for lying and cheating he had to choose a room with a temporary internal partition and a crack over the door and he had to cajole a member of the Laboratory staff into doing him the favor of turning cards in spite of the fact that the staff member protested that the procedure had nothing to do with any experiment. The staff member was not "completely mystified" at all. As soon as the 22 hits were scored from the record Hansel obtained by peeking, the staff member asked (laughing, and completely without surprise): "Okay, what did you do?" Professor Hansel answered immediately that he had peeked through the crack over the door.

Since Professor Hansel acknowledges in his book that I helped him in every possible way, why did he not duplicate Hubert Pearce's results with me in my old experimental room and working under the same conditions that existed for the Pearce-Pratt series? Why did he need instead to choose an entirely different room and to take as the intended victim of his effort at deception a staff member who had never completed an experiment and published a scientific report?

Mr. Gardner had nothing to say on the final (and most important) paragraph of my letter which dealt with his (and Professor Hansel's) abuse of the privilege of scientific criticism.

J. G. Pratt

To this I replied:

Pratt's reference to a "revised statement" by Hansel is a bit slippery. I obviously revised my own statement, in the light of a letter from Hansel. Evidently Pratt believes that, since Dr. Rhine hath spoken, the matter is now settled.

Pratt suggests that Pearce could have answered my 1953 letter but that his reply was lost in the mail, or perhaps I lost it. I can only say that if someone wrote to me and said, "You did not reply to my letter of . . . ," when in fact I had, I would, in my reply to the second letter, set him straight. Pratt is not being "scientific," just rhetorical.

When Pratt said in his previous *New York Review* letter that "a true floor plan would show . . . " I assumed he had such a plan. Apparently he does not. When Hansel was at Duke, he made a strenuous effort to obtain details about the extensive alterations that had been made in Pratt's old offices, and to learn who had requested them. Promises were made to send him those details. They never arrived, nor was Hansel's second letter, requesting them, answered. To Pratt's question, "Is Mr. Gardner quite sure that the door to my room had a transom?" I will adopt Pratt's technique of innuendo and counter with: "Is Mr. Pratt quite sure that it hadn't?" It has a clear-glass transom now. If Pratt has evidence, or even if he remembers, that it had no transom in 1933–34, it would have been helpful if he had said so outright, and not implied it by asking a question.

Actually, two separate offices were used by Pratt during his tests with Pearce. For more than half those tests, Pratt used an office (in the medical building) that now has a transom of ripple glass above the door. (See Hansel's book, page 76.) Is Mr. Pratt sure that this transom had ripple glass in 1933? If it did, was it a transom that could be opened? These are not trivial questions. A competent investigator would have considered them in his original reports. Since Pratt did not do so, and details of office alterations are not available, one can only guess, and marvel at Pratt's carelessness.

Pratt suggests that no one would have risked standing on a chair in a "busy corridor." He does not say how many tests were made in the evening (more than half, I believe), when corridors outside the two offices would have been virtually empty. Besides, both offices had corridor windows (alterations removed the window of one office; the other now has a window of ripple glass) through which any tall person could peek. The clear-glass window removed from one office had a lower edge that was five feet, ten inches, from the floor. It is hard to believe, but Pratt did not deem it necessary to cover this window while he was turning his ESP cards.

My reply to the last paragraphs of both Pratt's letters is that I do not concede that Hansel and I have abused the privileges of scientific criticism. Hansel has shown in his book that Pratt's experiments with Pearce were almost as amateurishly designed as Rhine's early test of Lady Wonder, the mind-reading horse, but Pratt lacks the courage to admit it. No one is now interested in what Pratt has to say about these old tests; it is a statement by Pearce that one longs to see. What, for example, was in that letter that Rhine speaks about so guardedly on page 98 of *New Frontiers of the Mind,* a letter received by Pearce which so distressed him that from that day forward he has been unable to demonstrate his former ESP talents? A full report by the Reverend Hubert Pearce, on his sensational, unrivaled ESP work when he was a student at Duke, would make a dramatic book. (Publishers take note. He can be reached at the First Methodist Church, Cameron, Missouri, where he is pastor.) Since scientific truth is also God's truth, it seems to me that such a report would serve both God and man. But my precognition tells me that Pearce will never write it.

Martin Gardner

Hansel's conjectures about how easily Pearce could have cheated provoked many angry rebuttals in addition to the Pratt letters reprinted above. Ian Stevenson, best known for his research on the ability of certain persons to recall past incarnations, strongly attacked Hansel's book in a review published in the *Journal of the American Society for Psychical Research* (vol. 61, 1967, pp. 254–267). Stevenson concedes that Hansel correctly pointed out glaring inconsistencies in the nine major published accounts of the Pearce-Pratt tests, but he feels that none of this careless reporting invalidates the experiments. In 1967 Stevenson obtained from Pearce a notarized statement which he gives as follows:

> In reference to the suggestions made concerning the experiments that Dr. Gaither Pratt and I did at Duke University, I do not hesitate to say that at no time did I leave my desk in the library during the tests, that neither I nor any person whom I know (other than experimenter or experimenters) had any knowledge of the order of the targets prior to my handing the list of calls to Dr. Pratt or Dr. Rhine, and that I certainly made no effort to obtain a normal knowledge by peeking through the window of Dr. Pratt's office—or by any other means.
>
> Hubert E. Pearce

Pearce died in 1973, and Pratt in 1979. I do not know whether Pearce cheated in his tests with Pratt, but in view of the discrepancies in early accounts, the fact that crucial details did not come to light until twenty years later, and, above all, the fact that no one bothered to check on Pearce's whereabouts during the tests, all add up to such amateurish controls that no one today should take the Pearce-Pratt tests as good evidence for ESP.

In *New Frontiers of the Mind* (1937) Rhine reports that a difficulty in working with Pearce was that he liked to arrange the details of a test and tended not to do well unless all his suggestions were followed. For this and other reasons, one of Rhine's assistants, Sara Ownbey, strongly suspected Pearce of cheating. (Later, Miss Ownbey and George Zirkle together achieved sensational ESP scores at Duke, but after they married each other their psi powers evaporated.)

Although it constitutes no proof of trickery, I think it worth noting that psychic charlatans are notorious in wanting control over protocols when they are tested. Harold Puthoff and Russell Targ had this difficulty in testing Uri Geller. Like most parapsychologists, who feel they must capitulate when testing temperamental self-declared psychics, Puthoff and Targ went along with many of Uri's unusual demands on the grounds that not to do so might upset his psi abilities. One reason they allowed Geller's friend Shipi Shtrang to hang around was their belief that Geller's powers might be enhanced by having his best friend nearby, as indeed they were!

Now that Pearce, Pratt, and Rhine are no longer living, we shall never know many of the relevant circumstances surrounding the controversial Pearce-Pratt tests, or precisely under what conditions Pearce made his famous 25 hits in guessing ESP cards. We are unlikely ever to know whether Pearce cheated. But in view of how easily he could have (I can list ten entirely different ways he could have known the 25 ESP cards, not one way ruled out by any of Rhine's brief descriptions of this miracle), it is hard to respect any parapsychologist today who argues that any of Pearce's achievements were adequately controlled.

A group of hard-hitting letters from parapsychologists severely critical of Hansel appeared in the *British Journal of Psychiatry* (vol. 114, 1968, pp. 1471-1480), all drubbing an earlier favorable review of Hansel's book by Eliot Slater. The letters were followed by sensible replies from both Hansel and Slater.

In his contribution to this debate Stevenson reports that, unlike Hansel, he had no difficulty at Duke University in obtaining plans of Pratt's office before alterations were made. I had the same difficulty Hansel had. Pratt wrote to ask me for copies of all my correspondence with Pearce, offering to "reciprocate" in some way for my cooperation. When I sent him the copies, I proposed that he in turn send me the plans. I never heard from him again.

My impression of Pratt has always been that he was an honest fellow whose incompetence was exceeded only by his gullibility. He accepted just about everything on the paranormal front—poltergeists, metal bending, thoughtography, even dogs with ESP. Among all the true believers who howled loudly in protest over Hansel's evidence that Soal had beefed up his data on Shackleton, no one howled louder than Pratt. In fact, he did not toss in the towel until statistician Betty Markwick, in her famous paper, "The Soal-Goldney Experiments with Basil Shackleton: New Evidence of Data Manipulation" (*Proceedings of the Society for Psychical Research,* vol. 56, May 1978, pages 250-277), gave incontrovertible evidence that Soal had indeed cheated.

Actually, Pratt tossed in only half a towel. Following Markwick's article is a statement by Pratt that I find one of the funniest in the entire annals of parapsychology. First he calls Markwick's research "exemplary" and regretfully agrees that "we must set aside, at least for the time being, all of the Soal experimental findings as lacking scientific validity." As for his own collaborations with Soal, "For the present I must put all of this work aside, marked to go to the dump heap." Then follow some remarks that I could hardly believe when I read them.

To understand Pratt's remarks, I must first explain that Markwick discovered that Soal, in preparing his list of random digits from 1 through 5 (to be used for randomizing the five animal pictures he used in his test with Shackleton), had periodically inserted dummy digits. About three-fourths of these fake digits corresponded to target hits. When the fake digits and their corresponding guesses are taken out of the data, Shackleton's scoring falls to chance levels. The odds against such high scoring on the fake digits are estimated by Markwick to be thousands of millions to one.

Does this mean that Soal cheated? According to Pratt, not necessarily! Pratt suggests that Soal, while preparing his list of random numbers, may have "used precognition when inserting digits into the columns of numbers he was copying down, unconsciously choosing numbers that would score hits on the calls the subject would make later. For me, this 'experimenter psi' explanation makes more sense, psychologically, than saying that Soal consciously falsified for his own records."

A revised and updated edition of Hansel's book, under the new title *ESP and Parapsychology: A Critical Re-Evaluation,* was published in 1980 by Prometheus Books.

20

"Ideas in Conflict"

Unorthodox science is like a spectrum. At one end is creative, path-breaking work that no one considers crackpot. Richard P. Feynman, for example, developed the highly unorthodox theory that a positron can be viewed as an electron moving backward in time. It led to his famous "space-time view" of quantum mechanics, for which he shared a Nobel Prize in physics. At the other end of the spectrum is ignorant, trivial, at times pathological work. Here we can put the flat-earthers, the hollow-earthers, the saucer nuts who make trips to Mars and Venus, the angle-trisectors, the perpetual motion inventors, and so on.

The middle of the spectrum is a vast, confused jumble of claims about which judgments are hard to make. Obviously there is no place where a sharp line can be drawn. Nevertheless, there is a qualitative difference between regions near the ends of any spectrum, and only confusion results when those regions are lumped together and talked about in the same terms.

Theodore J. Gordon, a young space engineer with Douglas Aircraft Company, has written *Ideas in Conflict* (St. Martin's, 1966) about nine areas of contemporary unorthodox science. The book is informative and amusing. It provides excellent, sympathetic introductions to the nine unorthodoxies. But it has one glaring defect: it treats with equal seriousness work that lies near both poles of the unorthodoxy spectrum.

There is a pleasant chapter on James McConnell, a University of Michigan psychologist who rocked the biological world a few years ago

Reprinted with permission from *Book Week*, October 2, 1966.

with his famous cannibalism experiments with planaria. Mr. McConnell and his associates trained a group of these tiny, cross-eyed flatworms to associate electrical shocks with sudden bursts of light. After a while, the light alone would produce a convulsive reaction. Trained worms were then chopped into fine bits and fed to untrained planaria. When the cannibals were taught the light reaction, they learned, insists Mr. McConnell, in a significantly shorter time. His theory is that a primitive memory is somehow stored in the body proteins of planaria; that a portion of this memory can be transmitted to other planaria by feeding.

Unfortunately, this and similar experiments are statistical in nature and it is easy for an experimenter's enthusiasms or prejudices to bias his observations. When the number of flatworms convulsed by light are being counted, two observers may not agree on whether a certain flatworm responded. Some independent researchers have confirmed Mr. McConnell's results. Others, including Nobel Prize winner Melvin Calvin, have tried to duplicate Mr. McConnell's results and failed. The work is going on and the issue is far from settled. Mr. McConnell may never win a Nobel Prize, but he probably belongs on the Feynman side of the spectrum.

To be sure, Mr. McConnell has run into opposition from the "establishment." In view of the revolutionary character of his claims, it could hardly have been otherwise. Most unorthodox ideas are false, and organized science would grind to a halt if it did not place the burden of truth on its revolutionaries. But Mr. Gordon should know better than to liken today's scientific establishments to the tribunals that persecuted men like Galileo and had Bruno and Servetus burned at the stake. All three were persecuted because they came in conflict with an established religion. Servetus was incinerated by a Protestant establishment for having attacked the doctrine of the Trinity. No church in the United States has pressured the University of Michigan to fire Mr. McConnell. No government bureau has decreed his views unsound. His articles have appeared in establishment journals and he himself edits a delightful periodical, *The Worm Runner's Digest,* to which I am both a subscriber and contributor. His claims have been taken with utmost seriousness and are now being tested around the world.

Had Mr. Gordon confined his book to the McConnells of unorthodoxy (you can find them in every field), he might have made a contribution to the psychology and sociology of modern science. Unfortunately, other chapters in his book approach, in the same language and spirit, men who are near the *other* end of the spectrum. There is, for instance, a serious chapter on the Rev. Franklin Loehr, whose book *The Power of Prayer on Plants* was published by Doubleday in 1959. There is no sense in which Mr. Loehr can be called any kind of scientist. His book makes clear that he has not the slightest notion of how to design an experiment. No one would have heard of him had not his publisher seen in his manuscript a chance to make a fast buck by hawking it to a gullible public, hungry for miracles.

The same applies to Mr. Gordon's chapter on R. C. W. Ettinger, who thinks everyone should arrange to have his body put in deep freeze after death, so that a century or so from now, when science has found a way to revive frozen corpses, they can rise like Lazarus from the grave. If a relative dies of an incurable disease, freeze the body anyway, says Mr. Ettinger. By the time it is revivified, medical science may have the cure. Another chapter deals with Immanuel Velikovsky, who continues to have defenders if not followers, and whose views, if correct, would require rewriting physics, astronomy, geology, and ancient history.

Somewhere near the middle of the spectrum are Mr. Gordon's other unorthodoxies: theories about how life first came to earth on meteorites, the psychedelic theories of Timothy Leary, the views of Albert Schatz, a soil chemist who is convinced that cavities in teeth are not caused by acids that are a by-product of microbes, but by the microbes directly attacking the enamel the way they attack soil. A chapter on the search for extraterrestrial life seems out of place, since it deals almost entirely with orthodox speculations. And there is the inevitable chapter on Dr. J. B. Rhine and ESP.

It is in the chapter on Dr. Rhine that Mr. Gordon's own occult beliefs are most evident. He takes for granted that Edgar Cayce, the Kentucky clairvoyant who diagnosed medical ills and prescribed bizarre remedies, was a genuine psychic, and Mr. Gordon is a bit miffed with Dr. Rhine for not agreeing. Mr. Gordon also takes seriously such prophetic writings as those of Nostradamus, and he recounts a number of personal anecdotes involving precognition.

"Next to Nostradamus ranks my mother-in-law, the prophet," he writes. "My mother-in-law is a witch. She reads tea leaves." He then describes several of her remarkable tea-leaf predictions that came true. When I first read this section I assumed that Mr. Gordon was merely trying to be funny. He wasn't. My own tea leaves tell me that the readers who will be most impressed by his book will consist mainly of those who are angry at the Air Force for withholding information on flying saucers, who are furious with the government for hounding Dr. Andrew Ivy, and who cannot understand why the State Department does not hire Jeane Dixon, the crystal gazer, as a foreign policy adviser.

21

"ESP, Seers and Psychics"

Our country, as everyone knows, is in the throes of a rousing revival of public interest in every aspect of the occult. At a time of mounting anxieties, coinciding with a rapid erosion of faith in traditional religions and a growing hostility toward science and technology, millions of people, hungry for miracles, are turning to astrology, spiritualism, extrasensory perception, and other psychic beliefs to find some sort of solace. The number of major publishing firms that have not pandered to this hunger by issuing at least one occult book can be counted on one hand.

In view of this trend it is like a breath of unpolluted air to come upon a work that surveys the occult scene with calm and amiable skepticism. Milbourne Christopher's *ESP, Seers and Psychics* (Crowell, 1970) will have no influence whatever on the true believers. It is not likely to sell a tenth as well as Jeane Dixon's autobiography. The believer in astrology and witches will be as hostile to Christopher's point of view as a flat-earther is to the views of establishment astronomers. But for those few who can still distinguish scientific evidence from emotional claptrap, Christopher's book is marvelously enlightening.

Christopher is a professional magician, extremely knowledgeable about methods of deception, and it is this insider's knowledge that gives his book an authenticity it could not otherwise have. The sad truth is that the history of occultism is peppered with bogus practitioners. One would suppose that distinguished philosophers and psychologists who develop a

This review appeared in a brochure for the Library of Science, 1970, and is reprinted with permission.

passion for psychic research would take a year or two off to study the curious art of deception (i.e., magic) before they appoint themselves authorities on matters in which fraud obviously can play an enormous role. It never happens. Again and again a well-meaning, well-educated investigator of the occult, but an ignoramus about magic, will be flim-flammed by the oldest and simplest of methods.

Consider Dr. Joseph B. Rhine. His first scholarly paper reported his investigation of Lady Wonder, a mind-reading horse near Richmond, Virginia. Rhine was completely convinced of Lady's psychic ability to guess numbers written on a pad. "The greatest thing since radio," he wrote the horse's owner. Christopher has an entertaining chapter on the history of famous calculating and mind-reading animals—not only horses but also dogs, pigs and even two learned London geese—and the subtle methods by which such animals are trained. Christopher himself, using a false name, had a session with Lady. A clever ruse proved conclu-sively that the horse's owner was "pencil reading"—the ancient art of determining what a person writes by the wigglings of a pencil—then sig-naling the mare by traditional methods. Has Rhine ever disavowed his youthful indiscretion? Not at all. He has admitted that Lady's owner later resorted to unscrupulous signaling, but only after the poor horse had mysteriously lost her previous psychic powers.

This is only one tidbit in Christopher's eye-opening book. You will learn about Jeane Dixon's colossal clinkers, prophecies that failed to come true but which she and her admirers conveniently forget. There is not a single written record of her much vaunted prediction of John Ken-nedy's assassination. (Exactly what she *did* predict is quoted by Christo-pher, and it is stupendously unimpressive.) *ESP, Seers and Psychics* con-tains excellent chapters on table tipping, the Ouija board (now earning millions for Parker Brothers), dowsing rods and pendulums, haunted houses, fire walking, astral projection, poltergeists (who, oddly, almost always work their mischief in houses occupied by a teen-ager), burials alive, and such mediums as Eusapia Palladino, Daniel Dunglas Home (who almost destroyed the marriage of Robert and Elizabeth Browning; she pro, he con), and other internationally renowned charlatans. The gruesome events surrounding the death of mind-reader Washington Irving Bishop, during a performance at a Lambs Club Gambol in Manhattan, will come as a shock even to many magicians.

A central theme that runs through Christopher's lurid pages is the unbelievable gullibility of intelligent men who know nothing about the art of deception. The roster includes philosophers and churchmen as eminent as Henry Sidgwick, William James, and Bishop James Pike; psychologists of the caliber of Gardner Murphy, writers as prominent as Sir Arthur Conan Doyle, and television personalities such as Merv Grif-fin and David Susskind. It is not without significance that the skeptics among TV talk-show moderators include Johnny Carson and Dick Cavett, both of whom once did professional magic, and Hugh Downs, a good friend of many prestidigitators. It was Susskind, for example, who produced "Maurice Woodruff Predicts," surely one of the silliest television series of 1969. Christopher has covered all this with fascinating detail and documentation, and just enough disclosure of techniques to give readers an insight into how things work without getting the author in trouble with his fellow conjurors.

It is a sobering thought that Nazi Germany was swept by just such an obsession with astrology and other occult nonsense, with Hitler the biggest nut of them all. It would be a sign of hope for America if half as many people read *ESP, Seers and Psychics* as, say, the number of scientific illiterates who devoured the recent spate of books about Edgar Cayce. At the moment, however, this seems as unlikely as the probability that General Motors will develop a clean engine or that a thousand newspapers will drop their astrology columns before the close of the crazy seventies.

Postscript

Milbourne Christopher, an old friend and fellow conjuror (he a professional; I, an amateur and a novice), has since written two other books on the occult that I heartily recommend: *Mediums, Mystics and the Occult* (Crowell, 1975) and *Search for the Soul* (Crowell, 1979).

22

"The Roots of Coincidence"

The curious thing about Arthur Koestler is that he believes in God. Not the transcendent, personalized deity of Moses and Jesus, but a deity more like the abstract God of Alfred North Whitehead. His faith is a kind of Neoplatonic pantheism. Behind the fragmented shadow universe, there is a vast, unthinkable Oneness, with laws so subtle that science has not yet formulated them, yet laws that are partly within the reach of science. In his final, most striking metaphor, Koestler speaks of scientists as "Peeping Toms at the keyhole of eternity." If they will only take the "stuffing"—their dogmatic prejudice against psychic research—out of the keyhole, they can touch off an intellectual revolution that will transform the world.

It is important to understand these metaphysical impulses behind Koestler's growing preoccupation with parapsychology. In all ages, religious believers have tried to bolster faith with material evidence. Saint Thomas Aquinas might not have been canonized had the Church not been convinced that one day he had floated in the air while praying. Today, when Christian miracles have dwindled to faith healing and glossolalia, Western theists who seek for signs (to use the condemnatory phrase of Jesus) are finding them in parapsychology. They are not trivial signs. If minds can influence the fall of dice, the growth of plants, and the healing of lesions in mice; if telepathy, clairvoyance, and precognition are genuine phenomena, then Koestler is right. Joseph Banks Rhine,

---

This review, which appeared in *World* magazine, August 1, 1972, and the letters quoted in the Postscript are reprinted with permission.

if not the prophet of a religious awakening, is at the least the Copernicus of a new rotation point in the history of science.

No one writing today is a more skillful polemicist than Koestler. His earlier books persuaded countless readers, including me, that things were far from what we thought they were in Stalin's Russia. For this I am grateful. How will he fare with his new, more positive rhetoric? In the light of current enthusiasm for astrology and the occult, I suspect he will fare extremely well.

The Roots of Coincidence (Random House, 1972) is a small book of five brisk, colorful chapters. The first, "The ABC of ESP," sketches some recent work in the field and suggests that in both the Russian and U.S. space programs telepathy will have "important strategic uses as a method of direct communication." Chapter 2, "The Perversity of Physics," argues that recent theories about space, time, and matter are so wild that the hypotheses of parapsychology pale by comparison. The third chapter, "Seriality and Synchronicity," outlines an approach toward coincidence, which I will return to in a moment.

Chapter 4, "Janus," introduces Koestler's concept of the "holon." Nature is a "multileveled, hierarchially organized" system of sub-wholes, nested together like Chinese boxes. Each part, or holon, is to a degree autonomous, yet subservient to higher holons. A cell in your heart is a thing in its own right, but dependent on the heart, in turn dependent on you, in turn dependent on your family, and so upward through the holons of city, state, nation, and humanity until one reaches the Atman, the great world-soul of Hindu philosophy. Like Janus, each holon has two faces: that of the "proud, self-assertive whole" and that of a "humble integrated part."

"The Country of the Blind," Koestler's final chapter, applies the title of H. G. Wells's greatest short story to all of modern civilization. Unseeing scientists continue to concentrate on the physical world, unaware of the powerful gusts of new knowledge blowing through the cracks pried open by a few courageous psychic researchers.

Readers with little knowledge of contemporary psychology, and how its experiments are designed, are likely to fall under the spell of Koestler's persuasive prose and finish the book without any inkling of its many shortcomings. The most glaring is his failure to give us the slightest information about painstaking research, underway for decades, which runs counter to the claims of leading parapsychologists. He is annoyed by the suggestion that early researchers made unconscious recording errors, but he does not cite a single paper reporting tests of ESP and PK made by skeptical psychologists which indicated the prevalence of such errors. Koestler is even more scornful of the charge of fraud. C. E. M. Hansel, whose book *ESP: A Scientific Evaluation* (1966) documents a strong case for the prevalence of ESP cheating, is called the "most bellicose" among hostile scientists, and his valuable book is waved aside as a "last-ditch stand." (See Chapter 19.)

Contemporary research in parapsychology exhibits no evidence of increasing rigor except in the keyhole-stuffed laboratories of the skeptics, where results are monotonously negative. The most recent big splash in the world of psi was made by Jule Eisenbud, a psychoanalyst at the University of Colorado, with his preposterous book *The World of Ted Serios* (1967). Serios, a Chicago bellhop, had a truly delightful talent. He would glance at, say, a photograph in *National Geographic*. Years later, after he had forgotten about the picture, someone could aim a Polaroid camera at him, snap a picture (using a wink-light and with the lens set at infinity), and ten seconds later, lo, there on the print would be a photograph, line for line like the one in *National Geographic*!

When *Life* wrote up Ted Serios for its September 22, 1967, issue, the article's author withheld one crucial bit of information. Nowhere did he disclose that, before a picture was taken, Ted always held a small one-inch-wide paper tube (which he called his "gismo") in front of the camera lens, presumably to focus psi-radiation from his skull. Photographers David B. Eisendrath and Charles Reynolds, both also amateur magicians, had no difficulty constructing a simple optical device which, secretly loaded into a gismo and later palmed out, could produce all the photographs in Eisenbud's book. The device is nothing more than a tiny cylinder with a positive transparency of a photograph at one end and a lens at the other. Light bouncing off the shirt and face of whoever holds the loaded gismo in front of the Polaroid camera is strong enough to produce excellent images on the film. Since their sensational exposé in *Popular Photography,* October 1967, Ted has softly and happily vanished from the psi scene.

Let me give another instance of the "rigor" of modern parapsychology. Perhaps the most respected of recent work is that being done by Stanley Krippner and Montague Ullman in the Dream Laboratory at the Maimonides Medical Center in Manhattan. (See their book, *Dream Studies and Telepathy,* 1970.) Koestler plugs their findings on pages 37–38, and Renée Haynes, in her postscript to *The Roots of Coincidence,* also praises the book on pages 146–147.

How trustowrthy is Krippner? To answer indirectly, let us now turn to "Parapsychology in the U.S.S.R.," a magazine article by Krippner and Richard Davidson (*Saturday Review,* March 18, 1972). On the first page is a photograph of Ninel Kulagina, identified as a "noted Russian sensitive," causing a "plastic sphere" to float in the air. (The sphere is a mere Ping-Pong ball, light enough to be levitated by a variety of techniques known to magicians.) According to the authors, "a heightened biological luminescence seems to radiate from her eyes while she is performing."

Krippner well knows that Mrs. Kulagina is a pretty, plump, dark-eyed little charlatan who took the stage name of Ninel because it is Lenin spelled backward. She is no more a sensitive than Kreskin, and like that amiable American humbug, she is pure show biz. In 1964 when there was great excitement in Russia over ladies who could read *Pravda* with their fingertips, Ninel became the country's second most publicized finger reader. Alas, Soviet establishment psychologists caught her cheating, using techniques familiar to all magicians, and familiar even to Dr. Rhine, who took a dim view of the practice. (See Chapter 6.)

On May 21, 1968, in a story from Moscow, the *New York Times* reported that Ninel — now using the pseudonym of Nelya Mikhailova — had been caught again. She was found employing concealed magnets to fool "Soviet scientists and newsmen into thinking she possessed the ability to move objects by staring at them." (Magnets are only part of the tale, but one hesitates to give away trade secrets.) Four years earlier, the same report revealed, Ninel had received a four-year prison sentence.

According to Sheila Ostrander and Lynn Schroeder, in their book *Psychic Discoveries Behind the Iron Curtain* (1970), Ninel's crime had been black-marketeering. The late Leonid L. Vasiliev, the Soviet's top parapsychologist (Koestler praises his work) had intervened on her behalf, so she went to a hospital instead of a jail. Vasiliev had personally tested her and pronounced her talents genuine.

Vasiliev spent his last years working on finger vision. The "star subject" of this field, as Krippner refers to her, is Rosa Kuleshova. She is a skilled performer of card tricks who likes to "show them to every comer"; so writes Ms. G. Bashkirova in an article about Kuleshova, reprinted in the *International Journal of Parapsychology,* Autumn 1965. Bashkirova admits that Rosa often cheats, apparently just for the hell of it. In one demonstration she correctly guessed the color of objects by *sitting* on them. "Of course," Bashkirova adds, "she peeped." (She doesn't say with what.)

Unfortunately, Krippner didn't get to see Rosa either. She had, he writes in his above-mentioned article, gone off and joined a circus. Krippner did not tell his readers what he surely knows — that the *New York Times* on October 11, 1970, disclosed that Rosa, too, had been caught cheating by Soviet scientists (no doubt scientists with keyholes as tightly stuffed as those who had so cleverly trapped Ninel). Koestler is filled with respect for the Soviet Union's pioneering work in parapsychology. It is hard to understand how, having read the hogwash in Ostrander's book, he could imagine that any of the Russian work deserved to be taken seriously.

But the oddest aspect of Koestler's book is his argument that "meaningful coincidences" (they provide the book's title) may have "a-causal" explanations beyond the known laws of physics and mathematics. He spends many pages quoting from an untranslated study of coincidences written by Paul Kammerer, the Austrian biologist who was such a passionate defender of Lamarckism, and about whose work Koestler has recently written a sympathetic book, *The Case of the Midwife Toad.* (See Chapter 10.)

Kammerer, it seems, spent enormous amounts of time keeping records of meaningful coincidences and trying to account for them on nonchance grounds. For example, he records a day in 1910 when his brother-in-law went to a concert, sat in seat 9, and was given a cloakroom check numbered 9. The next day the same fellow went to another concert where he had seat 21 and check 21. Kammerer calls this a "second-order series" because it repeated the previous coincidence.

Everyone has such experiences, and there is a simple explanation. The number of events in which you participate for a month, or even a week, is so huge that the probability of noticing a startling correlation is quite high, especially if you keep a sharp lookout. For the same reason a numerologist, with a large supply of words and numbers to play with, can turn up incredible correlations. Assume that the alphabet is a closed circle, *Z* joined to *A*. Shift each letter in OZ backward one step and you get NY, the abbreviation of the home state of L. Frank Baum, who originated the *Oz* series. For a second-order coincidence, shift each letter of OZ forward one step. You get PA, the abbreviation of the home state of Ruth Plumly Thompson, who continued the *Oz* series after Baum's death. Numerical and alphabetical coincidences of this sort have been taken with utmost seriousness by countless groups, from the Greek Pythagoreans and Hebrew cabalists to Christian sects (some still flourishing today) who found 666 (the "Number of the Beast") in the names of eminent adversaries.

Kammerer sat for hours in public parks jotting down information about passersby—age, dress, sex, what they carried, and so on. He found a strange tendency for like things to cluster. A woman would pass in a red dress and then, unaccountably, other red dresses would swish past him. Kammerer firmly believed that higher laws were operating, that there was a "world-mosaic . . . which, in spite of constant shufflings and rearrangements, also takes care of bringing like and like together."

It is hard to believe, but Koestler is impressed by this theory. He ties it to similar views advanced by Jung, and argues that paranormal phenomena also may be a series of "confluential events" that are "a-causal manifestations of the Integrative Tendency" (page 122). The falling dice in Rhine's PK tests thus only *seem* to be manipulated by the will of the experimenter. In old-fashioned religious language, a crapshooter's prayers do not influence the bones directly. They are answered by a divine intermediary.

I urge Koestler to make the following simple experiment. It was invented by A. D. Moore, professor emeritus of electrical engineering at the University of Michigan. He calls it the "nonpareil mosaic" because he uses large quantities of tiny colored balls of nonpareils, a sugar candy made in Milwaukee. Fill a beaker with thousands of the little beads, half of them red, half green. Shake thoroughly, then inspect the beaker's sides. Do you see an intimate, homogeneous mixture? You do not. You see a marvelous mosaic: irregular large clumps of red interspersed with similar clumps of green. The pattern is so unexpected that most physicists, when they see it, suspect an electrostatic effect, or, in Kammerer's words, a "quasi-gravitational attraction between like and like."

Nothing is operating here except elementary laws of chance. A less dramatic but simpler demonstration of what statisticians sometimes call the "bunch effect" can be made with a deck of playing cards. Arrange the cards alternately red and black. Shuffle the deck thoroughly, as many times as you wish, then spread the cards. The bunching will be obvious. Runs of four or five cards of the same color are very common, and even longer runs of seven or more cards occur more often than most persons would expect.

In spite of so many animadversions, I find Koestler's book far above the garbage in most popular books on parapsychology. He is particularly good at summarizing crisply some of the recent theories of the particle physicists, wild guesses about quarks, negative mass, anti-matter, backward-moving time, and so on. Modern science should indeed arouse in all of us a humility before the immensity of the unexplored and a tolerance for crazy hypotheses. But as for parapsychology, Koestler is a poor guide. He is too strongly biased by emotional commitments. He is too unaware of the queer sorts of controls necessary in a field in which deception, conscious and unconscious, is all too familiar.

Koestler obviously is convinced that paranormal phenomena lend credence to his pantheology, despite the number of atheists (e.g., Freud) who were and are on his side. I confess to having a different mind-set. I consider it a spiritual blessing that you and I have isolated brains. I am happy we cannot communicate by ESP, that we cannot see through walls or move objects by PK, that spirits do not return from the dead to haunt us, like the head with half a face that Jung saw on his pillow when he slept in a haunted house in Buckinghamshire. (See Koestler's book, page 93.)

I do not believe that integrative tendencies were at work when Adam Clayton Powell died on the anniversary of the murder of Martin Luther King, or when the *AP*ollo 16 crew left for the moon on *AP*ril 16 from *CAP*e Kennedy, there being exactly sixteen letters from *A* to *P* inclusive. Above all, I am grateful to whatever gods there be that the future is mercifully hidden from us.

Postscript

Arthur Koestler's letter about my review appeared in *World* magazine (October 10, 1972):

> May I ask for the hospitality of your columns for a reply to Martin Gardner's review of my book, *The Roots of Coincidence*?
> I am an old fan of Martin Gardner's; to read his monthly column in the *Scientific American* gives as much pleasure as replaying Bobby Fischer's games. I was all the more disappointed when I discovered, many years ago, that he had an almost obsessional prejudice against ESP—prejudice which he was, to be sure, candid enough to admit. In his book *In the Name of Science,* published in 1952, he wrote:
>
> > There is obviously an enormous, irrational prejudice on the part of most American psychologists—much greater than in England, for example—against even the possibility of extrasensory mental powers. It is a prejudice which I myself, to a certain degree, share. Just as Rhine's own strong beliefs must be taken into account when you read his highly persuasive books, so also must my own prejudice be taken into account when you read what follows. (pp. 299-300)
>
> What follows is a self-confessedly biased account of the pioneering work done over the previous twenty years at Duke University. Although he admitted at the outset that Rhine was "an intensely sincere man," Gardner's version of Rhine's experiments was a caricature that made Rhine appear a victim of scientific ineptitude combined with self-deception. All other chapters in Gardner's book dealt with charlatans and cranks; and in spite of Gardner's assertions that Rhine was an honorable man, his inclusion in this rogue's gallery had the effect on the reader of establishing guilt by association.
> I was sorry to see that in reviewing *The Roots of Coincidence* Gardner has followed the same method. About a third of the review is devoted to making fun of various suspected charlatans in America and Soviet Russia, thus conveying the impression that they play a part in the book under review. Otherwise why should the reviewer drag them in? The fact is that not a single one of these characters—Ted Serios, Ninel Kulagina, Rosel Kuleshova, etc.—is quoted or discussed or mentioned in my book. But since Gardner discusses them at such length in his review—though pointing out that Koestler himself is an honorable man—the impression given is again one of guilt by association with such dubious company.
> I was even more distressed by a grave misrepresentation. Gardner quotes me as saying that in future Russian and U.S. space programs, telepathy will have "'important strategic uses as a method of direct communication.'" I did not say that, and I do not believe that. The phrase in quotes appears on page 16 of my book and refers to a belief apparently held by certain people in Russian and American space agencies. The reason I do

not share it is the rare and capricious nature of ESP phenomena; and their dependence on emotional and motivational factors—which seems to exclude the possibility of their exploitation for utilitarian purposes.

Arthur Koestler

Koestler takes me to task for attributing to him the view that ESP may have important uses as a method of communication in space programs. He denies he said that, adding that the statement I quoted represented only the views of certain persons in space agencies.

On page 16 of his book Koestler tells us that Vasiliev had quoted a Soviet rocket pioneer as having said, "The phenomena of telepathy can no longer be called into question." Koestler then adds, in his own voice: "This conveyed to any Soviet scientist trained to read between the lines that ESP, once its technique has been mastered and made to function reliably, might have important strategic uses as a method of direct communication. This seemingly fantastic idea was confirmed as far back as 1963 by a high official of NASA."

A long quote follows in which this official speaks of the great potential value of ESP as a new technique for communication in flight systems. Nowhere does Koestler express doubts about such a possibility. I leave to any reader of pages 16–18 whether I misrepresented when I wrote that Koestler "suggests" the validity of using ESP in space communication. I was, of course, pleased to learn of his skepticism—a skepticism that clashes sharply with the views of many parapsychologists whose recent research on ESP communication has been funded by government agencies.

The reason I spent so much time on Ted Serios, Nina (as she is now called) Kulagina, and Rosa Kulashova was because a basic theme of Koestler's book is that parapsychology has finally become a "rigorous" science in contrast to its loosely controlled earlier days. Rhine and others did indeed learn how to tighten their controls (with an accompanying marked decline in sensational results), but in the past few decades the trend has gone the other way. It was to make this point that I introduced the extraordinary claims of Ted, Nina, and Rosa. (Uri Geller was not yet on the scene.)

As I said in my review, Koestler had singled out Stanley Krippner and Montague Ullman as examples of the new breed of parapsychologists who were doing "rigorous" work in the field. Yet both men were firmly convinced of the psi powers of Rosa, Nina, and Ted. The jacket of Eisenbud's *The World of Ted Serios* carries an enthusiastic endorsement by John Beloff, one of the most respected of Great Britain's parapsychologists. (In Chapter 12, I cite other top parapsychologists who to this day refuse to believe that Ted is a charlatan.)

At the time I wrote my review it is likely that news of Ted's ability and its acceptance in high psi circles had not yet reached Koestler. It would be interesting to know what he thinks now of Eisenbud's book. We do know that Koestler firmly believed in Uri's PK powers after witnessing them at Birkbeck College (see Chapter 27). It is not irrelevant to Koestler's claim that parapsychology has become a rigorous science to discuss the extent to which today's noted parapsychologists (and Koestler himself) were taken in by the simplest, most obvious flimflammery.

I owe Krippner an apology for having said that he well knew Nina to be a charlatan. My mistake was to overestimate his acumen. I assumed he knew this because news about Nina's cheating had been widely publicized, but Krippner has recently made it clear that he still does not "know" that Nina cheats. He certainly considered her powers genuine when he and Davidson wrote their naive article for the *Saturday Review*.

Koestler's concept of the "holon" may have boosted the current popularity of the word "holistic," especially in reference to "holistic medicine." There is nothing new about "holism." Its basic ideas were thoroughly explored by philosophers of the "emergent evolution" school— Henri Bergson, Samuel Alexander, C. Lloyd Morgan, and many others. See in particular the book *Holism and Evolution* by Jan Christian Smuts (1926) for metaphysical views almost the same as Koestler's.

Koestler says he is disappointed by my blindness toward the reality of psi, and I am flattered by this concern. I in turn am even more saddened by the fact that a man of Koestler's intelligence, who had the courage and vision to escape from one "country of the blind," has now himself become blind to the incredible shabbiness of parapsychological research around the world, and especially in the Soviet Union. One can only hope he will escape again.

For more on coincidences and their natural explanations, see my *Scientific American* column for October 1972, and the many number coincidences in my *Incredible Dr. Matrix* (Scribner's, 1976). Norman T. Gridgeman, reviewing Koestler's book in *Philosophy Forum* (vol. 14, 1975, pp. 307–316), closed by citing two favorite coincidences of his own:

> One is that William McDougall, the Lamarckian who ran a 14-year experiment on the inheritance of a learned behaviorism in rats—an experiment that has since been thoroughly discredited—was assisted in that enterprise by Dr. J. B. Rhine, the future doyen of parapsychology. And the other concerns an eighteenth-century Austrian bio-alchemist who created, nurtured, and exhibited ten homunculi. Sensation! But a cynical witness described them as "loathsome toads." The bio-alchemist's assistant and secretary, to whom we owe the story, was named Kammerer.

23

"Arthur Ford"

Modern Spiritualism began in 1848 in Hydesville, New York, when the Fox sisters discovered they could produce spirit raps by cracking their toe joints. The movement grew rapidly, peaking in the Reconstruction period and spreading to England, where it won such distinguished converts as Conan Doyle, Oliver Lodge, and William Crookes. By 1960 it had reached such a low ebb in the United States that it was almost impossible to find a medium willing to produce physical phenomena unless you went to a Spiritualist camp such as Lily Dale, in upstate New York.

Then suddenly, in 1967, Spiritualism began a comeback. It was, of course, part of the big Occult Explosion, but the strongest shove came from three men: the late Bishop James Pike, the later Reverend Arthur Ford, and Allen Spraggett, a Canadian fundamentalist preacher turned occult journalist.

To appreciate the significance of Spraggett's latest book, *Arthur Ford: The Man Who Talked with the Dead,* written with the help of William V. Rauscher (New American Library, 1973), it is necessary to give a quick sketch of Pike's sad, discombobulated life. After two years' study for the Roman Catholic priesthood, he lost his faith and dropped out of training to become a lawyer. He worked for the Securities and Exchange Commission, remarried (his first marriage had been annulled), regained his faith, and was ordained an Episcopalian priest. From a

This review, which appeared in the *New York Review of Books,* May 3, 1973, and the letters quoted in the Postscript are reprinted with permission from the New York Review of Books. © 1973 NYREV, Inc.

church in Poughkeepsie he moved to Columbia University, where he headed the department of religion until he became dean of that monstrous edifice near Columbia, the Cathedral of St. John the Divine. In 1958 he was made Bishop of California.

Back came the old doctrinal doubts, all proclaimed with great public fanfare. The Virgin Birth and the Trinity were the first to go. Then the Incarnation. Pike joined Alcoholics Anonymous. He entered Jungian analysis. When Spraggett first met him in 1963 (Pike was fifty) his impression was that of a man "incredibly old . . . either on the verge of utter exhaustion or afflicted with a terminal disease." In 1966 James, Jr., eldest of his four children, shot and killed himself at the age of twenty. Pike quit the ministry to join Robert Hutchins's Center for Democratic Institutions at Santa Barbara.

Like thousands of today's Protestant ministers, thirsting for signs, Pike grew increasingly obsessed with parapsychology. Although he no longer called himself a Christian (the church, he declared, was "sick and dying"), he retained a firm belief in God, the spiritual resurrection of Jesus, and life after death. Consumed with guilt over his son's suicide, he longed for hard evidence that Jim was happy on the Other Side.

He found it. Two weeks after his son's death, a series of poltergeist events took place in the apartment in Cambridge, England, where he was then living. He found that books had been mysteriously moved. A shaving mirror slid off the shelf. Milk soured. An alarm clock stopped at 8:19, the London hour when Jim might have killed himself in New York. He found safety pins open at the angle clock hands have at 8:19.

Convinced that Jim was trying to reach him, Pike sought the help of a London medium, Ena Twigg. In two séances with Mrs. Twigg, and several later ones in California with George Daisley, a London medium who had settled in Santa Barbara, Pike spoke to his discarnate son. His account of these séances, in his book *The Other Side,* was a Daisley bonanza. The medium moved from a tiny abode to a $70,000 house, now the headquarters of his Hallowed Grounds Fellowship for Spiritual Healing and Prayer. In 1972 his fee for a sitting was $30, and there was a six-month waiting list.

In 1967 Pike had his most dramatic encounter with Jim. At Lily Dale, Spraggett had met Arthur Ford, an almost forgotten American medium, and had been overwhelmed by Ford's clairvoyant powers. Why not bring Pike and Ford together for a séance and televise it? The sitting was video-taped in Toronto on September 3. Two weeks later, a half-hour portion of the two-hour session was shown on Canadian prime time. It was the biggest psychic news story since Bridey Murphy.

Compared to the great mediums of the past, with their jangling tambourines, floating trumpets, and glowing ectoplasm, Ford put on a dull performance. As was his custom, he covered his eyes with a black silk cloth, fell into a trance, and was immediately taken over by Fletcher, his spirit control since 1924. After introducing several discarnate churchmen Pike had known, Fletcher presented a "boy" who turned out to be Jim. "This boy," said Fletcher, ". . . wants you to understand that [neither] you nor any other member of your family have any right to feel that you failed him in any way."

Pike was deeply moved. "Thank you, Jim," he said. Three months later he had a private séance with Ford that moved him even more. In *The Other Side* he goes into detail about its "evidential" material. The ESP theory (so dear to parapsychologists who cannot buy the spirit world) that Ford was picking Pike's mind is ruled out; too much evidential data was unknown to Pike himself at the time.

In 1968, divorced from his second wife, Pike married Diane Kennedy, his secretary, who had helped him write *The Other Side*. The following year, when he and Diane were touring the Holy Land, they became lost in the Dead Sea desert and Pike died of a fall while Diane was seeking help. Spraggett lost no time dashing off *The Bishop Pike Story* (1970). Diane lost no time dashing off *Search* (1970), her account of Pike's death in the wilderness.

Ford died in Miami on January 4, 1971. On the day of his death, his spirit instantly began dictating a book on the afterlife to Ruth Montgomery. It was published later in the year as *A World Beyond*. Ford had willed his papers to the Reverend Canon William V. Rauscher, rector of Christ Episcopal Church, Woodbury, New Jersey, a convinced spiritualist and long-time friend. He proposed to Spraggett that the two collaborate on what the book's jacket calls the "authoritative biography of the world's greatest medium."

World's greatest? Perhaps Ford was the best-known native medium of recent decades, but modern mediums are a scrubby lot compared to the earlier giants. One thinks of the Davenport brothers, with their marvelous stage show; Daniel Dunglas Home, who floated out one window and back through another; Eusapia Palladino, the fat little Italian lady who hoisted heavy tables; Henry Slade, the slate writer; Margery, the "blonde witch of Boston"; and scores of others. Those were mediums who really *did* things!

Ford's dreary history begins with his birth in 1897 at Titusville, Florida. At seventeen he was studying to be a minister in the Disciples of Christ Church, at their Transylvania (shades of Count Dracula!) College, Lexington, Kentucky. He dropped out to enlist in the army. In his autobiography, *Nothing So Strange,* and his later book of memoirs, *Unknown but Known,* Ford said flatly that he never got overseas, but Spraggett cites four published interviews in which Ford spoke of his psychic experiences in the French trenches. (Ford later claimed to have had a son who died in World War II, but Spraggett and Rauscher, unable to find evidence that Ford even had a son, say nothing about this in their book.) After the war, Ford returned to Lexington, became ordained, but soon gave up his church to become a professional medium. It was in New York, 1924, that Fletcher became his lifelong control.

Spraggett does not conceal the fact that Ford gave conflicting accounts of who Fletcher was. In *Nothing So Strange* Ford said that Fletcher was a French Canadian he had met when they were both five, but had not seen since. The Canadian had been killed in the First World War. In 1928, in a psychic magazine, Ford gave a different story. He and Fletcher had been college chums. In another magazine article he called Fletcher "a Canadian chap with whom I went to school and college." Throughout his life Ford kept on the wall a framed photograph of Fletcher. The picture, that of a handsome youth with dark wavy hair, is reproduced in Spraggett's book.

A former male secretary of Ford's (Spraggett describes him as "middle-aged" and "effeminate") told Spraggett that Ford's trances were a put-on, and Fletcher a "pure invention." The photo of Fletcher, he said, "just happened along. It could have been anybody—a very, very good friend that Ford had many years ago, you know what I mean?" (The book has many of these sly, almost snickering hints about the homosexuality of Ford and others.)

Although prominent in Spiritualist circles during the Twenties, Ford was little known to the general public until 1928, when he became embroiled in a crazy controversy. The story is told in the final chapters of two biographies of Harry Houdini (by William Gresham and Milbourne Christopher), but in telling it again Spraggett has added interesting new details.

When Houdini died in 1926, his wife, Bess, announced that her husband had given her a secret message. Bess offered ten thousand dollars to any medium who could reveal it. After receiving thousands of incorrect guesses, she withdrew the offer.

In 1928, during a séance in Ford's apartment, Fletcher told Bess that Houdini's mother had asked him to give her the word "forgive." Bess was staggered. This was not the Houdini message, but it was a code word on which Houdini and his mother had agreed. No living person except herself, she told the press, knew that word. Alas, it was soon disclosed (by Conan Doyle, no less!) that a year earlier Bess had told a *Brooklyn Eagle* reporter all about "forgive." Doyle hastened to add that he did not believe Ford knew this! Still, the fact that the word had been published and Bess apparently had forgotten about it weakened the evidential value of Fletcher's disclosure.

Fletcher tried again. This time, speaking through Ford's lips in a series of eight sittings (none with Bess), he located Houdini himself. The message was "Rosabelle, believe." Bess requested a private sitting. In her apartment on January 7, 1929, Fletcher repeated the message. Bess signed a statement that it was correct.

On January 10 the *New York Evening Graphic*, a lurid tabloid, ran banner heads calling the revelation a hoax. Rea Jaure, a *Graphic* reporter, said Ford had told her that he and Bess had cooked up the whole thing to publicize a lecture tour the two of them had planned. Ford and Bess denied it. Walter Winchell published Mrs. Houdini's long, tearful letter defending herself and Ford. Joseph Dunninger (later to become a well-known television mentalist) got into the act. He declared that Houdini had been intimate with a red-headed magician's assistant, Daisy White, who knew the secret message and had passed it on to Ford. (Dunninger recently rehashed his version of the affair in *Fate*, November, 1971.) Bess never denied that Ford's message was correct, but in later years she insisted she had never received a spirit communication from her husband. My own opinion is that Bess, ill and drinking heavily in 1928, had blabbed the secret but was never able later to admit it.

From Houdini's peak to Pike's peak, Ford's career was undistinguished. He traveled about the world, lecturing to believers, giving public readings and private séances for the famous and not so famous. He lived for a time in Hollywood—a "lovely, mad city," he called it. He became an alcoholic. Just before his second divorce he got blind drunk in California, blacked out, and woke up in Florida. He joined AA. He popped pills. Fearful of the dark, he always slept with a light on.

Spraggett and Rauscher did not see all of Ford's papers because a secretary, on Ford's instructions, had destroyed an unknown amount. But there were piles of letters, diaries, and scrapbooks to go through. In checking this material Spraggett and Rauscher made a horrifying discovery.

Ford had thoroughly researched Pike in advance of the televised séance. During the sitting, Fletcher had introduced the Right Reverend Karl Morgan Block, Pike's predecessor as Bishop of California. (Fletcher: "The name is like Black, or something. . . . Charl. . ., Carl, Black, Block. . . .") Block mentioned several facts so obscure that Pike was certain no amount of research could have turned them up. But there among Ford's papers was a clipping—an obit on Block from the *New York Times*—containing just those facts. And there were other obits on other discarnates who had, through Fletcher, impressed Pike with their evidential data.

Shocked and outraged, Spraggett and Rauscher dug deeper. Ford kept obits on thousands. His file on Pike went back for years. There was a fragment from a destroyed notebook containing typed data on clients. There were handwritten notes from a psychic researcher and friend giving Ford facts about members of a church where the medium had been booked for a clairvoyant demonstration. There was Ford's note a week later: "Thanks for all you did for me. . . . You are forever putting me in your debt."

One of Ford's many male secretaries told Spraggett that Ford

. . . never went to a thing like the Pike sitting without untold research. He did the research himself. He showed me how to do it. He went to the library in Philadelphia. . . . School records, see? . . . One woman sent him five hundred dollars in advance for a sitting. I was with him when he did the research. . . .

One day I said to Arthur, "Are you reading your poems? . . ." That was the code name for his notes. . . . He kept his poems up to date by reading the papers constantly and cutting out obituaries from all over the United States.

He carried these poems of his in a gladstone suitcase and we'd hide it under the front seat of my car.

Do these disclosures, so common in the history of Spiritualism, shake the confidence of Spraggett and Rauscher in Ford? Not a bit. It only makes Ford seem to them more enigmatic. They conclude that he was "a genuinely gifted psychic who, for various reasons, scrutable and inscrutable, fell back on trickery when he had to." When he *had* to? The implication (it's a hoary one) is that clients force mediums to cheat by their incessant demands for evidential information.

The book does, however, have a villain, and you'd never guess who. It is the television entertainer Kreskin! The authors are apoplectic with indignation over this "late-adolescent" and "poor-man's Dunninger" for posing as a "sensitive" when all he does is magic tricks. It misleads the public. It casts doubt, you see, on genuine clairvoyants like Ford.

One of the strongest reasons for believing that mediums do not talk with the dead has always been that they never report the dead as saying anything much. Surely, if there is life beyond the grave, and God permits contacts with the Other Side, the dead do not become half-witted. Yet all they can say is that they are happy, that everything is peaceful and filled with light, and so on. They spout kindergarten metaphysics. As Spraggett reminds his readers, spirit controls are notoriously undecided about even so simple a question as reincarnation. And who should know better than they? Nor do they agree about the nature of Christ. Indeed, they don't agree about anything.

When Ford made a sensational platform appearance in London, after being introduced by Doyle, here are some of the marvelous messages he brought to those in the audience from departed loved ones. "They send greetings and love." "She says hold on and all will be well." "Yes, he [a dead child] is growing up nicely. He wanted to tell you so. And do not worry about your mother. She will be all right. I give you that."

Can anyone in his senses suppose that the dead would speak to the living in such puerile phrases? Perhaps one reason Spraggett is silent about the art of "cold reading"—the art of letting a listener unconsciously tell you, by his reactions, what to say next—is that Ford was never very good at it. I've heard palmists in carnival "mitt camps" who gave better readings.

The truth is that Ford was a mediocre mountebank. He would be forgotten today if it hadn't been for Bess Houdini and for his luck in catching poor Pike at a time of severe mental anguish. There is no deep mystery about Ford. He is easier to understand than Spraggett and Rauscher.

Postscript

Allen Spraggett's letter to the *NYR* was published in its October 11, 1973, issue:

> In regard to Martin Gardner's review of our book *Arthur Ford: The Man Who Talked With the Dead,* please let me reply for my collaborator, William Rauscher, and myself.
>
> Since Mr. Gardner, whose intellect has won him fame as an authority on games and puzzles, confesses (in one of his dozen or so pop science potboilers) to "an enormous, irrational prejudice against ESP," why, pray, is he reviewing the biography of a psychic? Isn't that a little like asking a member of the American Nazi Party to review *Fiddler on the Roof,* or a eunuch to do an in-depth critique of a sex manual?
>
> Mr. Gardner's menopausal outburst tells more about him and his "irrational prejudice," which has gotten worse with age, than about our book. We learn, for example, that he is bold in slithering to the attack when the victim is dead, as in his contemptuous remarks about Bishop James Pike.
>
> Mr. Gardner appears to hallucinate quotations, or maybe he has ESP. At any rate, my collaborator and I nowhere say that we found in Arthur Ford's files obituaries on "thousands." We nowhere say that we found in his files obituaries of any purported communicators in the Pike television séance but one, that of Rt. Rev. Karl Morgan Block.
>
> We nowhere say or imply that Bishop Pike had "a firm belief in life after death" at the time of his son's suicide and therefore that he may have expected to receive a message. In fact, Pike had renounced publicly any such belief some time before his son died. When Mr. Gardner describes the notoriously skeptical James Pike as a man "thirsting for signs," he is drawing on that vivid imagination evident throughout the review.
>
> We nowhere say that Canon Rauscher, a priest of the Episcopal Church, is or ever has been a "Spiritualist"—an error almost as quaint as the suggestion that Martin Gardner is to be taken seriously as a reviewer of books on ESP.
>
> We nowhere say or imply that the discovery of Arthur Ford's mediumistic fraud did "not shake" our confidence in him. The fact that our confidence was shattered is made plain enough in the book for even a prejudiced reviewer, however irrational, to have noticed. We go on to explain, however, that a qualified belief in Ford's extrasensory powers was rebuilt on the basis of hard evidence that he was—in spite of the sometime cheating—genuinely gifted psychically.

This evidence, Mr. Gardner ignores, of course. Obviously the psychic who impressed a' writer such as Aldous Huxley, a psychologist of the stature of William McDougall, a Pulitzer Price-winning novelist such as Upton Sinclair, and an astronaut, Edgar Mitchell, cannot pull the wool over the eyes of a wide-awake expert on games and puzzles.

Mr. Gardner is an expert in the art of the smear. He tortures prose to create the impression that my collaborator and I grudgingly disclose the unhappy facts about Ford's cheating. We do "not conceal the facts" is the weasel way he puts it.

The truth is that Rauscher and I are the people who discovered the evidence of Ford's cheating and we freely, without constraint, reveal it in the book. For your reviewer to have said simply that, however, apparently was too much for his "irrational prejudice." Poor embittered old man.

Allen Spraggett

To the above letter I replied:

It is typical of Spraggett that he would begin his attack with a fake quotation. A paragraph in my 1952 book, *Fads and Fallacies in the Name of Science,* begins: "There is obviously an enormous irrational prejudice on the part of most American psychologists—much greater than in England, for example—against even the possibility of extrasensory mental powers. It is a prejudice which I myself, to a certain degree, share." The degree to which I shared it then, as now, is slight. ESP is, obviously, a possibility. Like most psychologists, I consider it an unlikely possibility.

Spraggett correctly says that he did not claim to have found thousands of obituaries in Ford's files. I never said he did. On the contrary, I reported Spraggett's statement that an unknown portion of Ford's files had been destroyed by a secretary after Ford's death. My assertion that "Ford kept obits on thousands" was based partly (not entirely) on Spraggett's book. A former male secretary of Ford's told Spraggett that Ford used the code name "poems" for his obits, carried them with him in a suitcase, and "kept his poems up to date by reading the papers constantly and cutting out obituaries from all over the United States." Spraggett himself writes (p. 248): "Arthur Ford's private files revealed that he had a marked propensity for clipping obituaries. . . ." I do stand corrected on a trivial point. Only one obit (not several) contained evidential information on which Ford drew in his famous séance with Bishop Pike.

It is true that for a brief period before his son's death Pike voiced doubts about immortality. They were short-lived. Two weeks after his son's suicide, Pike was convinced that his son was trying to reach him from beyond the grave. In view of Pike's long Christian ministry, his temporary doubts only deepened his hunger for hard evidence that his son's soul had not utterly perished.

Now about the word "spiritualist." Canon Rauscher does not like to be called a spiritualist because it implies that he belongs to a spiritualist church. That I did not use the word in this sense is evident from the fact that I called Rauscher a "convinced spiritualist" immediately after identifying him as an Episcopalian priest. The latest *Webster's New Collegiate Dictionary* defines spiritualism as "a belief that spirits of the dead communicate with the living, usually through a medium." Canon Rauscher closes his introduction to Spraggett's book by writing: "If, after reading this book . . . you were to ask me, 'Do you believe that Arthur Ford talked with the dead?' my answer would be yes." I apologize if capitalizing "Spiritualist" misled some readers into supposing it meant anything more than Rauscher's long-standing conviction that mediums do indeed talk with the dead.

Spraggett's pity for those who do not share his adolescent enthusiasm for Protestant occultism is touching. In a way, one must envy him his ability to believe almost anything. In his book *The Unexplained* (Bishop Pike, in his preface, calls it a "thoughtful book," and Norman Vincent Peale, on the jacket, says it is "just about the best book on the phenomenon of ESP that has appeared in many a day") you will find Spraggett believing in astrology, teleportation, haunted houses, helping plants with prayer, Kathryn Kuhlman's miraculous healing of a cancer victim (Spraggett later wrote an entire book about this faith-healer), Ted Serios's ability to project thought pictures onto Polaroid film, and scores of even wilder miracles about which so many of us embittered old skeptics have our doubts. [For the record, I was 59 when Spraggett called me a "poor embittered old man."]

One final point. Of the four men cited by Spraggett as "impressed" by Ford, only McDougall was a scientist. Although Spraggett quotes McDougall as saying that Ford had "supernormal powers," he does not give the source of this quote, and on page 226 he states that the psychologist was "not unduly impressed" by a Ford séance. As for the other three, their capacity for uncritical belief is exceeded only by Spraggett's. Huxley wrote an entire book to promote the worthless views of Dr. William ("throw away your glasses") Bates (see Chapter 19 of my *Fads and Fallacies*), and until the day he died, Sinclair defended the nutty theories of Dr. Albert Abrams, this country's funniest medical quack (see Chapter 17 of *Fads and Fallacies*). We'll learn all about Mitchell's views when Putnam publishes the book he is writing about psychic phenomena. Meanwhile, he has been appearing on television shows testifying to his faith in Israeli magician Uri Geller's ability to bend iron spikes by psychokinesis.

Martin Gardner

For the best account of Bishop Pike's encounter with that holy three — Ford, Spraggett, and Rauscher — see pages 207–240 of *The Death and Life of Bishop Pike* (Doubleday, 1976) by two of Pike's friends and admirers, William Stringfellow and Anthony Towne. Pike's widow, Diane, authorized the biography and wrote its introduction. I also recommend an excellent review of this book by Raymond Schroth in the *New York Times Book Review,* August 1, 1976.

The book is filled with hitherto undisclosed details about Pike and Ford. You will learn how Ford kept his traveling bag (containing his "poems") closed with leather straps, each with a padlock. You will learn about the suicide attempt of Pike's daughter. You will learn about the suicide of Pike's mistress, Maren Bergrud. After a stormy argument at 2 A.M., Pike handed her a bottle of sleeping pills and said, "Take your pills and go." She went, took fifty-five pills, and died the following morning. Her suicide note said that, although she had finally accepted the fact that Pike did not love her, her will had left everything to him, and it concluded: "I needed hope. You never offered it—never once offered it." Pike found the note, tore off the part about himself, and kept it from the police. He later gave the note to Diane, and she in turn allowed Stringfellow and Towne to print it. The authors make a strong case for their belief that Maren Bergrud, ill, neurotic, and on drugs, was in cahoots with the medium George Daisley and that it was she who faked the poltergeist phenomena in Pike's Cambridge apartment.

You will also learn from the book the real cause of the death of Pike's son. It was not so much drugs as it was intense anxiety and depression over his discovery that he was homosexual. Pike himself had had a homosexual experience when he was a law student at Yale. He spoke about it to Diane, and he told a homosexual friend of long standing that he had not found the episode "unpleasant or distasteful." To the authors' regret, Pike apparently never told his son about this experience. They feel that Pike's church, with its official condemnation of homosexuality, was "cruelly complicit" in the young man's self-murder.

Rauscher impressed Stringfellow and Towne as having a passion for spiritualism that "was genuine. . . . He had a long, close personal association with Arthur Ford. The two men were good friends. He officiated at Ford's burial service. Father Rauscher seems to us to be ingenuous to a fault. He is the sort of fellow who is easily used by disingenuous connivers."

Rauscher was president and leader of Spiritual Frontiers Fellowship, Inc., founded in 1956 as a quasi-corporation for which Ford was forever seeking donations. Rauscher collaborated with Spraggett on *The Spiritual Frontier* (Doubleday, 1975), and wrote the introduction to a book by a reformed fraudulent medium, M. Lamar Keene (as told to Spraggett), published in 1976 by St. Martin's Press under the title *The Psychic Mafia*. Rauscher is an amateur conjuror who gives occasional stage shows. I have met him briefly through our mutual hobby of magic, and my opinion of him coincides with that expressed by Stringfellow and Towne. I find him sincerely persuaded that Christianity should be wedded to spiritualism and to the claims of parapsychology, but (like his friend Spraggett) a man of boundless gullibility and scientific ignorance.

Stringfellow and Towne are less kind to Spraggett. They describe his book on Pike as a "potboiler" in which he violated many confidences. Much of it, they say, is "coarsely cribbed" from their own earlier book, *The Bishop Pike Affair*. I am also in agreement with them in the picture they draw of Spraggett. He impresses me as a hyperactive ham, an opportunist with an enormous ego and a strong drive for fame and fortune.

Spraggett was born in Toronto in 1932, and educated at Queen's University at Kingston. He was a hell-fire, Bible-pummeling fundamentalist preacher until 1962, when he became the religion editor of the *Toronto Daily Star*. He left the *Star* six years later to write a newspaper column on the occult, called "The Unexplained," that was widely syndicated in Canada and the United States. His popular potboilers on occult topics, his radio and television shows, and his rousing lectures soon made him Canada's best-known drum-beater for astrology and all things paranormal.

I mentioned in my review that homosexuality runs like a subtle fugue through the book by Spraggett and Rauscher (see, for example, their cruel footnote about Kreskin on page 171). They tell about Ford's alcoholism and his manic-depressive moods. There are many hints about Ford's sexual preferences, but they do not say outright that Ford was homosexual—or perhaps bisexual is a better word.

Ironically, in 1979 Spraggett, along with six other Canadian men, was served with a warrant by the Winnipeg police and charged with having paid for homosexual acts with two boys, ages 14 and 15, in a Winnipeg hotel in 1978. Spraggett vigorously denied knowing the boys or having ever been involved with homosexuality. In 1980, five months after the ending of his sensational trial, a judge found him not guilty. But the damaging publicity ended his radio show on astrology and his popular weekly CBC television show, *Beyond Reason,* on which three psychics tried to guess the profession of two hidden guests. Spraggett, now fully vindicated, has announced his intention of writing a book about what he calls his crucifixion and resurrection.

The stirring together of parapsychology with Christology reached towering heights of vapidity in 1965, when Jesus himself began to dictate to a woman known only as Helen, who at that time was an assistant professor of medical psychology at Columbia University. The result, ten years later, was a half-million-word Revelation called *A Course in Miracles,* published in 1975 by Judy Skutch, a wealthy patron of Uri Geller, Dean Kraft, and other psychics and psychic causes. You can read all about it in Brian Van der Horst's article, "Simple, Dumb, Boring—and a Course in Miracles" (*New Realities,* vol. 1, no. 1, 1977); the same author's follow-up piece, "Miracles Come of Age" (vol. 3, no. 1, 1979); and "The Gospel According to Helen," by John Koffend in *Psychology Today,* September 1980.

The "Jesus" who dictated this mammoth clunk of a work makes L. Ron Hubbard and Werner Erhard sound like profound theologians. The *Course*'s introduction sums itself up in two gaseous lines:

> Nothing real can be threatened.
> Nothing unreal exists.

24

"The Preachers"

A group of students on the campus of the University of California at Berkeley are in animated discussion. Are they debating subtle differences between Maoism and Castroism? Between Therevada and Mahayana Buddhism? Listen carefully. They are talking about the "rapture." Will the saved be caught up in the air with Jesus at the time of his Second Coming, or will the rapture take place *before* the Second Coming?

This unexpected revival of Protestant fundamentalism among the young (what sociologist predicted it?) is one of the craziest aspects of the current American scene. "Demythologizing"—purging Christianity of the historicity of its great myths—was supposed to keep the young people in the liberal churches. It had a reverse effect. Mythology was what they wanted, not do-good sermonizing that put their heads to sleep. They *wanted* to be told about heaven and hell, God and Satan, sin and redemption.

The liberal churches are now half-filled on Sunday mornings, mostly with sad-faced elders who are there largely from habit. They sing tuneless hymns with vacuous phrases. They recite dreary creeds they no longer believe. They drink a communion wine that has lost even its symbolic savor. On the other side of town, Pentecostal churches are jammed with bright-eyed youngsters who are belting out the old melodious songs about the Cross, shouting "Thank you, Jesus!" and having a marvelous time. Pentecostalism is the belief that the "gifts" of Pentecost, especially

Reprinted from the *New York Review of Books,* February 21, 1974, with permission from the New York Review of Books. © 1974 NYREV, Inc.

faith-healing and glossolalia, were not restricted to the early Church but are still in effect as signs of the power of the Holy Spirit. Modern Pentecostalism was confined to Protestant fundamentalist sects until a few years ago when it suddenly became fashionable among conservative Roman Catholics and Episcopalians.

Riding the crest of this new wave, in part fomenting it, are the great evangelists. Who are they? What are they like? James Morris, who grew up in Tulsa, has written *The Preachers* (St. Martin's, 1973), a frightening, funny account of nine of the biggies. "One-man denominations," he calls them. Their simple-minded books are selling by the millions. Their colorful magazines have larger circulations than *Playboy*. More psychosomatic ills are being banished in one day by Bible-thumping faith-healers than by all the psychiatrists in a year, and the cures are probably just as lasting.

Consider Tulsa, once the proud "oil capital of the world." Today it is the "fundamentalist capital of the world." Oral Roberts and Billy James Hargis, two of the country's most successful one-man denominations, make their homes there. Roberts, the more flamboyant of the two, is now second only to Billy Graham in fame, fortune, and adulation. Morris's chapter on him, painstakingly documented, tells a remarkable story.

The story begins in 1918 on a farm near Ada, Oklahoma. That was the year both Roberts and Graham were born. Although Oral's parents were devout members of the Pentecostal Holiness Church, Oral did not take his religion seriously until one day when he was playing high school basketball and collapsed on the floor with blood running from his nostrils. Doctors pronounced it advanced tuberculosis of both lungs. Some time later, at a local tent revival, the evangelist touched Oral's head. A blinding flash engulfed him. He leaped from his chair, shouting, "I am healed!"[1]

Roberts became a Pentecostal minister at seventeen. Eleven years later, when he was pastoring a church in Georgia, he discovered that he, too, had the healing power. A heavy motor had dropped on the foot of one of his deacons. In one of his many autobiographies, Roberts describes how the blood streamed from the man's shoe, "which was nearly cut from his foot," Roberts touched the shoe and cried, "Jesus, heal!" "In amazement," Roberts writes, "I saw him take his shoe off, stamp his foot. . . . I saw with my own eyes that his foot had been instantly restored."

It was the first of Roberts's many miracles. The divine power seemed to flow through his body only when he established what he called a "point of contact" with the sufferer. If present in the flesh, Roberts touched the person with his hand. If there only on radio or television, listeners were asked to touch the loudspeaker or screen. Brother Roberts also became a skillful exorcist. "First, I feel God's presence, usually through my hand," he told an interviewer. "Then I catch the breath of a [demon-possessed] person—it will have a stench as a of a body that has been decayed. Then I notice the eyes. They're—they're like snake eyes."

In the late forties, Roberts established Healing Waters, Inc., in Tulsa. By the mid-fifties he had far outdistanced his nearest rival, the Dallas healer Jack Coe, in the size of his tent crowds, sales of literature, number of radio shows, and the amount of money pouring in. In 1956 Healing Waters employed 287 workers, mainly to open envelopes and count the cash. Last year a Tulsa banker estimated the annual cash flow to Roberts as fifteen million dollars.

There were the inevitable tragedies. A diabetic woman, healed by Roberts, threw away her insulin and promptly died. A lady with cancer of the spine expired a few days after testifying about her cure. Roberts came under increasing criticism by doctors and non-Pentecostal clerics.

Slowly Brother Roberts began to change. He phased out his television healing. He collapsed his big tent. Rumpled suits gave way to tailored pinstripes. A Tulsa bank made him a director. So did the Chamber of Commerce. He joined Rotary. He joined the city's most exclusive country club.

He established ORU (Oral Roberts University) on a large tract of land in suburban Tulsa. Billy Graham was the dedication speaker. The university's dazzling modern architecture is dominated by a 200-foot-high Prayer Tower with a "crown of thorns" around its observation deck and a perpetual flame on top to signify the Holy Ghost. All day and all night, seven days a week, "prayer partners" in the tower pray for and counsel those who telephone from all parts of the world.

Three years ago, to the horror of his old Pentecostal associates, Roberts—now "Dr." Roberts—became an ordained Methodist minister and joined Boston Avenue, Tulsa's wealthy, fashionable, middle-of-the-road Methodist church. Does Dr. Roberts, Tulsa's most distinguished millionaire, have some long-range plan up his natty coat sleeve? No one knows. His major interest at the moment seems to be his TV musical spectaculars, featuring prominent guests and the "now" sound of his singing son, Richard. No one on these shows speaks in the Unknown Tongue.

Billy James Hargis, Tulsa's number two windbag, had more formal training for the cloth than Roberts; he graduated from Ozark Bible College, Bentonville, Arkansas, to become a Christian Church minister at eighteen. Hargis doesn't practice faith-healing. His forte is fighting communism. The Christian Crusade, which he oversees, is the largest Christ-centered anti-Red movement in the land. Indeed, beside it the American Communist Party pales into insignificance. The beautiful cathedral in Tulsa draws almost as many visitors as Roberts's Prayer Tower. Billy has his own university in Tulsa: American Christian College. The Crusade also runs colleges in Colorado and Maine. New Yorkers may regard this pudgy, moon-faced Tulsan as a bigger buffoon than Roberts, but in the Middle West he is a greatly admired celebrity. The half-million-dollar mansion where he lives, on a hill in Tulsa, has ninety telephone outlets.[2]

Kathryn Kuhlman, Pittsburgh's aging evangelist (in her sixties, though she doesn't look it), is the only lady preacher in Morris's book. Although faint thunder compared to Aimee Semple McPherson, Miss Kuhlman is unquestionably the most renowned faith-healer in America. Some of Morris's best writing is his description of her triumphant appearance in Tulsa in 1971, with Dr. Roberts on the platform, and the deaf, halt, and blind being "slain by the Lord" when Miss Kuhlman touches them.

Allen Spraggett, a Canadian fundamentalist who gave up preaching to write preposterous potboilers about the occult, is Miss Kuhlman's leading biographer. Spraggett believes that mediums talk with the dead, and just about every other aspect of the current psychic scene. His 1970 paperback, *Kathryn Kuhlman: The Woman Who Believes in Miracles,* is of special interest, Morris suggests, because it approaches Miss Kuhlman's miracles from the standpoint of parapsychology.

Spraggett has a low opinion of Roberts, but considers Miss Kuhlman a saint. What impresses him most is that so many of those she heals are unbelievers. Could it be, he asks, that what Miss Kuhlman calls the Holy Spirit is really a "field phenomenon," that her healing is less an act of God than a parapsychological event? This mixing of Christianity with the occult is a rapidly growing trend, not only among professional psychics like Jeane Dixon but also among liberal Protestant leaders such as Norman Vincent Peale and the late Bishop James Pike. It has, of course, long been the stock-in-trade of mediums such as Arthur Ford, about whom Spraggett has also written a book.[3]

The funniest chapter in Morris's book concerns the eighty-one-year-old Pasadena prophet, Herbert W. Armstrong. It seems unlikely that a single man, with undistinguished oratory and a patchwork of antique heresies, could build a fundamentalist empire that now has operating funds even larger than Roberts's. Somehow the old man did it, and he did it all on the radio without a single appeal for funds. He simply drones on about biblical prophecy and the "wonderful world of tomorrow," then offers subscriptions to his magazine, *Plain Truth,* and booklets attacking evolution and other false doctrines—all "absolutely free." This absence of requests for money is one of his cleverest innovations. He waits patiently until converts are solidly hooked on his Worldwide Church of God, then subjects them to rigid tithing.

Armstrong's revelations are incredibly old hat. The Anglo-Saxon people are the true descendants of the lost tribes of Israel. There is no Trinity. The dead are truly dead until Judgment Day when God will restore them to life with a new body. (This doctrine of "conditional immortality" has had many distinguished advocates: John Milton and Karl Barth, to name only two. It is a central dogma of Seventh Day Adventism, Jehovah's Witnesses, and other sects.) Old Testament food regulations must be observed. "Pagan holidays" such as Sunday, Christmas, and Easter are taboo.

Listening to Armstrong explicate Daniel and Revelation, one finds it hard to conceive how anyone could suppose that this smiling little man, with white hair and large eyes, is the only person on earth with a direct pipeline to the Almighty. Yet his Church of God has millions of converts. I used to think Bobby Fischer refused to play chess on Saturday because he is Jewish. Not so. As an Armstrong follower, Bobby shares with Seventh Day Adventists the conviction that God never authorized a Sunday sabbath.

What is one to make of Herbert's handsome son, Garner Ted, whose voice sounds so curiously like his dad's? Does he really believe the nonsense he talks, or is he just concerned about the multi-million-dollar empire he will soon inherit if he pretends to keep the faith and Jesus delays His coming for a few more decades?

There are hilarious pictures in *The Preachers* of other top Bible pounders. A. A. Allen had to abandon plans to raise the dead when it became apparent that the faithful would start shipping him corpses. C. W. Burpo once devoted an entire radio program to revealing that H. Rap Brown was none other than John Green, an undercover agent of the Senate Appropriations Committee. (Burpo had read this in Russell Baker's column and didn't know it was a joke.) And there is the Right Reverend Father-in-God, His Divine Eminence, Dr. Frederick J. Eikerenkoetter II, better known as Reverend Ike. Ike is a black evangelist from South Carolina whose major revelation is that God wants you to make lots of money. Ike has even found support for this in the Good Book: "A feast is made for laughter, and wine maketh merry: but money answereth all things" (Ecclesiastes 10:19). A chapter is devoted to Carl McIntire, the irascible New Jersey fundamentalist who has been in the news recently because the government is giving him such a hard time about his pirate radio station.

Billy Graham, the biggest of the biggies, is the only preacher about whom Morris has nothing unkind to disclose. Graham really does practice what he preaches. Others attack him from left and right, but he turns the other cheek and goes his own way, firmly persuaded by his vast pride and ignorance that he, Billy Graham, like Billy Sunday before him, is in firm possession of God's truth. If he were a clever charlatan he would be more interesting. But the awful truth is that Graham is not a charlatan. He may even be convinced, God help us all, that his friend Richard Nixon golfs with him not to boost his political image but because he, Nixon, actually shares Billy's evangelical faith.

In spite of its comic scenes, *The Preachers* is a sad, nostalgic book, and when one finishes it large questions loom in the mind. Why has it happened? How is it that today, when science and medicine are advancing on a thousand spectacular fronts, people seem caught up in every conceivable variety of irrationalism? This new irrationalism seems to be worldwide. It can be found in England, Japan, France, Germany — even in Russia. There is an astonishing article in the recent issue of *Time* (October 8, page 102) about how Pentecostalism is "spreading like a spiritual wildfire" around the globe, especially in Korea!

One view, ably defended by Father Andrew Greeley in his recent book *Unsecular Man,* is that the percentage of persons, at any one time or place, who cannot live without faith in the supernatural is a relatively fixed constant. Only the myths change. What we are now witnessing is not an increase of belief in the supernatural; just noisy alterations in the content of such belief. As liberal Protestantism rolls toward humanism, and Catholicism rolls toward Protestantism, unsecular Western souls start looking elsewhere — to the past and to the East — for gods and magic.

Another view is that superstition does indeed fluctuate in intensity, like the business cycle; that a combination of social forces—the bomb, wars, science, future shock, deteriorating education, decline of established churches, and so on—have produced a genuine upsurge of uncritical faith. Whatever is the true state of affairs and the reasons for it, strange things are happening to American Protestantism, and its future now seems as wildly unpredictable as the future of America itself.

NOTES

1. There is no shred of evidence that Oral Roberts ever had tuberculosis aside from his own statements that doctors said he had it. His several autobiographies give many details about doctors and hospitals who reported, *after* Oral was instantly healed, that he did *not* have TB. You will look in vain for the name of a physician who said Oral had TB, or the name of a hospital where it was diagnosed.

2. For another good account of Billy Hargis and his activities when he was riding high, see the chapter on him in *Danger on the Right,* by Arnold Foster and Benjamin Epstein (Random House, 1964).

3. See Chapter 23.

Postscript

Since I reviewed *The Preachers,* the born-again fundamentalist revival has grown with surprising rapidity in both stridency and political power. If Morris were to write a sequel he'd have to devote chapters to Pat

Robertson and his 700 Club, to Jim Bakker and his PTL Club, to Rex Humbard, James Robison, Robert Schuler, Jerry Falwell, and the up-and-coming piano-pounding Jimmy Swaggart — to mention only some of the better-known "electronic church" revivalists. Charles Colson and Jeb Stuart Magruder, of Watergate notoriety, have been born again. Eldridge Cleaver has seen the light. Mark Hatfield and Harold Hughes continue to be the nation's leading evangelical congressmen.

Fundamentalist books are selling faster than pornography. Recent jackpot-hitters include Billy Graham's monograph on angels, Hal Lindsey's potboilers about the Second Coming, Charles Colson's *Born Again,* Marabel Morgan's *The Total Woman,* and Johnny Cash's *Man in Black.*

Jimmy Carter's 1976 election was strongly influenced by the growing born-again vote, especially among blacks. But in 1980, disenchanted by Jimmy's attitudes toward such things as abortion and school prayers, the fundamentalists flocked to Ronald Reagan, who did his best to appear even more born-again than Carter. Addressing a gathering of fundamentalists in August 1980 he announced that the theory of evolution was "just a theory" and was marred by "great flaws." He said he favored the teaching of creationism in the schools as an alternative theory. John Anderson disagreed. As a born-again Christian, Anderson acknowledged that the Bible was the word of God, but he pointed out that the "days" of Genesis can be taken to mean long periods of time.

Jimmy Carter, responding to a query from *Scientific American,* gave the most enlightened comment on evolution. I quote it from *Scientific American,* November 1980, page 80:

> The scientific evidence that the earth was formed about four-and-a-half billion years ago and that life developed over this period of time is convincing. I believe that responsible science and religion work hand in hand to provide important answers concerning our existence on the earth. My own personal faith leads me to believe that God is in control of the ongoing processes of creation. Insofar as the school curriculum is concerned, state and local school boards should exercise that responsibility in a manner consistent with the Constitutional mandate of separation of church and state.

Carter's statement suggests how far his religious convictions are from those of the fundamentalists who make up the new "Christian right." It is a strange commentary on our times that the three presidential candidates of 1980 all professed to be evangelical Christians and that the man who won the election not only doubts the theory of evolution but also believes in astrology (Reagan and Nancy are good friends of astrologers Carroll Righter and Jeane Dixon) and the results of parapsychology. In an interview by Angela Fox Dunn (see the *Washington Post,* July 13, 1980) Reagan said he consulted Righter's horoscope column every day. "I believe you'll find," he told her, "that 80 percent of the people in New York's Hall of Fame are Aquarians" (Reagan, born February 6, 1911, is an Aquarian). In short, our country has elected as president a man who holds Protestant fundamentalist opinions and believes in astrology and ESP.

Jerry Sholes, a former associate of Oral Roberts, blasted Roberts in an eye-opening book called *Give Me That Prime Time Religion* (1979). It was first privately published in Tulsa as a paperback, later as a hardcover Hawthorn book. Since 1977 Oral has been battling with Tulsa civic leaders over his plans to build a mammoth $100 million medical complex opposite ORU (Oral Roberts University). He wants the hospital to have 777 rooms. Donations were to be in repetitions of seven: $7.77, $77.77, $777.77, and so on. When the ground was broken in '77, 77 white doves were released. The hospital will emphasize "holistic medicine," by which Oral means a combination of traditional medicine and divine healing. Tulsa doctors claim that the one thing Tulsa doesn't need is a new hospital, but Oral is proceeding anyway because, as he says, God told him to build the hospital just as God told Noah to build the Ark.

Oral has a thing about the number 7. Is it because 777 has been a traditional numerological symbol of Christ, in contrast to 666, the Number of the Beast, or because "Roberts" has seven letters? The address of ORU is 7777 South Lewis Avenue, and the phone number of "The Prayer Tower" (2×7 letters) is 492-7777. "The City of Faith," Oral's name for his medical complex, also has 2×7 letters, and so does Oral's favorite phrase "Expect a Miracle." His 1961 autobiography *My Story* has seven letters, and also his 1972 autobiography *The Call.* Many of his book titles spell with multiples of seven, such as *The Miracle of Seed-Faith* and *A Daily Guide to Miracles.*

For a good account of Oral's hospital troubles see "And God said to Oral: Build a Hospital" (*Science,* April 18, 1980, pp. 267 ff.). Officials of the National Health Planning and Resource Development Act (Roberts thinks they are doing the work of the Devil) are trying to force Oral to reduce the number of beds in his City of Faith from 777 to 294. As Christopher Nelson Sinback pointed out in a letter (*Science,* May 16, 1980), 294 is equal to $(7 \times 7 \times 7) - (7 \times 7)$, making it two sevens better than 777, though Oral is unlikely to be impressed.

In September 1980 the wire services reported that an unknown gambler, in jeans and cowboy boots, stalked into the Horseshoe Club at Las Vegas, placed a single bet of $777 thousand on the craps table, won, and left with more than $1.5 million. I suspect this was a publicity event staged by the casino, but if not, could it have been Oral making a bet God told him to make?

A month later the *New York Times* reported (October 17) that in his latest appeal for funds Oral told his financial "partners" in a letter that at precisely 7 P.M., on May 25 ($5 + 2 = 7$), as he stood silently praying in his City of Faith, a 900-foot-tall Jesus appeared before him:

"I felt an overwhelming presence all around me. When I opened my eyes, there He stood, some 900 feet tall, looking at me. His eyes—Oh! His eyes! He stood a full 300 feet taller than the 600-foot-tall City of Faith. There I was, face to face with Jesus Christ, the Son of the Living God."[2]

And God said to Oral: "I told you that I would speak to your partners and, through them, I would build it." According to an Associated Press story, Roberts added: "If you will obey, it will not be difficult to finish the second half of the City of Faith." I have been told by an ORU official that the millions are pouring in.

Oral was not available to the *New York Times* or to the Associated Press, but George Stovall, executive vice-president of ORU, said, "What he said he saw, he saw."

I believe it. Many of Tulsa's medical officials are convinced that Oral is a total fraud. All things are possible, but in my opinion Oral is not a total fraud. I think he saw what he said he saw, just as I think Paul saw what he said he saw on the road to Damascus, and that Joan of Arc saw and heard what she said she saw and heard.

Billy Hargis met his Waterloo in 1974 when five students at his college, four of them men, accused him of having sexual relations with them. One student, as reported in *Time* (February 16, 1976, p. 52), said Billy justified his behavior by citing the Old Testament friendship of David and Jonathan. After a sad farewell sermon Hargis turned the college over to his vice-president, but he was soon back in Tulsa to continue his Christian Crusade and attack sexual laxity in America. He has a wife and four children. "I have made more than my share of mistakes," he said through a lawyer. "I'm not proud of them. Even the Apostle Paul said, 'Christ died to save sinners, of whom I am chief.'"

Kathryn Kuhlman's troubles began in 1975 when Paul Bartholomew, her former TV agent and personal administrator, sued her for about a half-million dollars, charging her with diverting funds from her tax-exempt foundation to private use. Bartholomew's brother-in-law, Dino Karsonakis, who had been Kathryn's pianist and close companion, had earlier broken with her, contending that she owned a million dollars worth of jewelry and that her art and antiques were worth another million. Bartholomew's suit was settled out of court for an undisclosed sum. A few months passed before Kathryn entered a Tulsa hospital for open-heart surgery. She died in Tulsa in 1976, two months later.

Herbert Armstrong began to have problems in 1977 when some disillusioned Armstrongites published *Ambassador Report,* an 89-page hatchet job in magazine format. It featured an interview with chess grand-master Bobby Fischer in which he told of his disenchantment with Herbert after giving him about $100,000 over a period of fifteen years. Bobby was so enraged by the publication of his tape-recorded remarks that he looked up one of the ladies involved and allegedly struck her twice. She filed assault charges, which were dropped after a financial settlement.

The Fischer interview took second place in *Ambassador Report* to a more sensational article, "In Bed with Garner Ted." A cover photo shows Ted smiling and saying, "And greetings suckers around the world. . . ." The article contained such irrefutable evidence of Ted's philandering that even his father couldn't disbelieve it. He excommunicated Ted, who vanished into Texas where he has been struggling to get his rival Church of God, International, off the ground. Herbert has denounced Ted's supporters for following a "mere man" instead of God.

Armstrong's multi-million-dollar empire (which includes odd little possessions like *Quest* magazine[1]) is in the hands of a recent (1975) convert, Stanley R. Rader, Herbert's former accountant and lawyer. In 1977, Herbert, in his mid-eighties, married Rader's 40-year-old secretary. They live in Tucson, Arizona. If Jesus doesn't come soon, Rader is the most likely person to inherit the empire, unless Ted has some legal tricks up his soiled sacerdotal sleeves.

To add to Herbert's misfortunes, the attorney general of California is currently battling the Worldwide Church of God, claiming that Armstrong and Rader have been selling property, thereby draining the church of millions for their private use. To give his side (God's side of course) of the story, Herbert has been taking full-page ads in the *New York Times*, the incredible text surrounding either his own face or the face of Rader. Rader has written a book about it all, called *Against the Gates of Hell* (Everest House, 1980).

It used to be Herbert and Garner Ted on the airways. Now it is only Herbert, with treasurer Rader waiting patiently in the wings. Night after night on the radio, and weekly on his TV show, Herbert keeps spouting the same dreary doctrines, telling his listeners to "wake up and blow the dust off your Bibles" and to realize that he and he alone is preaching the true gospel. Herbert never says a word about Seventh Day Adventism, with which he was once involved, and which has long preached such fundamental Armstrong doctrines as a Saturday sabbath, the soon return of Jesus, soul-sleeping until Judgment Day, and the permanent annihilation of the wicked.

For whatever it's worth, I think the old man believes every word he says. He is a type all too common in the history of Christian sects. As for Garner Ted and Stanley Rader—well, a lot of money is involved! And a lot of money will also descend like manna on whoever inherits Oral Roberts's empire. I predict that "something good is going to happen" to Oral's heir-apparent, his singing son Richard (seven letters). Oral is going to excommunicate him one of these days.

Billy Graham is rolling along accustomed grooves, still looking handsomer than all his rivals and hitting harder than most of them on the soon appearance of the anti-Christ, to be followed by the Second Coming and the final defeat of Satan. Johnny Cash has written a stirring song about it that I heard him sing at a recent Graham crusade: "Matthew 24 is Knocking at the Door." The title was the text of Billy's fire-and-brimstone sermon.

Billy was profoundly shocked by Watergate—not so much by his friend Nixon's behavior as by those four-letter words he heard on the tapes. Dick had never talked that way on the golf course! But the old born-again Nixon is back, Billy now assures us, and repentant of his sins.

Then there's the Reverend Sun Moon, the great new messiah from Korea, on the U.S. scene, surrounded by his smiling, empty-headed children. But that's another story, even funnier than the sagas of Oral Roberts and the Armstrongs.

NOTES

1. In January 1981 the inevitable happened. Herbert Armstrong wanted to publish in *Quest* an article of his own. It had been understood that the magazine was to be independent of the Worldwide Church of God; so when Stanley Rader demanded that the article be published the six top editors of *Quest* resigned. It was high time. For four years their efforts had accomplished little more than to dignify one of the wealthiest boobs on the Protestant landscape.

2. How did Oral know that Jesus was 900 feet tall? For various hypotheses, see Robert McAfee Brown's analysis, "Oral Roberts and the 900-foot Jesus: Investigating the Credibility of a Claim from the Oral Tradition," in *The Christian Century,* April 22, 1981, pp. 450–452. Perhaps, Brown suggests, Oral confused 300 yards with 300 cubits, which would give Jesus a height exactly equal to the length of Noah's Ark. I am indebted to the statistician William Kruskal for calling this important mathematical paper to my attention.

25

"Uri" and "Arigo"

It's hard to say which of these two books—Andrija Puharich's *Uri: A Journal of the Mystery of Uri Geller* (Doubleday, 1974) and John G. Fuller's *Arigo: Surgeon of the Rusty Knife* (Crowell, 1974)—is the nuttier, but there's no denying that Puharich's book is the funnier. The comedy begins in 1952. Andrija Henry Puharich, an ex-Catholic with an MD degree from Northwestern, is starting his career as a distinguished parapsychologist. Through a medium from Poona, India, he makes his first contact with the Nine. The Nine are the highest minds in the universe, roughly equivalent to what everybody else calls God. Their messages are partly confirmed by another medium who reports to Puharich through one Dr. Laughead (*sic*) of Whipple, Arizona.

Nineteen years go by. Puharich has brought a Dutch psychic to America, studied hallucinogenic mushrooms, patented fifty devices for hearing aids, investigated Brazilian psychic-healer Arigo, photographed UFOs, and written two Doubleday books: *Beyond Telepathy* and *The Sacred Mushroom*. In 1971 comes the climacteric of his career. He meets Uri Geller.

Has any reader *not* seen Uri Geller on television? His most sensational appearance last year was on "Not For Women Only," when he caused a spoon, held by Barbara Walters, to bend at right angles. On the "Mike Douglas Show" he bent a nail held by Tony Curtis (ironically the man who played Houdini) and Hugh Downs displayed his car key, which

This review, which appeared in the *New York Review of Books,* May 16, 1974, and the letters quoted in the Postscript are reprinted with permission from the New York Review of Books. © 1974 NYREV, Inc.

Geller had bent backstage. On Merv Griffin's show he wowed Merv by picking the one nonempty film can out of ten. His only flopperoo was on the "Tonight Show." Johnny Carson, ex-magician, had taken the trouble to establish a few simple backstage controls.

Last fall, on a triumphal tour of England, Uri got more favorable publicity than any sensitive since the medium D. D. Home. Geller's picture had made the cover of *Psychic, Fate, It Is Divine* (official organ of Guru Maharaj Ji), and Germany's *Der Spiegel*. He might have made the cover of *Time* had it not been for the sanity of science editor Leon Jaroff. Parapsychologist Gertrude Schmeidler, speaking last March at New York University, closed her lecture on psychokinesis by saying that the most exciting new development in this field is the appearance of sensitives like Geller who can demonstrate PK at will. Geller is far and away the hottest figure today on the world psi scene.

To return to our story: Puharich is quickly persuaded that Uri Geller is a genuine "sensitive" even though Geller is then earning his living in Israel by doing a nightclub act featuring magic tricks similar to those of Dunninger and Kreskin. Uri and Puharich become pals. Under hypnosis, strange voices issue from Geller and are tape-recorded. The tapes later mysteriously self-destruct. What was on these lost tapes? Nothing less than a new Revelation from the Nine!

Subservient to the Nine, Puharich learns, are the Controllers who supervise countless planetary civilizations. Earth's Controller is Hoova. Every 6,000 years Hoova intervenes in earth's history. The last time was 6,000 years ago. Now a new contact has been made. The revelation is by way of a spacecraft called Spectra that has been stationed over earth for 800 years. It is the size of a city and is occupied by supercomputers. Vast intelligences of the future, under Hoova's control, have gone millions of "light-years" back in time to enter Spectra's computers. Uri has been chosen to be the bearer of Hoova's new message to mankind. Puharich has been chosen to be Uri's witness, keeper, and scribe. "There is no other on earth that we will use for the next fifty years but you and Uri," the computers tell Puharich.

The book swarms with UFO sightings, taped messages from Spectra, and accounts of hundreds of miracles that Geller performs with the aid of computer power. One is stunned by the smallness of these wonders. Compared to walking on water and rousing people from the grave, Uri's feats have a picayune, slapstick quality more in keeping with a clever charlatan than a messiah. A cartridge filler vanishes from Puharich's fountain pen. Three days later it appears in Geller's hand while he is watching a UFO in a field outside Tel Aviv. On Uri's twenty-fifth birthday, at Abraham's Oak at Mamre, Uri holds aloft Puharich's watch. Puharich asks Spectra for a sign. The watch hands are found to have moved ahead twenty-nine minutes. Eggs miraculously become hardboiled.

A camera case left by Puharich at his home in Ossining, New York, turns up on Uri's bed in Israel. A belt massager that Uri wants, to take off his excess weight, is delivered to him even though no one ordered it. Puharich believes his watch to be on a dresser. Uri screams. The watch is on Uri's wrist.

Uri materializes a crystal ball, levitates a camera, turns lead wire to gold. A conch shell on a shelf near Uri falls "slowly" to the floor. A cable car is stalled by Uri's power. In Munich he stops an escalator. Uri's Derringer pistol vanishes and is found back in its case. Puharich's tape recorder and a lady's scarf are translocated to the seat of a locked car. After a verbal fight with Uri—Uri is in a "shouting, towering, abusive rage"—Puharich finds that nine pens have arranged themselves on his desk top to spell WHY. Puharich calls Geller into his study to show him this poignant message from the Nine. "We both looked at each other with brotherly love and understanding, and we wept." Soon thereafter a heavy cabinet vanishes from Puharich's study and is discovered in a bedroom. The nine pens then arrange themselves cryptically into six upright lines and a triangle.

This is only a tiny sample of the miracles that accompany Geller everywhere. On stage he duplicates drawings in sealed envelopes, causes nails and silverware to bend, starts watches that have stopped, makes their hands change positions, guesses numbers, colors, and cities thought of by someone in the audience, and so on. His methods are familiar to magicians who specialize in a branch of magic known as "mentalism."

Off stage Uri's stock-in-trade is translocating objects. My favorite Geller miracle occurred in Ossining when Uri translocated Wellington, Puharich's black Labrador retriever. Puharich, Uri, Uri's Israeli friend Shipi Shtrang, and Puharich's assistant, Melanie Toyofuko, were having breakfast. When Puharich left to answer a kitchen phone, he noticed that Wellington was no longer in the doorway to the kitchen. A moment later Geller spotted the dog outside the house. "This demonstration of the power of Hoova made a lasting impression on all of us. We now fully realized that it was possible to translocate a living thing with safety."

One dares to wonder: While Puharich was in the kitchen, could Geller have snatched up Wellington and heaved him out the front door or through the window? A few days later Wellington, for no apparent reason, gave Uri a nasty bite on the wrist.

When the book ends, Uri's immediate future is far from clear. Hoova has ordered that a documentary film be made of Uri by Ms. Toyofuko. A *Knowledge Book* is soon to be revealed, first to Shipi, then to Uri, then to Puharich. Mass landings of UFOs will soon take place, although they may be invisible. A computer called Rhombus 4D has given Puharich permission to write his book. Two physicists at the Stanford Research Institute have investigated Uri and are much impressed. One of them,

Harold Puthoff, has been a strong supporter of Scientology, but can L. Ron Hubbard bend car keys? The soundtrack of SRI's film on Uri is given as an appendix in Puharich's book. Astronaut Edgar Mitchell has become one of Uri's most enthusiastic defenders.

Events since the book was written cloud the Advent of Uri even more. Early this year a British periodical, the *New Scientist,* arranged with Geller for a careful testing by a committee which included a professional magician. At the last hour, Geller refused to show. Ominous death threats, he told the press, made it necessary for him to go into hiding. Heinemann, an English publisher, has announced that Geller is writing an autobiography for them. Doubleday is angry, claiming that Geller is obliged to promote *their* book.

No one knows how Geller will play it. Will he repudiate Puharich, risking the disenchantment of his old buddy? Will he endorse Puharich, announce that Hoova has ordered him to stop the trivial trickery (after all, his powers have been validated by the prestigious SRI), take up psychic healing, and start a new religion? He is certainly better looking, better muscled, and cleverer than Guru Maharaj Ji. On the other hand, he has a monstrous ego, a low boiling point, and a consuming passion for fame and money that could propel him into disaster. If he were wiser he would pull a Marjoe and collaborate with Ms. Toyofuko on a documentary exposé of himself, but I wouldn't bet on it.

John Grant Fuller, who wrote *Arigo,* is a former *Saturday Review* columnist, author of two profitable books on flying saucers and the recently published *Fever!*. When he saw the color films Puharich arranged to have made of Arigo, he realized that here was something more mindblowing than extraterrestrial spacecraft, perhaps linked to them. "Was he [Arigo] able to tap a computer-like energy field beyond himself . . . ?" Fuller asks. We all know now, of course, where that energy field is located. Puharich's afterword, in Fuller's book, closes with a tribute to Geller and a plea to mankind to stop persecuting these two messengers "from the higher powers of the universe." Thus the two books interlock and reinforce one another.

José Pedro de Freitas, popularly known as Ze Arigo (meaning jovial country bumpkin), was a Brazilian peasant who lived in the village of Congonhas de Campso in the mountains north of Rio de Janeiro. For twenty years this man, whose education stopped at the third grade, practiced illegal medicine, treating an average of 300 patients a day, diagnosing, prescribing drugs, and performing major surgery. He was a legend throughout Brazil when in 1971 his blue Opala, which he was driving through a heavy rain, crashed into a truck. In *Uri,* Puharich quotes a communication from Spectra revealing that Arigo had experienced no pain. He left his body just before the crash. News of Arigo's death reached Puharich by a mysterious phone call that came *fifteen minutes before the accident.*

Arigo was a stocky, black-mustached man of boundless energy and gall. Although he and his wife, and their five children, were devout Catholics, Arigo was also a follower of Kardecism, an occult sect popular among the less ignorant classes of Brazil. Allen Kardec, the founder, was a mid-nineteenth-century Frenchman who combined spiritualism, or Spiritism as he preferred to call it, with reincarnation. The movement that bears his adopted name (his real name was Denizard Rivail) made little headway in England or the U.S., but has always been the dominant spiritualist sect of France.

Arigo's spirit control was Adolphe Fritz, a German doctor who died in 1918. Arigo insisted that when he worked on patients he was always in a trance state, unable to recall afterward what he did. While possessed by Dr. Fritz his voice had a gutteral German accent. Fritz always spoke through Arigo's left ear. At times Arigo would cup a hand around his ear, the better to hear Fritz's medical commands. Arigo did not understand German. Fortunately, Fritz spoke to him in Portuguese.

To service the hundreds of patients who flocked daily to his Spiritist clinic, Arigo worked at lightning speed diagnosing at a glance, scribbling prescriptions instantly, and performing operations in less than a minute. His instruments were unsterilized pen-knives, kitchen knives, and scissors. No anesthetics. No tying of blood vessels, no pain, little bleeding. When bleeding did occur, Arigo would stop it with a "sharp verbal command." No antiseptics. Once, Fuller tells us, when Arigo wiped a bloody knife on a patient's blouse, it miraculously *left no stain*. He accepted no money. There were persistent rumors of kickbacks from local druggists, but Fuller assures us they were never verified.

Arigo loved to demonstrate his powers by giving patients an "eye checkup" regardless of their complaint. This consisted of inserting a four-inch blade into the person's eye cavity and levering the eyeball until it protruded from the socket. On one occasion, he allowed Puharich to do this, guiding Puharich's hand. Puharich felt an inexplicable repellent force. Puharich believes that Arigo cut flesh with this force because he sometimes used the *dull* side of a knife or even just a finger.

Let's take a look at some of Arigo's major operations. Brazilian senator Lucio Bittencourt has lung cancer. He and Arigo are at the same hotel. After a night of heavy beering, the senator tries to sleep. The door opens and Arigo bursts in, eyes glazed, holding a razor. Bittencourt blacks out. Next morning he finds blood on the sheets and an incision in his back.[1] In Rio, an unnamed "doctor" says that the senator's lung tumor has been removed "by a very unusual surgical technique that did not seem to be known in Brazil."

A woman with cancer of the uterus has received last rites. Arigo gets a knife from the kitchen. "Pulling down the sheets, [Arigo] spread the woman's legs and plunged the knife directly into the vagina, probing violently. . . . Then he removed the blade, forced his hand into the opening. . . . In a matter of seconds he withdrew his hand, yanking out an enormous bloody uterine tumor. . . ."

Sonja has cancer of the liver. Arigo cuts her abdomen, inserts scissors, removes his hand. *The scissors move by themselves.* He reaches in, takes out the tumor. When he wipes the incision, its edges instantly adhere. Arigo puts a crucifix on the wound. Another woman has ovarian cancer. Arigo shoves three scissors and two knives up her vagina. Inside, the scissors become animated. Arigo stops the bleeding by saying, "Lord, let there be no more blood." He removes from the opening a piece of tissue thirty-one by fifteen inches. The operation is over in a few minutes.

Arigo also specialized in eye surgery. The clouded lenses of cataract patients were snipped out with nail scissors. One man, Fuller tells us, did not need glasses for his operated eye because the eye "accommodated" without the lens. Accommodation is what the lens does when the ciliary muscle alters the lens's convexity. To say that an eye accommodates without a lens is like saying that a hand from which fingers have been amputated has no trouble fingering a flute.

Arigo sometimes vomited between operations. On one occasion, after he had finished retching, Arigo explains that the man who had approached him had been possessed of evil spirits and he, Arigo, had taken the demons into himself. The man thanks Arigo profusely. Arigo, by the way, once believed himself to be demon-possessed and actually underwent a formal Catholic exorcism.

As a documented study, Fuller's book is worthless. In most cases he gives no names of doctors who diagnose, or names of doctors who check a patient afterward. We are merely told that "doctors found" or "X-rays showed." He favors the reader with a few snippets from Puharich's study of 1,000 Arigo patients, but after reading *Uri* who can take Puharich seriously? Although Arigo was violently opposed by the Catholic Church and by Brazilian doctors, not a paragraph in the book presents the views of any informed person who did not buy the Arigo mystique. Who in America will bother going to Brazil to interview doctors and check the court records of Arigo's two prison sentences? What publisher would publish such a book?

And this brings us to a serious moral question. Fuller's book is just persuasive enough to convince some fuzzy-minded readers, with curable ailments, to stop seeing their physicians and fly to Brazil or to the Philippines to be mangled by psychic quacks. The Kardecists in Brazil control several hospitals, Fuller tells us, where "doctors" operate on the "etheric body." They go through a pantomime of surgery, moving their hands several inches above the skin. "An enormous number of successful cases"—I am quoting Fuller—"have been verified by responsible doctors." Eli, Arigo's younger brother, has announced that the spirits of both Arigo and Dr. Fritz have visited him and urged him to carry on their work. More than one reader of Fuller's book may die a needless death because he read it.

Let me be clear. As a believer in democratic freedoms I oppose any federal or state legislation to prevent horrors like *Arigo* from being published. But Thomas Y. Crowell is an old and distinguished house. Unlike Doubleday, it has made its first venture into despicable medical journalism. The book may make lots of money, but one can feel only contempt for those at Crowell who are not ashamed of this shoddy book.

NOTE

1. In Chapter 3 Fuller tells us that Senator Bittencourt was suffering from cancer of the lung. Arigo operated on his back ("a clean, neat incision in the dorsal area of the rib cage"), removing a cancerous tumor from the lung. On the Long John Nebel radio show (October 24, 1974) I heard Fuller several times say that the scar was on the senator's chest. Anne Dooley, in the book cited in my postscript, tells us that the senator suffered from cancer of the colon. Arigo operated on his stomach, and left the blood-stained tumor on the bed. Dooley says her "researched account" is based on a lecture she heard in London and is "vouched for by Dr. Andrija Puharich, the well-known New York neurologist." Conflicts like this are typical of the kind of irresponsible, "gee whiz," anecdotal reporting of miracles in which Fuller and Dooley specialize.

Postscript

John Fuller, understandably furious over my review, sent the following letter, which the *NYR* published in their July 18, 1974, issue:

> I have heard somewhere in the past that the only answer to calumny is silence. But your review of my forthcoming book, *Arigo: Surgeon of the Rusty Knife* . . . by Martin Gardner has gone so far beyond calumny that he cannot remain unanswered.
>
> There is a world of difference between healthy criticism and vicious, scurrilous attacks. In effect, he "reviewed" the book before he even read it, by writing my publishers months ago a long condemnatory letter which paralleled his present review, simply because he heard the book was coming out.
>
> In my note at the beginning of *Arigo* I mention that the story is strange and incredible, but that there are undisputed facts that cannot be altered even by the most obdurate skeptic. Then the introductory note goes on to say:
>
>> It is an established *fact* that Ze Arigo, the peasant Brazilian surgeon-healer, could cut through the flesh and viscera with an unclean kitchen- or pocketknife

and there would be no pain, no hemostasis—the tying off of blood vessels—and no need for stitches. It is a *fact* that he could stop the flow of blood with a sharp verbal command. It is a *fact* that there would be no ensuing infection, even though no antisepsis was used.

It is a fact that he could write swiftly some of the most sophisticated prescriptions in modern pharmacology, yet he never went beyond third grade and never studied the subject. It is a fact that he could almost instantly make clear, accurate, and confirmable diagnoses or blood pressure readings with scarcely a glance at the patient.

It is a fact that both Brazilian and American doctors have verified Arigo's healings and have taken explicit color motion pictures of his work and operations. It is a fact that Arigo treated over three hundred patients a day for nearly two decades and never charged for his services.

It is a fact that among his patients were leading executives, statesmen, lawyers, scientists, doctors, aristocrats from many countries, as well as the poor and desolate. It is a fact Brazil's former President, Juscelino Kubitschek, the creator of the capital city of Brasilia and himself a physician, brought his daughter to Arigo for successful treatment. It is a fact that Arigo brought about medically confirmed cures in cases of cancer and other fatal diseases that had been given up as hopeless by leading doctors and hospitals in some of the most advanced countries in the Western world.

But none of these facts, all carefully brought together and examined, can add up to an explanation. And it is for this reason that this story is so difficult to write. . . .

It *was* difficult to write. But not half so difficult as encountering an hysterical diatribe (certainly not a critique) by a person who is so afraid to face facts and history that he descends to unprecedented levels of calumny.

This reviewer seems to want to rewrite Brazilian history, and to pretend that Arigo never existed. He ignores the conservatism with which the book is written, and the central thrust of the entire story: *There exist certain phenomena which have yet to be explained. Science is just beginning to explore these by-passed pockets. They should be cautiously explored under meticulous controls.*

This is exactly what the book says. Yet your reviewer utilizes two-and-a-half columns of your space to vilify documented records which are plain and simple facts. I went to great pains not to extrapolate.

There are two independent advance reviews from highly-respected publishing trade journals which seem to take a totally different point of view in reviewing the book:

1. *Publisher's Weekly*: "How Arigo did what he did remains a subject for wonder and conjecture. That he was a psychic healer of phenomenal powers the reader of this well-documented study will hardly doubt."

2. *The Kirkus Reviews*: ". . . his immaculately objective report will unnerve the most stalwart skeptic. . . . Fuller does not proselytize. . . ."

But it is not simply a divergence of opinion that surfaces here. Your reviewer has made statements which go so far beyond the bounds of decency that I am stunned that anyone could stoop to such a level.

I will quote one sentence from your reviewer which is so appalling that I cannot believe you permitted it to appear. It states: "More than one reader of Fuller's book may die a needless death because he read it."

When I saw that in print I can only say that I was in a state of shock. How could any person, even with the most distorted mind, make such a statement if he had read the book?

Your reviewer precedes this statement with the following sentence: "Fuller's book is just persuasive enough to convince some fuzzy-minded readers, with curable ailments, to stop seeing their physicians and fly to Brazil or to the Philippines to be mangled by psychic quacks. . . ."

In stating this, your reviewer has made it a point to ignore the following statements in my book:

1. Many reports have come from the Philippines about feats of surgery by untutored and untrained psychics there, but there has been a constant exposure of trickery in their work. Further, their lack of cooperation with medical researchers has made their case untenable. [Page viii]

2. There had been psychics in the Philippines who claimed to do surgery similar to Arigo's, but they had been easily exposed as fakes, and had refused direct observation in full day-light. [Page 249]

All through this review, your reviewer implies that the Arigo book is irresponsible and undocumented; that it gives a blanket support to unregulated practice of medicine; that it recommends that people stop seeing their own doctors in favor of such practices that Arigo engaged in.

In implying this, your reviewer is flatly libelous. The book *Arigo* is the record of a factual, historical phenomenon. It is written in a low key, and with little or no extrapolation. It states very simply: *Arigo should have had, before his death, a long, careful, objective medical study by highly qualified medical scientists.*

Your reviewer ignores statement after statement in the book which carefully qualify the material in it. It is important that some of these be listed here:

"I thought about the story for a long time. It was obvious that, in spite of the considerable amount of medical records available, I would have to go to Brazil and check the story in detail. Nothing could be secondhand in a story like this. It was a chronicle that would have to be verified in every aspect.

"But there was another problem. If the story *did* check out, and the book drew wide readership, what would it do to Arigo and his work? Would people who were desperately ill in the United States spend large sums of money to go to Brazil—only to find Arigo so flooded and exhausted with additional patients that he could not handle them all?" (Page 242).

It was for this reason that I tabled the idea of doing the book at all.

It was only after Arigo was killed that I felt the story should be told, so that there would be absolutely no chance of anyone forsaking conventional medicine. Further, Arigo worked *with* doctors, not against them. I simply would not have written the book if he were alive, and I said this repeatedly to editors during the two years I knew about the story and held back from writing it. This was another reason that the shock was so great when I read your reviewer's vicious and unprincipled comment about a reader suffering a "needless death" because of the book. Over and over again, it goes through my mind: How could a man write this about an author? What possible motive does he have? This isn't literary criticism, this is intolerable cruelty.

In combing the voluminous trial records, I found that they recorded time after time that there was no testimony that Arigo had harmed anyone in his quarter of a century practice. Furthermore, your reviewer ignores the fact that I have and quote the opinions of many doctors and scientists who observed and investigated Arigo.

Further statements in the book refute your reviewer's unprincipled accusations:

"Even those who were convinced of Arigo's validity felt he could not go on with his practice in an open, uncontrolled situation. Unscrupulous charlatans, inspired by Arigo, would proliferate throughout the country, with utterly disastrous results for the public" (page 122).

"Khater [a Brazilian newsman] was convinced that instead of being prosecuted, Arigo should be supported by funds for a special scientific study. In this way, Arigo would be placed under the control of licensed doctors — *a necessary step* [italics added] to prevent uncontrolled charlatanism from proliferating. . . ." (page 133).

"The only possible track to take would be to ask for a suspended sentence, with a court order placing Arigo in the custody of a group of competent medical doctors who would work with him in trying to unveil the mystery of his powers. . . ." (page 142).

"Plans were already taking shape under the aegis of several Brazilian doctors to build a hospital in Conghonas where permission would be obtained to have Arigo continue to work under the direct supervision of Brazilian medical men. If there had been some way to foster this type of project at the time of the first court process, perhaps some real clue to Arigo's rare effectiveness would have already been available. . . ." (page 209).

"Perhaps Dr. Oswaldo Conrado, the cardiology specialist from Sao Paulo, summed up the most interesting attitude from the point of view of the medical profession when he said: 'If doctors were able to open up new hope for patients, it would be a wonderful experience. When I find that I am directly confronted with a hopeless case, and when every possible medical avenue is closed, I see no reason not to look for other means. We wouldn't be human if we didn't.'

"The facts about Arigo exist. They have happened, simply and naturally. A commission of scientists, free from preconceived ideas *must* study him, and study him thoroughly. We might be on the edge of discovering entirely new and extremely beneficial therapeutic resources" (page 219).

Do these typical statements sound like a book which would cause a reader to "die a needless death because he read it"? Do they indicate that the book recommends patients give up their own doctors to go to Arigo, if he were alive? Do they indicate "despicable medical journalism," as your reviewer has written and would have his readers believe?

John G. Fuller

To the above I replied:

There is no way to reply adequately to Mr. Fuller short of writing a book on scientific method, the ethics of medical journalism, and how to distinguish anecdotes from facts. If Mr. Fuller can believe he is stating a "plain and simple fact" when he describes an operation during which Arigo slices open a woman, drops in a pair of scissors, then watches while the scissors, animated by a mysterious force unknown to science, move by themselves until a malignant tumor is cut out; if he can believe that his collection of miracle tales, unrelieved by a single note of humor or skepticism, is "well-documented" and "immaculately objective," then his mind-set is so different from mine (or from that of anyone I know) that communication is impossible.

It is true that no reader of Mr. Fuller's book can go to Brazil to be treated by Arigo in the flesh, because Arigo is dead. And it is true that Mr. Fuller has a low opinion of psychic surgeons in the Philippines. But Mr. Fuller speaks highly in his book of other Brazilian "doctors," carrying on Arigo's Spiritist surgery, and any reader impressed by Mr. Fuller's book will be just as impressed by the equally fulsome accounts, by equally respectable journalists, of the Philippine healers. (See Tom Valentine's *Psychic Surgery,* published by Regnery, or Harold Sherman's report on "Psychic Surgery in the Philippines" in Martin Ebon's anthology, *The Psychic Reader.*) Both Brazilian and Philippine psychic surgeons are among the many charlatans whom some poor reader of Mr. Fuller's book, suffering from an affliction, will try to locate.

They are not hard to find. Indeed, Arigo's rival, the Spiritist surgeon Lourival de Freitas, is regarded by many students of the occult as even more sensational a healer than Arigo. Anne Dooley, writing on "Psychic Surgery in Brazil" in *Psychic,* January, 1973, devotes most of her article to de Freitas. Like Arigo, he gives his patients "eye checkups" (extruding the eyeball with a knife). Like Arigo he operates under the direction of discarnate spirits. Like Arigo, he accepts no payments. Like Arigo, he is a heavy drinker. If Mr. Fuller believes he has written an objective account of Arigo, one would suppose that he would feel obligated to call attention to this healer's similar abilities. Yet nowhere in *Arigo* is Lourival de Freitas even mentioned. If Mr. Fuller considers this man a fraud, it would be interesting to know what criteria Mr. Fuller employs for distinguishing between the miracle tales that are told about both men.

Martin Gardner

Fuller correctly says that I wrote to T. Y. Crowell before his book was published. It was a letter to Robert Crowell, chairman of the company, whom I had met, explaining why I was canceling a contract. Crowell had just offered me a contract for a book, and I had only to sign it to obtain the advance. I was so appalled by *Arigo* that I returned the unsigned contract as a personal protest. Mr. Crowell limited his reply to two short sentences. He said he understood, and that he admired my principles.

Cynthia Vartan, Fuller's editor at Crowell, was not so polite. She fired off a vitriolic letter to *NYR,* which they did not print because Fuller had said the same things at greater length. Most of the editors at Crowell were amazed that anyone could suppose a moral issue was involved in publishing a book that was an insult to the medical community. However, at least one editor was on my side, and I later learned that the president of Crowell regretted they had accepted the book. *Reader's Digest,* by the way, condensed *Arigo* in its March 1975 issue.

Anne Dooley's fullest account of the psychic surgery of Arigo's rival, Lourival, is in her book *Every Wall a Door* (Dutton, 1974). Other chapters in the same book are devoted to Arigo and to the spirit hospitals of Brazil. Neither Dooley nor Fuller mention how Arigo made his money. Contrary to Fuller's picture of Arigo as a poor, devout Catholic who "never charged for his services," Arigo died a wealthy man and owned considerable property in the area. His brother ran the village pharmacy where Arigo's useless drug prescriptions, some thousand a week, were filled. The brother also owned the most expensive hotel in the village, where wealthy out-of-town patients stayed. Did Fuller know all this? If so, it was conscious deception to leave it out of his book. If he did not know it, it is evidence of the low quality of his "research."

Uri Geller's autobiography, *My Story,* ghosted by Fuller, was published by Praeger in 1975. Fuller's next book was *We Almost Lost Detroit.* The book got mixed reviews. For strong criticism see *We Did Not Almost Lose Detroit,* a booklet by Earl M. Page, published by Detroit Edison in 1976.

Fuller has since become an enthusiastic spiritualist. His *Ghost of Flight 401* (Putnam's, 1976) is about the haunting of several Eastern Airlines jets by the spirits of the pilot and flight engineer of Eastern Flight 401 after it crashed in 1972. In 1979 Putnam issued Fuller's book *The Airmen Who Would Not Die.* I have not read this great scientific treatise, but according to Putnam's full-page ad in the *New York Times Book Review* it deals with the 1930 crash of the British dirigible R101. "Drawing on fully documented evidence," says the ad, "Fuller reveals the amazing supernatural phenomena surrounding the disaster—from the pre-vision of famous medium Eileen Garrett, in which she predicted the calamity in full technical detail, to the séance held two days after the crash, when the R101's dead commander recreated the crew's final, agonizing moments. . . . It's a riveting, thoroughly factual report of human tragedy and superhuman prophecy. . . ." *Reader's Digest* condensed the book in June 1979.

In March 1979 Fuller and his new wife, Elizabeth, then a 33-year-old ex-airline stewardess who had helped Fuller research his last book, were in the Himalayas on an expedition. Benjamin Franklin, during one icy night, suddenly began to dictate to her a series of new proverbs. Some she scribbled down by "automatic writing" (as if a spirit were guiding her hand); others she recorded on tape. Now 124 of them are collected in a book called *Poor Elizabeth's Almanac.* In 1980 Elizabeth was touring the United States to promote the volume.

The aphorisms are clever, such as: "No one religion has a corner on God. He owns the whole block." And "Step down an inch to help a friend and raise yourself a foot." An Associated Press wire story of October 13, 1980, quoted Elizabeth as saying she had never made up a proverb until these new sayings of Ben began to pop into her brain.

I have not tried to keep pace with Puharich's latest psychic adventures. In my files is a report by Marc Seifer (from the *Journal of Occult Studies,* August 1977) of the Toward a Physics of Consciousness Symposium held at the Harvard Science Center, in Cambridge, Massachusetts, May 6–8, 1977. The symposium was coordinated by Ira Einhorn, who had lived at Puharich's house in Ossining, New York, during the period when Uri was there and getting his extraterrestrial messages from Hoova. In his acknowledgements at the front of *Uri,* Puharich thanks Einhorn, whose "imagination helped to formulate this book and to get it to the attention of publishers," although he adds that his "greatest debt is to Uri for giving me the privilege of being his scribe." Einhorn wrote the introduction to Puharich's *Beyond Telepathy* (Doubleday, 1962) and is the author of a book of his own titled *78-187880* (the title is its Library of Congress catalog number), published by Doubleday in 1972 in both hard and soft covers.

Ira opened the conference with a keynote speech in which he stressed the Western world's need for a spiritual rebirth based on Eastern religion and parapsychology. Two years later Einhorn was arrested in Philadelphia for the alleged murder of his girlfriend, whose body was found in a trunk in Einhorn's closet. (See the cover story by Albert Robbins, "Blinded by the Light," in the *Village Voice,* July 23, 1979. As I write, Einhorn has jumped bail and his present whereabouts are unknown.)

Puharich spoke at the conference about his recent investigations of "space kids" who are able to bend metal by PK. They have other powers. One of them, Puharich said, had materialized a tree. Six of them were teleported to his home (which, by the way, later burned down for undisclosed reasons) from as far away as Switzerland. Puharich said he believed the children were in psychic communication, like Geller, with higher-order civilizations, and he had "mapped out about 30 different civilizations, besides Hoova. . . ." He estimated the number of "Geller children" on earth to be 1.5 million. They all chose before birth to come here from other dimensions to fulfil special missions.

That ridiculous set of psychic entities, the Nine, are back on the psi scene, stronger than ever. A young medium from London, Jenny O'Connor, claims to be in touch with the Nine. They are eight-million-year-old chuckleheads from the star Sirius. In 1979 Jenny took up residence at Esalen, where she quickly became a celebrity. Esalen staff members take the Nine very Siriusly. According to Jeffrey Klein, in his article "Esalen Slides Off the Cliff" (*Mother Jones,* December 1979):

> Under the Nine's guidance, the institute's board of directors fired its chief financial officer and reorganized its entire management structure. The Nine have also acted as co-leader both in open meetings and in Gestalt workshops. They have even provided advice about Esalen's physical plant. . . .
>
> People at Esalen drop the Nine's name in conversation in the same intimate phony way one might mention a famous person who'd recently married into one's family. "Well, you know, the *Nine* say that. . . ."

Puharich and Fuller, like Esalen, are also sliding off the cliff, because they believe, they really do believe, the crap they write about. Not Uri. How he must roar with laughter over the fools who suppose, or once supposed, his claims to be genuine!

26

"The New Nonsense" and "Supersenses"

Our nation is now in the midst of unprecedented enthusiasm for beliefs that medieval astrologers would have considered insane. The enthusiasm is more frightening than funny. When I wrote *Fads and Fallacies,* in 1952, I likened this zeal to German crackpottery before Hitler. Charles Fair, neuroscientist and poet, finds even stronger similarities with pre-Napoleon France, when everybody was mesmerized by Mesmerism, and with Rome, swept by outlandish mystery cults before its fall. "It is hard to imagine a Napoleon or a Hitler arising in this country," Fair writes in *The New Nonsense* (Simon & Schuster, 1974), "but the psychological preconditions . . . clearly seem to exist. . . ."

Although Fair's book teems with good sense and sound scholarship, it will have no effect on the rising tide. Indeed, many will read it only to discover what new cults are "in." For every reader who agrees with Fair, two will be so captivated by his attacks on, say, Silva Mind Control that they will rush out to enroll in a Silva course.

Why is it happening? Surely Fair is right in focusing on two major causes: the decay of religious orthodoxy and the disenchantment with science. After decades of the demythologizing of Christianity, is it so surprising that the populace hankers for new myths? Objects for new faith are all over the landscape: Guru Maharaj Ji, Hare Krishna, transcendental meditation, parapsychology, encounter therapy, scream therapy, Rolfing, *I Ching*—Fair's list is endless. The cults flutter (in Fair's words)

Reprinted with permission from *Book World,* January 12, 1975.

289

like gaudy autumnal butterflies, "springing up as the old orthodoxies die but before those to come have yet taken shape."

In almost any month, a dozen hardcore examples of the new nonsense spew forth from otherwise reputable houses. A prize specimen is Charles Panati's *Supersenses* (Quadrangle, 1974), published, heaven help us, by the *New York Times*. Its theme: parapsychology is moving away from street superstition into rigidly controlled, reputable laboratories.

It is true that parapsychologists now play with electronic gear, Faraday cages, and computers, but their experimental designs are becoming shakier than a house of ESP cards. Panati, former physicist and now a science writer for *Newsweek*, has a timid way of dealing with this deterioration. He simply ignores it.

For example, he writes not a line about the sensational work of Dr. Jule Eisenbud with Ted Serios, the Chicago bellhop who could project thought pictures onto Polaroid film until two magicians explained how he did it. Panati has a low opinion of Eisenbud, so he just leaves him out. How will readers know that Panati's chief heroes—Thelma Moss, Stanley Krippner, Edgar Mitchell, and others—are firmly on record in support of Ted's powers?

Take the curious case of Dr. Walter J. Levy. On pages 220–221, 246–248, Panati describes the great work by Levy and his colleagues on the precognitive ability of jirds and baby chicks. All footnote references are to papers by Levy, but Levy is not mentioned in text or index. Why? Because just before Panati's book went to press, Levy, caught beefing up computer records, was fired by his boss, Dr. J. B. Rhine.

What Panati does include is wild enough. Several pages are devoted to Russian psychic Nelya Mikhailova's psychokinetic ability to separate yolk from albumen in raw eggs. You'll never learn from Panati that Nelya, under her former stage name of Ninel Kulagina, was caught (by USSR establishment scientists) using secret magnets and invisible thread to move and levitate objects, served a prison term for black-marketeering, and before that had been caught cheating in tests of her ability to read *Pravda* with her fingertips.

Uri Geller, Ninel's Western counterpart, is also in the book. Panati admits that Uri is a magician. Nevertheless Uri has a "real and sizable" psi gift. This has been proved, says Panati, by "rigorous tests" at Stanford Research Institute. (Panati calls tests "rigorous" whenever he describes reports that would flunk an undergraduate psychology major.) You won't learn, from Panati's laundered account, about Andrija Puharich's sad book on Geller, although poor Puharich appears elsewhere in *Supersenses* as a distinguished expert. You won't learn that Dr. Harold Puthoff, who tested Geller at SRI, wrote a preface to L. Ron Hubbard's *Scientology: A Religion,* and in 1970 described Scientology as "a highly sophisticated and highly technological system . . . characteristic of the best of modern corporate planning . . ."

Indeed, you won't learn anything from *Supersenses* that you wouldn't have known had you been a charter subscriber to *Psychic* magazine.

Postscript

Charles Panati is young, good-looking, likable, sincere, bright, and enormously naive. Before he left *Newsweek* to free-lance, he telephoned me to say he was about to meet Uri Geller for the first time. What advice could I give him? I urged him to take along a magician. He didn't, and a day or so later he got back to me to say how tremendously impressed he had been. Indeed, he was so impressed that he was soon editing *The Geller Papers* (Houghton Mifflin, 1976), a ridiculous anthology that I discuss in Chapter 15.

The Geller Papers was followed by Panati's psychic horror novel, *Links* (Houghton Mifflin, 1978). His latest book, *Breakthroughs* (Houghton Mifflin, 1980), is happily not about psi but about new developments in medicine. Panati is now considerably disenchanted about Uri, but like John Taylor (see Chapters 16 and 27) it will be a long time before he will be able to recover his reputation as a science writer after taking the "Geller effect" seriously.

27

"Superminds" and
"The Magic of Uri Geller"

You suspect someone of habitually cheating at cards. Whom would you hire as a secret observer to settle the matter? A physicist?

A self-proclaimed psychic goes about performing miracles exactly like the feats of magicians who specialize in what the trade calls "mentalism." You suspect the psychic of cheating. Whom do you call upon as an expert witness? A physicist?

One of the saddest, most persistent aspects of the history of alleged psychic phenomena is that there always has been a small, noisy group of scientists who, combining enormous egotism with even greater gullibility, actually imagine that *they* are competent to detect psychic fraud. Let's take a quick look at a prize specimen: Johann Zöllner, an Austrian professor of astrophysics.

In the 1870s Zöllner was bowled over by the miracles of an American medium, Henry Slade. Slade was a handsome scoundrel whose most unusual flimflam was causing knots to appear on closed loops of cord. He also was a virtuoso in producing, on blank slates, insipid chalked messages from discarnates in the "other world." He could wave his hand over a compass and make the needle gyrate. Small objects, sometimes water, had a way of falling out of the air near him.

Not once was Zöllner capable of seriously entertaining the hypothesis that so charming a gentleman as Slade could be a fraud. Indeed, Zöllner rushed into print an entire book about Slade, *Transcendental Physics,* in

---

Reprinted from the *New York Review of Books,* October 30, 1975, with permission from the New York Review of Books. © 1975 NYREV, Inc.

which he argued that physical space has four dimensions, and Slade had a supermind capable of moving test materials in and out of four-space.

The book is a classic case, written by an honest but stupid pedagogue. Magicians read it today with hilarity, because Slade's methods are well known and by reading between the lines they can reconstruct what Slade was doing. I myself, a lifelong student of conjuring, never expected to see a later volume that would demonstrate, with so much unconscious humor, how easily a scientist can be hornswoggled by the simplest of deceptions.

Until I read John Taylor's *Superminds* (Viking, 1979). This big, glossy book, plastered with sensational photographs of levitated tables and mediums, and forks twisted by the superminds of grinning children and pretty ladies, must be seen to be believed. That it should be written at all, by a man who now runs the risk of being remembered only as the British boob of the century, is more improbable than any "miracle" it describes. The book is subtitled, "A scientist looks at the paranormal." It should have been subtitled, "A scientist gapes at Uri Geller," because Geller, the young and handsome Israeli prestidigitator who insists he never presti-digitates, is both the book's immediate cause as well as its superstar.

It all began in November, 1973, when John Taylor, then forty-two, appeared on a BBC television show with Uri. Taylor, a respected mathe-matical physicist at Kings College, London, had expected to be unim-pressed. Instead, he became bug-eyed with astonishment. Uri performed his now familiar, increasingly tiresome, little bag of tricks. He made a fork bend. He started a stopped watch. He duplicated a drawing inside sealed envelopes.

Magicians watching the show were singularly unimpressed. But Dr. Taylor and his friends, who sometimes call themselves "counterculture physicists," suffer from a peculiar syndrome which I call PPE or "pre-mature psi ejaculation." Instead of waiting for a psychic to be tested by skeptical psychologists, aided by competent magicians, whenever a CCP sees a self-dubbed psychic do a few magic tricks he instantly pronounces the feats genuine and fires off a press release, article, or book. This is how Taylor describes his emotions after testing Uri in his laboratory:

> One clear observation of Geller in action had an overpowering effect on me. I felt as if the whole framework with which I viewed the world had sud-denly been destroyed. I seemed very naked and vulnerable, surrounded by a hostile, incomprehensible universe. It was many days before I was able to come to terms with this sensation.

After Taylor regained his composure, he discovered that he had become a true Gellerite. Not only does Uri possess a supermind (the best), but hundreds of British children (Taylor believes) also can produce the "Geller effect"—making objects bend and break by a power of the mind. Like his CCP friends, Taylor is less interested in proving the Geller effect exists (after all, has he not *personally* witnessed it?) than in measuring it and developing a theory. What force of nature is responsible? Gravity? The weak force? Neutrinos? Conjectured particles such as tachyons, intermediate bosons, magnetic monopoles, quarks? He finally settles on electromagnetism as the most likely candidate.

Taylor's ignorance of conjuring is almost total. Describing Uri's reproduction of a drawing inside envelopes, he writes, "No methods known to science can explain his revelation of that drawing. . . ." Well, what about methods *not* known to science? I can assure Professor Taylor that there are more than thirty distinct techniques by which a mentalist can accomplish just such a feat, the methods varying with conditions under which the performer is restrained.[1]

After watching Geller make a compass needle rotate by waving his hand above it, Taylor attempted to imitate Uri's movements. He even tried stamping on the ground. Nothing happened. "Nor could Geller have been using a magnet," he writes with incredible presumption, "unless he could palm it with consummate skill. . . ."

It never occurred to Taylor—why should it?—that a magnet need not be palmed to make a compass needle move. When Slade performed this old trick he had the magnet in the tip of a shoe. He crossed his legs, and when he wanted the needle to turn he simply raised the tip of his shoe to the table's underside.[2] Another handy spot for a magnet is under the trousers on the knee. Today one can obtain powerful, flexible little magnets that are wafer-thin and easily concealed in the mouth between lower teeth and cheek. When Uri moved the compass needle on Tom Snyder's "Tomorrow" show (August 14, 1975), he had the magnet either in his mouth or sewn in his shirt collar. This was evident from the fact that, each time the needle deflected, Uri's head darted close to the compass. By vigorously milking his left fist over the compass, Uri misdirected attention from his head. Then he put the compass on the floor, the camera glided in for a close-up, and no viewer could see *where* Uri's head was.

But I'm already in trouble with some of my magic friends for revealing too much. On page 52 Taylor describes a closely related miracle. Geller produced bursts of noise in a Geiger counter. Did Taylor or any of the other brilliant CCPs present think of inspecting Geller afterward to see if he had a bit of harmless radioactive substance concealed on his person? It never occurred to them!

Magicians are understandably reluctant to expose Geller's methods. There is even a small group of mentalists who regard Uri as "one of the boys," no different from Kreskin and other pros, except that by acting the role of a psychic more convincingly, and introducing innovations such as bending car keys and silverware, Uri is making more money than they are.

I disagree. There is, in my opinion, a qualitative difference between Uri's career and the careers of the great mentalists of the past. Uri is not in the tradition of such entertainers as Anna Eva Fay and Joseph Dunninger. Uri is in the tradition of the great physical mediums of the nineteenth century. True, the spirits of the departed have departed from most of today's psi laboratories, but in their places are those mysterious energies "unknown to science," possibly emanating from superminds in outer space. There is the same damage to science and to individual lives.

Consider some of the effects of Uri's ruthless pursuit of money and adulation. Andrija Puharich, at one time respected by his colleagues in parapsychology, has had his career reduced to shambles. Prominent journalists and scientists, and one brave astronaut who walked on the moon, have been made to appear almost feeble-minded. Thousands of others, especially among the young, have been so dazzled by Uri that they believe him to be in the vanguard of a new revolution in human consciousness.

For these and other reasons I cannot agree with those magicians who snipe at James Randi, professional conjuror, for his just published paperback, *The Magic of Uri Geller* (Ballantine, 1975). It is an excellent book, sensible, well-informed, witty, compassionate, and utterly devastating. Uri has often hinted that anyone who attacks him is in danger of being whammied by his great powers. Indeed, in Puharich's addlepated book, *Uri,* Geller professes to be dismayed over an occasion on which he became angry at someone, wished him harm, and the poor man obediently died. But Randi doesn't scare easy.

The Magic of Uri Geller is a collection of eye-opening articles by skeptics, mixed with amusing and acid commentary by the Amazing Randi. Joe Hanlon, who wrote the *New Scientist's* excellent special issue on Geller (October 17, 1974), gives his reasons for believing Uri to be a mountebank. Andrew Weil's two-part article in *Psychology Today* is here: Part 1 telling how Uri bamboozled him, Part 2 recording his disenchantment after seeing Randi bend keys faster and under better controls. The inside details of Uri's failure to impress Leon Jaroff and other editors of *Time* is faithfully told in contrast to Puharich's muddied account.

Did Uri once teleport himself from a sofa in Puharich's home in Ossining to the streets of Rio, then return with a thousand cruzeiro note? Uri says yes, but read Randi's chapter 8! Yale Joel, a *Life* photographer, discloses how Uri was caught faking a photo supposedly taken of himself without removing the camera's lens cap. How does Uri repair "broken" watches? See chapter 12. Why did Uri bomb on the Johnny Carson show? Was it because Uri was nervous, or because Johnny, an ex-magician, never allowed Uri or his friends access to the test materials before the show began? How did Uri manage to "melt" a spoon so convincingly for Barbara Walters that for a year her outlook on life was changed?

One of Uri's most sensational achievements in clairvoyance, when he was tested at the Stanford Research Institute, was guessing eight times in a row the top face of a die that had been shaken in a metal file box by "one of the experimenters." Is there a simple way Uri could have cheated? There is indeed. Joaquin Argamasilla, a young Spanish magician, convinced many CCPs that his "X-ray eyes" could see through steel and silver boxes. Houdini exposed the Spaniard's technique in a rare pamphlet that Randi reprints; then Randi makes a shrewd guess about how Uri could have used the same basic method.

Perhaps the most damaging chapter in Randi's explosive book is the translation of an article about Uri that appeared last year in a Tel Aviv weekly paper. Itzhaak Saban, a former friend of Uri who served as his chaffeur in Israel, told the reporter how he once "stooged" for Uri by secretly signaling information from a front seat during Uri's stage performances. Hannah Shtrang, older sister of Uri's inseparable companion, Shipi, tells how Uri and Shipi first became interested in conjuring, how they developed Uri's stage show, and how Shipi became Uri's number one confederate. In those days Uri introduced Shipi as his brother. He would refuse to perform unless "little brother" had a front row seat. Hannah herself sometimes took over signaling chores.

"Uri and Shipi used to train for long hours together," said Hannah, "even after he was already famous, drilling together on the drawing and reproduction of certain objects after they cast only a quick glance at them." Saban showed the reporter how Uri uses sleight of hand to change the hands on a watch, apparently "without touching it in any way." When anyone mentions the scientists who have certified Geller, the reporter writes, Saban "reacts with a wide know-it-all grin."

The most hilarious sections in Randi's book are those in which he tells of his adventures in England early this year while gathering material for his book. Disguised as James Zwinge (his real name), a supermind from Canada, Randi visited the offices of *Psychic News,* London's leading psychic periodical. The office was soon in an uproar as Zwinge leaped about "Gellerizing" object after object. The result: a picture of Zwinge on the front page of the July 26 issue, and a big story about the awesome powers of this strange man "with a gray beard and intense eyes" who "seemed to radiate a magnetic aura."

A spoon bent and broke. A cabinet key twisted. A paper knife bent 45 degrees. "All could vouch Zwinge had not been near it," said *Psychic News.* A clock on the wall suddenly gained two hours. Another office clock gained two and a half hours. "Certainly Zwinge had no opportunity to interfere with them. He had been under constant surveillance. . . . I was fully alerted to any suspicious moves. But Zwinge made none."

Randi telephoned Taylor. The professor cut him off with "I have all the evidence I need" and hung up. But Taylor didn't know what Randi looked like, and Zwinge's face had not yet graced the London news-stands. Posing as a reporter from *Time,* Randi had no difficulty getting the interview he wanted. What happened in Taylor's office is one of the book's funniest highlights.

Take a look at the photograph on page 159 of *Superminds.* It shows Taylor's holiest relic—a plastic tube, corked and sealed at both ends. Inside is an originally straight strip of aluminum, now bent into an S shape by the supermind of a teen-age boy. Did Taylor actually *see* it bend? Well, no. There is, you see, what the professor calls (so help me) the "shyness effect." Things bend only when one is not looking at them. Superboy took the tube home, came back with the strip bent. Now, students, let's turn to Randi's account of what happened when he surreptitiously examined this tube. While pretending to admire the sacred relic, Randi tugged on one of the rubber corks. It came right out in his hand! Taylor didn't notice, so Randi quickly jammed the cork back in. "The screw-and-sealing wax precaution hadn't mattered a bit," comments Randi. "It was a very poor piece of preparation. . . ."

Since Randi wrote his book, two researchers at Bath University had no difficulty designing a simple test for six young superminds. The observer was instructed to relax vigilance after twenty minutes. Rods and spoons Gellerized beautifully while the unsuspecting children were secretly videotaped through one-way mirrors. "*A* put the rod under her feet to bend it; *B, E* and *F* used two hands to bend a spoon. . . . We can assert that in no case did we observe a rod or spoon bent other than by palpably normal means" (*Nature,* September 4, 1975, page 8).[3]

Taylor is only the latest of the many casualties that psychic hustlers leave in their luminous wakes. Will Randi's book result in fewer future casualties? I doubt it. The mind-sets of true believers go so deep that, even if Uri were to confess all, they wouldn't believe him. Remember Margaret Fox, who started modern Spiritualism by cracking her toes? (Her facial features, by the way, resembled Uri's so closely as to suggest a reincarnation.) None of Margaret's supporters believed *her* when she confessed.

Uri once said to Stefan Kanfer, of *Time*: "Randi is jealous of me because I'm young and good-looking, and have nice wavy hair."

"Well," Randi concludes his book, "I'm no longer as young as I'd prefer to be, and most of the hair has departed during the years, that's true. But I sleep well, Uri."

NOTES

1. For details about some of Uri's methods see *Confessions of a Psychic* and *Further Confessions of a Psychic,* two anonymous booklets that purport to be written by Uri's chief

rival, Uriah Fuller. I assume that "Uriah" is from Uriah Heep, the hypocrite in *David Copperfield*, but "Fuller" could be any of many different Fullers on today's psi scene: John G. Fuller, the psychic journalist; Curtis Fuller, publisher of *Fate*; and Brother Willard Fuller, an evangelist in Jacksonville, Florida, who specializes in using prayer to miraculously put gold, silver, and porcelain fillings in teeth.

For details about Brother Fuller see James Cranshaw's two-part article, "Reverend Fuller's Ministry of Dental Healing," in *Fate,* March and April 1975. Brother Fuller also uses divine power to replace missing teeth, fix malocclusions and heal gum diseases. In *Fate*, March 1975, Brother Fuller's pretty wife, Amelia, advertised for $2 a book by Daniel Fry entitled *Can God Fill Teeth? The Real Facts About the Miracle Ministry of Willard Fuller* (CSA Press, Lakemont, Georgia). I doubt if the author of the Uriah Fuller booklets had R. Buckminster Fuller in mind, even though in recent years he has been giving lectures in praise of est. (On est, see Chapter 28.)

Both booklets are intended only for sale to magicians, but I am told that interested readers can obtain them from the publisher, Karl Fulves, Box 433, Teaneck, New Jersey 07666.

2. Slade was actually caught using this method (well known to magicians of his day) by an astute investigator who observed Slade's foot in a mirror he had secretly placed on his lap. See the *Proceedings of the American Society for Psychical Research,* vol. 15 (1921), page 556. CCPs seldom adopt such sneaky stratagems because: (1) they don't know enough about magic to think of them; (2) if they did, they would hesitate to set a trap because it would indicate a lack of trust and that could disturb the delicate operation of psi.

3. Before this experiment was done, one of the researchers, Harry Collins, actually believed the children could bend metal by PK. He may even still think the Geller effect is genuine. In any case, he was surprised and disappointed to discover that the children cheated.

Postscript

Since I reviewed *Superminds,* Professor John Taylor has turned one of the fastest and funniest flipflops in the history of psi research. Not only does he now consider Uri a fake, but he has decided that ESP and PK do not even exist! See Chapter 16 for my appraisal of the book in which Taylor makes these retractions, though with a notable absence of apologies or credits to those skeptics who did their best to keep him from making an ass of himself.

James Randi's new book, *Flim-Flam!* (Lippincott & Crowell, 1980) goes into more details about Uri's career and the careers of other fraudulent or self-deceived psychics.

28

"Powers of Mind"

Today's pop counterculture, especially among the young, is an awesome mix of maximum mindlessness, minimum historical awareness, and a pathetic yearning for (to quote Chico Marx) strawberry shortcut. To hell with established religions, with science, with philosophy, with economics and politics, with the liberal arts—with anything that demands time and effort. Dig the rock beat, kink up your sex life, meditate, tack a photo of Squeaky Fromme on the wall.

George Jerome Waldo Goodman, alias Adam Smith, has a nimble mind, a quick wit, and a sharp nose for the latest fashionable crap. His earlier best-seller, *The Money Game,* was snapped up by middle-classers eager to make a fast buck on the rising stock market. His new best seller, *Powers of Mind* (Random House, 1975), will be snapped up by middle-classers eager to find instant health and happiness.

Of course it's not called happiness. You raise your consciousness, expand your inner space, increase your aliveness. To give fake credibility to his short-cut tour of what he calls the "consciousness circuit," Smith practices the old technique of first making a quick tour himself. George Plimpton at least spent considerable time making friends with top athletes and playing their game before he wrote a book about it, but Smith is in more of a hurry. A day here, a day there, skim the references, do what you can by phone, punch up the stories, make up some new ones. Did a

This review, which appeared in the *New York Review of Books,* December 11, 1975, and the letters quoted in the Postscript are reprinted with permission from the New York Review of Books. © 1975–1976 NYREV, Inc.

"Crazy Indian" give Smith some flowers at Pennsylvania Station, and did Smith then walk through the coach, handing everybody a flower and saying "Namaste," which the Indian told him meant "I salute the light within you"? Forgive me, Smith, but I doubt that.

Smith's machine-gun style is exactly right for the short attention span of his readers. *Black Mask* pulp. Dizzy-paced, hard-boiled, wise-cracky. Lots of one-word sentences: "Wow!," "Wiggy," "Yep." Names of "in" thinkers pepper the pages: Wittgenstein, Heidegger, Jung, Gurdjieff, Huxley (Aldous, of course), Chomsky, Thomas Kuhn, Robert Ornstein, Teilhard de Chardin. . . . I'd forgotten about Count Korzybski, but Smith must admire him because he's there too. When I finished the book I wondered why Karl Popper's name hadn't popped up, but no—at the back of the book, footnote 21 (on Kuhn's "paradigm") ends with, "No thumbnail reference to the history of science should leave out Karl Popper and Michael Polanyi, especially the latter."

Smith's anecdotal wonders chase one another like one-liners in a Henny Youngman routine. "A funny thing happened to me on my way to the Meditation Center. . . ." For openers, Stewart Alsop has an inexplicable remission of cancer after a strange dream about refusing to get off a train at Baltimore. Norman Cousins cures himself of a mystery malady by watching old Marx Brothers movies. A ten-year-old black girl's palms start to bleed after she reads about the Crucifixion. A man on LSD, who thinks he's arguing with Socrates, speaks *"in classic Greek, which he did not understand!"* (Italics and exclamation in the original.)

Doctors give Smith little lectures on placebos, on drugs, on the Rumpelstiltskin effect (naming an ailment makes a patient get better), on split-brain research. After instruction by an *I Ching* master, Smith asks the book's advice on stock investments. He visits Esalen. He studies Arica. His body is pummeled in a Rolfing session. Later, Ida Rolf herself tells him she can't stand osteopaths and chiropractors and that in the entire land there are only two competent Rolfers: Ida Rolf and her son. He takes a biofeedback course. He tries Yoga. He does the sliced-ping-pong-balls-over-the-eyes bit with Montague Ullman at the Maimonides Dream Lab. He floats in John Lilly's sensory deprivation tank. He half-practices TM and discloses (shame!) his secret mantra.

Several chapters cover Zen sports: Zen football, Zen golf, Zen tennis (no Zen bowling?). The Guru Maharaj Ji and the Reverend Sun Moon are passed over lightly because Smith failed to contact them, but he did meet Uri Geller, and he *thinks* he met the elusive Carlos Castaneda. He's not sure because the man said he was Carlos's double. This doesn't mean, you understand, that the man merely *resembled* Carlos. The "double," to quote from a recent Doubleday book on OOBE (out of body experiences), "is the solid second body—a living, breathing organism identical in appearance and behavior to the physical body." You can photograph it.

Baba Ram Dass gets a big play. Ram Dass is Richard Alpert, Tim Leary's old sidekick at Harvard before the two were sidekicked off the Yard. Alpert went to India, came back a guru. He is now much admired on the college consciousness circuit, even though his father (president of the New Haven Railroad) calls him Rum Dum and his older brother calls him Rammed Ass.

Ingo Swann, the New York Scientologist who, like Geller, has been pronounced a genuine psychic by Puthoff and Targ (the two laser physicists who do psychic research for the Stanford Research Institute), tells Smith about how he projected his consciousness to Jupiter and Mercury. When Smith phoned a NASA official to ask if they knew about Ingo's Mercury trip, the reply was: "No, I didn't know, I don't want to know, and please don't tell me. We didn't sponsor that. All our probes use regular old rockets."

Ingo assured Smith that 95 percent of all so-called psychics are frauds. That does not apply, of course, to him or to Harold Sherman, the Arkansas seer whose consciousness accompanied Ingo's on both space probes. You won't hear much about their Jupiter probe because the only things they saw that weren't in every elementary astronomy text were enormous mountains on the planet's surface. That was just before a Pioneer flyby revealed that Jupiter doesn't have a surface. Swann has recently been hired to dowse for oil. He doesn't use a forked stick—justs walks around and feels the vibes.

The books' saddEST chapter, "The High Value of Nothing," is about the latEST and hottEST of the new strawberry short cuts. EST is not T. S. Eliot's initials backward. It is the acronym for Erhard Seminars Training. Erhard? In the beginning he was Jack Rosenberg who grew up near Philadelphia and used to be a manager for *Parents' Magazine's* door-to-door encyclopedia salesmen. For a while he was into Scientology, but this distinguished church expelled him and he went to work for Mind Dynamics, a California outfit now OOB (out of business). Then he thought of EST.

The name is interESTing. It looks like ESP, and it rhymes with REST and ZEST. Above all it sounds like MEST, Scientology's great acronym for Matter, Energy, Space, and Time. According to L. Ron Hubbard, the original Thetans, omnipotent and immortal, became bored with eternity. To amuse themselves they began to create universes out of MEST. Slowly, over trillions of years, they became enmeshed in one of their worlds. And who are these fallen gods who have forgotten who they are? They are ourselves!

"But something has happened," writes Christopher Evans in his eye-opening account of Scientology in *Cults of Unreason.* "One man, Lafayette Ronald Hubbard, has stumbled on the secret, has remembered what it's all about and will lead us back until we cease to be pawns and return to our heritage as players."

Scientology is complicated. You have to read books to grasp the grand design. Erhard has a shorter short cut. EST is Latin for "is." What is, is. What isn't, isn't. The universe is what it is. It can't be anything else. It's perfect. You are one of its machines. You are what you are. You, too, are perfect. You have "free will" but in a paradoxical sense. You have to choose what you choose. The secret of satori is to relax and enjoy. "The whole idea of making it," Erhard told Smith, "is bullshit." In fact, everything is bullshit, including EST. Once you recognize this great truth, and that there is nothing to get, you "get it." You lose, of course, your $250 initiation fee. That's what EST gets.

The notion that peace comes with acceptance is one of the oldest ideas in religion and philosophy. A thousand eminent atheists, pantheists, and theists have said it much better than Erhard. Spinoza, for instance, wrote eloquently about how true freedom comes only to the person who knows he has no freedom. "In His will is our peace," wrote Dante.

Yet there are thousands of poor souls, eager to be privy to the latest shortcut, who are paying money to be told this. And they are told it in ways cribbed from a dozen other cults and carefully calculated to produce maximum shock, schlock, and publicity. Once you've paid the fee you are locked into a room with all the other shortcutters. You can't smoke, eat, or go to the washroom.

You're a tube. Food and liquids go in one end and out the other. "We make them look at their tube-ness," says Erhard to Smith. "At least if you don't let them pee, you begin to get their attention."

You are lower than a tube. You're a mechanical asshole. A lady with gray hair raises her hand to say she thinks the instructor could make his point with less vulgarity. Smith is scribbling notes.

"Fern, honey," says the instructor, "these are only *words*. . . . Why do you grant the words the power to make you an effect, Fern, there is no difference between fuck and spaghetti!"

Fern is stunned.

"And by the time this course is through," the instructor goes on, "you will be able to go to Mamma Leone's and order a plate of fuck! Or sing this whole bunch a dirty song!"

Next morning Fern stands up and sings the only dirty song she knows. Applause. Fern giggles. She's beginning to get it.

"I guess EST will keep growing because the demand is more than the supply," Erhard tells Smith. "But don't get me wrong. I don't think the world needs EST, I don't think the world needs anything, the world already is, and that's perfect."

"If nobody needs it, why do you do it?"

"I do it because I do it, because that's what I do."

And because, like Hubbard before him, Erhard (note the names' last syllables) is getting rich. And why not? He's packaging instant insight.

No fuss. No exercises. No need to seek wisdom from the past, not even from Hubbard. No written instructions. Just pay your fee, "get it," tell your friends. There are, naturally, advanced seminars. They cost more. Last month I read that Erhard had given a large sum of money to a group of California counterculture physicists who want to investigate the natural laws behind such things as Uri's spoon bending. Look for EST to get more and more into PK. It's good PR.

How shall I sum up my reaction to this flashy, unprofound book? It's impossible to guess what Smith's own views are. The over-all impression he leaves is that strange things are happening outside the range of established science. He has nothing serious or interesting to say about any of them. Some people may want to read the book to learn about the traps their relatives and friends may drop into. But they might as well wait for the paperback.

Postscript

Adam Smith did not, of course, care for my review. He sent me an angry personal letter and wrote a much milder one for the *NYR* which they printed in their January 22, 1976, issue:

> Martin Gardner implies, in his review of *Powers of Mind* that I wrote a quick book on a fashionable subject. When I began working in the area, the subject was utterly unfashionable. *Powers of Mind* took more than three years of research, some of it in psychology and physiology, which Mr. Gardner did not mention. The three years were necessary to cover many courses, more than three-hundred interviews, and upwards of a thousand references. In the interim, wavelets of books appeared on transcendental meditation, the "relaxation response," the application of Zen and martial-arts techniques to sports, and so on—each one chapter, or part of a chapter, in *Powers of Mind*. Had the object been to be quick and fashionable, there was certainly material enough for half a dozen books, each not much different from some current best-sellers. I do not object to the selling of books by publishers—that is supposed to happen—but obviously this was not my only intent.
>
> Adam Smith

I replied as follows:

> When Adam Smith says the subject matter of his book was "utterly unfashionable" three years before he finished it, I can only wonder where he was living. The consciousness-raising trend (roughly paralleling the occult and the back-to-fundamentalism revolutions) got underway in the mid-sixties

and was going full blast when Smith started his research. Consider one tiny datum: a special issue of *Cosmopolitan* devoted to this country's growing preoccupation with probing the unknown. One article is called "Drugs and the Mind's Hidden Powers." The date? January 1960. Zen and yoga had, of course, become fashionable long before that.

Almost every cult in Smith's book (except EST) was the talk of cocktail parties in Manhattan in the late sixties, especially in theatrical, art, and literary circles. They just hadn't yet spread to spots like Houston and Omaha. As for the "quickness" of Smith's research, I will mention only that his lengthy bibliography contains no books or articles critical of any of the movements or scientific claims about which he writes.

Martin Gardner

Since my review, so many books and articles have been written attacking and defending est (now usually written all in lower case) that I forgo even attempting to give a bibliography. The best book defending Erhard is, I regret to say, written by William Warren Bartley III, a philosopher who should know better. He maintains that est cured his insomnia. I know Bartley as a fellow Carrollian and author of *Lewis Carroll's Symbolic Logic,* published by the same house (Clarkson Potter) that later published his *Werner Erhard—The Transformation of a Man: The Founding of* est (1978). I once reminded Bartley that Bishop Berkeley had written a book praising the great medicinal virtues of tar water, but he could see no resemblance between that book and his tribute to Erhard and est.

For readers who have spent good money on est, or TM, or any other of the new cults that are tinged with Eastern mysticism, and have failed to find peace and happiness, I recommend the following exercise. For twenty minutes every morning and evening, assume the lotus position, close your eyes, let an image of your favorite guru float into your consciousness, and chant over and over again that ancient Hindu mantra: *Owah—Tanah—Siam.* Come to think of it, Erhard might even approve.

29

"The Gift of Inner Healing"

A year ago, at one of Ruth Carter Stapleton's evangelistic meetings, not many in the audience would have known she was Jimmy Carter's sister. Now everybody knows that it was Mrs. Stapleton who opened her brother's eyes to his need for rebirth.

It was 1967. Depressed over his failure to defeat Lester Maddox, Jimmy Carter was walking with Ruth in the pine woods near his farm. Would he, Ruth asked, give up anything for Christ, including politics? No, said Jimmy. Then he buried his face in his hands and wept. The episode led to his second birth — that peak experience which, for all evangelicals, means a recognition that one is a miserable sinner saved by the grace of Christ's atoning death.

Like all charismatic faith-healers, Mrs. Stapleton does not like to be called a faith-healer. Only Jesus, through the Holy Spirit, heals. Her first book, *The Gift of Inner Healing* (Word Books, 1976), is a gracefully written defense of her unusual technique for helping Jesus bring peace to Christians who suffer mental anguish.

The book opens with personal testimony. When Ruth married at 19 she was, she says, totally unprepared for life. Her father's love and protectiveness had not been "altogether healthy." As his favorite she had been taught to believe she was the most beautiful, talented, gifted person ever born. Because her mother treated all her children alike, Ruth felt rejected by her.

From the *New York Times Book Review*, August 22, 1976. © 1976 by the New York Times Company. Reprinted with permission.

The crisis came after she had her first child, followed in rapid succession by three others. A car accident plunged her into black despair. When an unnamed friend prayed for her at the hospital, repressed anger against her parents surfaced and she began to stop blaming them. Her cure was completed at a "Christ-centered camp" where she received a Pentecostal baptism of the Spirit. Later she spoke the Unknown Tongue.

After her rebirth she found she had the ability to help others in severe mental distress. Strongly influenced by Dr. W. Hugh Missildine's pop therapy book, *Your Inner Child of the Past,* she became convinced that most mental ills are caused by painful repressed memories of the unconscious mind's "inner child." To reawaken and heal these crippling memories she uses a technique she calls "faith imagination."

It begins with the patient closing his or her eyes. Then in a soft hypnotic voice Mrs. Stapleton takes the patient back in dreamy imagination to childbirth. She pictures the person's life as a narrow staircase lit by the light of Jesus. The patient is asked to imagine the Saviour carrying him or her up this stairway. Each step is a year. When they reach the year that a traumatic event took place, Jesus forgives the offending parties. The patient in turn is asked to love and forgive.

Consider Mrs. Z. She had been in and out of mental institutions for 30 years. When Ruth brought her to step 12 she began to scream. She had just recalled being raped by her father. ("She had never told any therapist.") Mrs. Stapleton describes Jesus as resting his hand on the shoulder of Mrs. Z's poor father. "Jesus is forgiving your father. Do you?" Mrs. Z nods and whispers, "Oh, Daddy, I forgive you too."

And that's all! The cure is complete. Mrs. Z's husband can't believe his wife's story. He wants to take her back to her psychiatrists, but she refuses.

Jody, a handsome young homosexual, requires many sessions of faith imagination to start him on the right road. Aided by Ruth's soothing monotone, he visualizes himself as a child and Jesus supplying the masculine experiences he had missed. Jesus plays baseball with him. Jesus takes him fishing. "Naturally, he [Jody] caught the biggest fish, and the most, which pleased his fishing partner as it would any father." (Mrs. Stapleton takes for granted that homosexuality is always an illness. "Show me a happy homosexual," she quotes someone, "and I'll show you a gay corpse.")

She also takes for granted that persons can be possessed by fallen angels, but she argues that symptoms of genuine demon possession are usually easy to distinguish from those of mental illness. There is no antagonism between the two kinds of healing. Did not Jesus practice both?

Mrs. Stapleton's mental healing is often accompanied by the cure of psychosomatic ills, and sometimes ills she believes are not psychosomatic. Elsewhere she has spoken of seeing a person born blind being healed. On a TV interview last May she told of a boy born with no inner organs of hearing. He began to hear after she had prayed with him.

Her book is one of the saddest I have ever read. Sad because the author is a woman of striking beauty and intelligence, vivacious charm, and deep spiritual commitment. None of the clouds of flimflammery that hung over Aimee Semple McPherson and Kathryn Kuhlman, and the early Oral Roberts, darken her brow. Then why sad? Because Mrs. Stapleton, untrained in psychiatry and with minimal knowledge of the field, is practicing a psychotherapy of unbelievable naïveté.

She sees vividly how she has helped those who can be helped, but a curious spiritual fog blinds her to those twin evils that dog the steps of all Pentecostal healers. Always there are those trusting souls who, in temporary ecstasy, avoid medical help until too late. And always there are those who, when the help from on high does not come, or does not last, believe it is because they lack sufficient faith. It is a conviction that can only deepen despair.

Nowhere does Mrs. Stapleton betray the slightest interest in confirming what patients tell her. How does she know that Mrs. Z was actually raped by her father? (As Freud discovered, such fantasies are common among the mentally ill.) How scrupulously does she check, years later, on the completeness of earlier cures? Her patients often believe that Jesus is actually coming to them in their daydreams. Is Mrs. Stapleton never troubled by the thought that Jesus may resent being used in this fashion?

Postscript

The *New York Times* inadvertently lost two lines of type in my review, which are restored above, and advertently chopped off my last paragraph, which read as follows:

> Surely God must want Mrs. Stapleton to obtain all the medical knowledge she can. As her fame grows, the number of disturbed persons seeking her help will increase, and her responsibility will be awesome. The temptation to stop learning, to bask in the adulation, to expand her program of Behold, Inc. — in short, the temptation to become another Kuhlman — will not be easy to resist.

Since my review appeared, Ruth Carter has written two more books. *The Experience of Inner Healing* was published in 1977 by the publisher of her first book, and the next year Harper and Row brought out *Brother Billy*. Her fame as a charismatic healer continues to grow. Numerous magazines have done cover stories about her, of which the most notable have been Dotson Rader's "First Sister" (*New York* magazine, March 27, 1978), and Rudy Maxa's "Ruth" (*Washington Post Magazine,* October 8, 1978). Maxa's earlier article about Ruth, "Hustling for the Lord," appeared in the *Washington Post Magazine* on January 8, 1978. It featured a photograph of *Hustler* publisher Larry Flynt, his fly bulging, standing by a painting of a laughing Jesus. The nation's leading psychic journal, *New Realities,* devoted an issue to "Holistic Health," with Ruth's face on the cover to herald the article called "Ruth Carter Stapleton, Spiritual Therapist."

Ruth's nonprofit Behold, Inc., continues to publish its newspaper *Behold . . . and Be Whole,* and it is said that more and more volunteers are needed each year to handle the volume of mail. Robert Stapleton, Ruth's husband, keeps the books. He is a tall, subdued retired veterinarian from Fayetteville, North Carolina. The Stapletons have purchased thirty acres of land at Argyle, Texas, near Dallas, where they are building an inner-healing retreat called Holovita, which means "whole life." Behold, Inc., finances Ruth's many trips around the world, where she lectures and meets with the mentally disturbed.

The most sensational news story about Ruth broke in 1977, when it was revealed that she and Larry Flynt had become friends. Larry announced that he was a born-again Christian. After Flynt was shot by some unknown would-be assassin, Ruth rushed to his hospital bedside. Cynics are not sure whether Larry's conversion was the real thing or just a ploy to try to beat the rap in Cincinnati where he was on trial for promoting pornography. In any case, the pages of *Hustler* have not reflected any changes in Larry's primitive attitudes toward women, sex, and Jesus.

Details about Ruth's life are filtering out. The Christian retreat where the cure for her depression was completed was Camp Farthest Out, in North Carolina, and the young psychologist there who gave her so much help (telling her she was a beautiful person after she stayed up late one night to detail her sins) was Norman Elliott. Ruth speaks often in tongues and has it under such control that she once considered giving her platform prayers in the Unknown Tongue. Her husband, too, spoke in tongues when he was reborn, and their son Michael broke into the tongue of the angels when he was nine, just after getting a mini-bike, for which he had been praying, on his birthday.

Ruth is aging marvelously. She is a strikingly beautiful woman, with a well-shaped body, lovely blonde hair, skillfully applied make-up, and a winning Carter smile less toothy than brother Jimmy's. The smile fans pleasant creases from the corners of sea-green eyes. Add to this her soft, breathy, intimate voice, and she comes across on the platform as a woman of considerable sex appeal.

On January 13, 1977, I happened to hear her interviewed on New York City's WOR-AM radio station by Patricia McCann. My name was not mentioned, but Patricia produced a copy of my review and read the passage in which I said that Ruth was practicing psychotherapy on the mentally ill even though she was "untrained in psychiatry."

Ruth responded by first saying that mine was the only negative review her book had received and that to her knowledge not a single doctor or psychologist had criticized the book. As for her lack of training in psychiatry, it simply wasn't true. She had taken psychology courses at college! (Ruth got a master of arts degree at the University of North Carolina, and for a while taught high school English.)

"So the man who wrote the article was misinformed," said Patricia. "Yes," Ruth replied with a small laugh. Then she added, "But its okay."

I took that "okay" to mean Ruth was forgiving me for my terrible ignorance. At one point during the interview Ruth mentioned that someone had been in an "insane asylum." Have you heard of a psychiatrist using that phrase in the past fifty years? I, too, took psychology courses at college, but I would never consider myself trained in psychiatry. At another point Patricia broke off to do a commercial for a book by Gaylord Hauser, a food faddist I had written about in my *Fads and Fallacies.* Ruth said she had read and admired all his books. I was startled also when Ruth said it had been a "big shock" to brother Jimmy when he first learned from newspaper accounts that she was doing spiritual healing. I got the impression that she and Jimmy are rather distant in their religious opinions.

This impression was strengthened by Ruth's book on Billy. Brother Billy comes through as a lovable, warm-hearted, generous, funny, anger-prone rogue. Jimmy comes through as a tough older brother who never gave Billy the breaks he deserved. Jimmy, by the way, is on record as having said he "can't recall" weeping when Ruth spoke to him in the woods. Does Ruth have a tendency to exaggerate? Some of her anecdotes about Billy seem to stretch facts to the point of incredulity.

There is such a thing as spiritual hubris, and in reading Ruth's first two books, which are supposed to be Jesus-centered, I found them embarrassingly Ruth-centered. Try the following experiment. Go through either book and draw a circle around every personal pronoun. Total the circled words and divide by the number of pages to get the page average. You'll be surprised. The two instruction manuals for the second book have identical introductions that open: "When I first became aware . . ." and the first nine lines contain nine personal pronouns. Her 30-page booklet *Power Through Release* (Macalester Park Publishing Company, 1968) is dedicated to "my husband and to my children whose love for me has been evidenced in kindness, patience and under-standing." Even her first book's dedication is as much to herself as to others: "To all of those whose ministry has helped me unlock doors of meaning in my own life."

Ruth Carter Stapleton

Holovita

Summer N' Senton, Texas 76201

October 1980

Dear One,

    During the past year much has happened at Holovita inspiring us to expand our program at home and abroad. This letter is to especially thank you for your support in making this all possible. Hundreds have been helped through Inner Healing at Holovita, and thousands more as I have traveled across the country and the world spreading the good news of the healing Love of Christ.

    New health and physical fitness capabilities in the form of a pool and spa have been added; an old house to be refurbished promises to be a great practical addition as a guest house for retreat participants; an underground chapel, which will provide a place for quiet, inspiring meditation, is under construction: the small lake near the entrance has been cleaned, and with a functional windmill used in conjunction with the lake, we will have the start of an irrigation system; the new fruit orchard, with a variety of trees, has survived the intensive Texas dry spell this past summer; and new shrubbery has been added to the property.

    Our monthly retreats have been inspiring to all who have attended. These experiences are rapidly being expanded to include international as well as national and local interest. Participants from Japan, India, England and Ireland have provided important exchange and communication in expansion of the International program.

    I reiterate and, again, thank you most gratefully for your past support in making all this possible. Your prayers and donations have encouraged us in all that has been accomplished.

    This is an urgent plea for you to support BEHOLD, INCORPORATED through Holovita, and to provide for its continued expansion. Will you please acknowledge this letter with a pledge for 1981, and be a partner in this work which is so important to all peoples of the nation and the world.

    A copy of the 1981 budget is yours for the asking.

One with you in Christ,

Ruth

Ruth Stapleton

Ruth's lovely face opens the first edition of her first book and also graces one of its jacket flaps. A color photo of a smiling Ruth is on the back of the jacket of the first edition of book number two. Her picture is on the Bantam paperback reprints of both books. The jacket of *Brother Billy* features photos of Ruth with Billy and Ruth with Jimmy on the front, and Ruth alone on the back.

Anxious to see a copy of Ruth's periodical, *Behold . . . and Be Whole,* I wrote to Holovita to ask how much it would cost for a subscription. The reply was the form letter reproduced on page 312. Note how cleverly "Dear One" makes it possible to send the letter to a person of any sex. Since it failed to answer my question I tried again on October 21 with the following short request:

> I would like to obtain a copy of the latest issue of your newspaper, *Behold . . . and Be Whole.* A contribution of ten dollars is enclosed.
>
> Sincerely,
> Martin Gardner

This produced another form letter, only now it began "Dear Martin." Ruth thanked me profusely for my gift, invoked a blessing, but again made no mention of my request for her newspaper. (I assume it exists, otherwise Ruth would not talk about it in her interviews.) In eleven lines I counted eleven Ruth pronouns.

But it's okay.

30

"Mind-Reach" and "The Search for Superman"

Mind-Reach (Delacorte, 1976) is the latest and most sensational of a spate of new books by "paraphysicists," a fast-growing breed of trained physicists who have taken up psychic research. Margaret Mead writes the enthusiastic introduction. Richard Bach, author of *Jonathan Livingston Seagull,* provides an equally ecstatic foreword. Eleanor Friede, editor and co-publisher of *Mind-Reach,* is the former Macmillan editor who launched Bach's book.

In case you ever wondered why Bach's ridiculous story became a best-seller, the answer is now clear. It is about a bird who raises his consciousness until he can perform such paranormal feats as flying through solid rock. The supergull caught the fancy of millions of supergullibles of the "me generation," eager to expand their inner space. Bach was inspired to write his story after hearing a "clear, deep voice" call out the bird's name when he was alone on a beach. He has since contributed tens of thousands of dollars to Harold Puthoff and Russell Targ, authors of *Mind-Reach,* and has become one of their most talented psychic subjects.

The fame of P and T, as they are often called, rests mainly on their validation at Stanford Research Institute of Uri Geller's ESP powers. This dismays them. Their work with Uri, they say several times, was only 3 percent of their psi research. Most of the book is about what they consider far more revolutionary: their experiments on "remote viewing," the clairvoyant perception of distant targets. They are convinced that everyone

---

Reprinted from the *New York Review of Books,* March 17, 1977, with permission of the New York Review of Books. © 1977 NYREV, Inc.

has this ability. "So far we have not found a single person who could not do remote viewing to satisfaction."

The protocols are simple. A subject sits in the laboratory with one experimenter while one or more other experimenters visit a sequence of randomly chosen spots within a half-hour's drive. The subject and the inside experimenter are kept ignorant of these targets. When the outside experimenters are at spot *A,* the subject tape records his impressions and makes rough sketches of what he "sees." These reports, unedited and un-labeled, are shuffled in random order and given to a "judge," usually an SRI research analyst who is a friend and booster of P and T. The judge is taken to spots *A, B, C, . . .* where he does his best to correlate reports with targets. He also weights each match with a number from one through nine to indicate how closely he thinks the report and target correspond. A statistical analysis is then made of the "blind matching."

P and T say they have tested more than twenty subjects and in every case the judges matched reports and targets to a degree that violated chance. Moreover, when a report is correctly matched, P and T contend that the accuracy of the sketches far exceeds what could be expected from chance. Their book is filled with drawings alongside target photos. The correspondences are indeed striking.

A question at once arises. Were the photos taken before or after the sketches were made? It is typical of the book's exasperating vagueness that nowhere can you find the answer. Targets are large: "Airport, Palo Alto," "Miniature golf course, Menlo Park," "Marina, Redwood City," and so on. By zeroing in on one aspect of a complicated scene, and photographing it from the best angle, it is possible to obtain photos that are extremely misleading.

Consider the three sketches on page 83 by Duane Elgin, an SRI ana-lyst who strongly believes he has psychic powers. His fifteen-minute taped description of one target makes clear that he thought the outside experi-menters were in a museum. His third sketch shows a stick-figure person in a large circular hall surrounded by four blobs that could be exhibit cases. Curved rectangles in the background are labeled "windows." The actual target, however, was a tennis court. P and T reproduce this sketch below one of the courts, the rectangled fences in back matching Elgin's windows.

Elgin's second drawing shows two people labeled H and P (initials, one assumes, of two experimenters) on either side of what was intended to be an exhibit panel. But the panel becomes a net when placed below a photograph of tennis courts. The first sketch shows a figure holding what looks like a gigantic tuning fork. By printing this below a tennis player with upraised racket one's mind instantly closes the top of the "fork" and interprets the ambiguous object as a racket.

A marvelous example of how easy it is to find something in a wide target area that will match almost any drawing is provided by the top sketch on page 9—a dome with two arches. P and T put it alongside a photograph of a small merry-go-round in a playground, taken at a camera angle such that curved bars on the merry-go-round match the two arches of the sketch. Now turn to pictures of two other targets (pages 49 and 85) and note in each the top of Hoover Tower at Stanford University. It is a much closer match of the sketch. With a magnifying glass you can read what the subject, a visiting scientist, scribbled beside his sketches. His comments fit the tower, not the merry-go-round.

"My God, it really works!" the unnamed scientist exclaimed when taken to the target area. One suspects he would have had the same reaction had he been taken to any spot where there are many kinds of structures (miniature golf course!), but that's not the sort of experiment P and T like to try.

Perhaps photos were made by experimenters when they first went to the target areas. If so, many pictures would have had to be taken, otherwise a subject might respond to a photograph instead of the entire scene. A selection of which photo to print beside a sketch would then be the same sort of fudging. This does not, of course, explain the nonchance results of blind matching. In one test with Hella Hammid, a psychic friend of Targ's, the odds against matching are given as 500,000 to 1. The point is that not saying when the photos were made is the kind of omission which, to skeptical psychologists (i.e., most professional psychologists), makes P and T reports seem amateurish.

P and T claim that their work strongly supports the prevailing view that ESP does not decline with distance. One successful test concerned target areas in Costa Rica, visited by Puthoff while three subjects responded in California. Another series of tests, with New York psychic Ingo Swann, involved randomly picking coordinates on a world map and having Swann describe what he "saw" at each location. P and T regard the results as overwhelmingly positive. Moving farther out, P and T monitored a space probe in which Swann (at SRI) and the Arkansas psychic Harold Sherman (in Arkansas) simultaneously viewed Jupiter. Results were disappointing but a later probe of Mercury by the same two men impresses P and T as "intriguing." Since there is "excellent statistical data" that psychics can remote view any spot on the globe, P and T see no reason why it cannot be applied usefully to space exploration.

Remote viewing is also independent of time. A chapter on this opens with Targ recounting four of his precognitive dreams. "Cheered on by these firsthand experiences," P and T repeated their remote viewing test four times with Ms. Hammid, using exactly the same protocols except for one change. Ms. Hammid was asked to describe each of the four targets twenty minutes *before* it was randomly selected. Three judges independently matched all four targets correctly. P and T regard this as "one of the most successful experiments we have done to date."

If the description by P and T can be trusted, this test is certainly impressive. However, as we have seen and shall see, P and T have a facility for writing elliptical and deceptive accounts. The "professional engineering consultant" brought in to "independently observe and record the events" was David B. Hurt, a true believer who had designed Targ's first electronic machine for teaching ESP—hardly an unbiased assistant. None of the three judges is identified. Although the experiment, if valid, is one of the greatest scientific breakthroughs of the century, it is tossed off so carelessly, and with such scant information, as to be impossible to evaluate.

In a chapter on Geller, P and T regret that they were unable to confirm his metal-bending powers. Contrary to what Uri says, they never saw anything bend except when Uri was allowed to touch it. "We saw many wild and wonderful things. We shot thirty thousand feet of 16 mm. film and thirty hours of videotape, although we did not ever photograph what we set out to observe."

Accounts of the "wild and wonderful things" leave no doubt about the authors' firm belief in Uri's psychokinetic powers. "We have seen dozens of objects move, bend and break," they write, but for some reason they couldn't capture those miracles on camera. They remind the reader that in quantum mechanics the "observer" affects what is being observed. Could this be responsible for what John Taylor, the eminent British Gellerite, calls the "shyness effect"? Metal bends only when persons (and cameras) are not looking!

The fact that magicians can duplicate a psychic's tricks (the Amazing Randi now bends keys better than Geller) naturally does not prove that psychics do them the same way. When Margaret Mead repeats this in her introduction, I wonder if she realizes what a bromide it is. Every time Randi bends a key some Gellerite uncorks this stale remark. Every time Houdini exposed a phony medium spiritualists said the same thing. Over and over again history has demonstrated that there is a type of neurotic personality who thrives on being admired for psychic powers and who will go to incredible lengths to perfect methods of deception. (Recommended reading for Ms. Mead: "Alexander the Oracle Monger" by Lucian.) If magicians can reproduce a psychic's bag of tricks it does not prove him a charlatan, but it enormously increases the probability that he is, and it makes mandatory the presence of a knowledgeable magician in any laboratory test of the psychic that can be taken seriously.

Uri's precognitive powers fail him at roulette, he says, but Swann and Puthoff have no such blocks. P and T tell us that Swann, using a home roulette wheel, "slowly built up his average scoring rate from 50 percent (chance) to around 80 percent." Swann and Puthoff then went to Lake Tahoe where Ingo quickly netted several hundred dollars. Puthoff, betting more conservatively, won only about a hundred.

P and T are persuaded that predictions of the future can be amplified by a technique similar to those used for computer enhancement of signals from space probes. The idea is to repeat a signal many times until the "noise washes out and the signal emerges." Sitting in a hotel room, Puthoff and his wife, and a sponsor and his wife, applied this technique to the hotel's roulette wheel by precognizing a red-black run that would begin after the first double zero to turn up when they entered the room. The four cast votes, over and over again, until they obtained an amplified prediction of eleven marble falls.

"To the table we went." All but two of their predicted colors were correct. "The two incorrect ones, the seventh and ninth, straddle a double zero . . . perhaps a confusion factor?" Such casino exploits, P and T continue, "have in fact stood up to scientific investigation and have resulted in published papers. For those interested, we include here the description of a proven and published strategy . . . it has provided a number of individuals known to us an opportunity to succeed at the casino and come away with money in their pockets as testimony to their psychic prowess."

Purchasers of *Mind-Reach* thus obtain a gold mine: a "proven" system of winning at roulette. And these are the two men to whom the Naval Electronics Systems Command recently gave, as reported in Wilhelm's book, $47,000 for more research on psi! It is not hard to understand why the government is interested. Amplify the powers of a group of psychics and you have an unparalleled espionage technique. And if one psychic can bend a spoon, perhaps a group of psychics could trigger a nuclear explosion in a warhead. Julius Caesar had his oracles. Hitler had his astrologers. Our military complex has SRI.

Businessmen, too, can profit from psi training. P and T devote many pages to "executive ESP." This suggests the "possibility of the prediction of future economic, social and political trends." Sure enough, Swann and Dr. Willis W. Harmon, of SRI's Educational Policy Research Center, actually did a pilot study on this in 1973. Ingo and two anonymous psychics put their psi forces together to concoct an oracular forecast that ends with a mysterious cataclysm in 1985 that will "bring current concepts of man to an end." (The Second Coming?) For details, see Swann's autobiography, *To Kiss Earth Good-bye.*

The best documentation yet of the incompetence of P and T is in *The Search for Superman.* The author, John Wilhelm, is a former science correspondent for *Time.* He is inclined to believe in psi. After extensive conversations with P and T and several of their associates, he set about to record as objectively as possible what he had learned. His book is extremely valuable for many reasons. It gives details about government funding of P and T projects that had hitherto been hush-hush. It is filled with eye-opening revelations about facts that should have been in *Mind-Reach* but are not.

Example. Years ago magicians explained how easily Geller's constant companion, Shipi Shtrang, could have obtained information about targets and communicated it to Uri. Astronaut Edgar Mitchell, who helped finance SRI tests of Geller, once complained that whenever you try to do something with Geller "Shipi is underfoot." True, admit P and T, but this has no bearing on their tests because Shipi was carefully "excluded from the target area."

Anyone would take this to mean that Shipi was not in the laboratory when tests were made. But P and T have a special way of using words. They mean that Shipi was kept out of the *room* in which target pictures were being "sent." He was very much underfoot. During some tests he was even locked inside the shielded room with Uri. "Part of our secret design," Puthoff told Wilhelm, "was to see if Geller did better work when Shipi was around. We wanted to know if Shipi worked as a psychotronic amplifier."

During some tests Shipi was directly outside the room where Targ and his assistant, Jean Mayo, prepared the target drawings. Targ and Mayo often conversed about them. At Uri's insistence the drawings were taped on a wall. (It is not clear what wall.) "Shipi was just seated at a desk by himself," Puthoff told Wilhelm. He was not even watched. "When Geller was inside the second-floor Faraday cage," Wilhelm goes on, ". . . Shipi was inside with him, guarded by Puthoff. Where, then, was Hannah?"

Hannah? Hannah Shtrang, Shipi's sister and Uri's former girlfriend, was also at SRI. Later she broke with Uri and told an Israeli journalist how she and Shipi used to aid Uri in Israel by secretly signaling. (A translation of the Hebrew article is in Randi's Ballantine paperback, *The Magic of Uri Geller*.)

It is hard to believe, but in the famous *Nature* report by P and T, on their work with Geller, there is no mention of the many others who participated. Actually, the tests were like stage performances. In Wilhelm's words: "On many occasions people not associated with the experiments clustered around, kibitzing the procedures. Sometimes members of the SRI management hierarchy were there to observe. . . . At other times the audience simply was a group of psi boosters." Swann summed it up: "It was like a monkey cage."

An astonishing number of lab assistants, Wilhelm reveals, were Scientologists. Puthoff himself is a dedicated follower of L. Ron Hubbard. He first reached the status of "clear"—a person free of all "engrams." (Engrams are neurosis-causing patterns imprinted on one's unconscious mind by what one hears as an embryo.) Later he advanced to Class-III Operational Thetan. He wrote the preface to Hubbard's *Scientology: A Religion.* He was married in a Scientology church. Eli Primrose, who assisted in the Geller tests, is a Scientologist married to a Scientology minister. George W. Church, Jr., whose Science Unlimited

Research Foundation (SURF) provided early financial backing for P and T, is a Scientologist.

These facts are not irrelevant because Scientologists passionately believe in all forms of psi. Puthoff's work thus strongly supports his and Church's doctrines. Jean Mayo, who did many of the target drawings, is not a Scientologist but a self-styled psychic and a devout Gellerite. She told Wilhelm her responsibility was to help "send" the targets to Uri.

Ingo Swann, the first superpsychic discovered by P and T, also is an active Scientologist. He's a Class-VII Operational Thetan. (There are fourteen "clears" working at SRI, Swann told Wilhelm.) The second SRI superstar was Pat Price, a businessman who died in 1975. He, too, was a Scientologist—a Class-IV Operational Thetan.

In Wilhelm's hilarious book, P and T emerge as a pair of bumbling Keystone cops, fervently believing, proud of their ability to detect any kind of fraud, yet over and over again violating the simplest canons of sound experimental design. Their accounts of remote viewing (on which Wilhelm touches only briefly) raise dozens of questions. Exactly how many people (including typists) knew each target selection? It is not just the superstar who has to be monitored, but also his friends. It is not enough to keep selections in a safe. One must guard against the selection process itself being overheard, against such things as carbon paper being retrieved from wastebaskets. Electrons and rats don't cheat. Professional psychics do.

The "judges" in the "remote viewing" experiments worked with unedited typescripts. If a typist copied the reports before they were shuffled, decreasing blackness of letters and increasing typing errors could provide cues to the original time sequence. Only short excerpts from a few reports have been published. Can we be sure that the subject, rambling on about each target, did not provide subtle time-ordering clues?

Ray Hyman, a psychologist who along with Randi, me, and others is severely attacked by P and T in a distortion-packed chapter on the "loyal opposition," has pointed out another possible source of bias. Targets for an experiment are chosen to have minimum scenic overlap. Suppose target *A* is a tennis court. The subject, usually taken to the site immediately after making his report, to give him immediate reinforcement, now knows that on the remaining targets he can avoid "seeing" a tennis court. Target *B* is, say, Hoover Tower. He now knows it is a good bet to avoid seeing both tennis courts and tall towers. After all, the number of basic gestalts in natural scenery is not a large number. To forestall bias from this elimination process, subjects should not have been taken to targets until an entire series of tests was completed.

When a judge was taken to target sites to match reports to targets, were visits made in random order? And did P and T make sure that whoever took him here and there had no knowledge of the time ordering of either targets or reports? Otherwise a judge could unconsciously pick up cues. Can you imagine a believer completely concealing his feelings each time a judge announced a tentative decision? One must assume that P and T took all these precautions into account, but that is not the point. Extraordinary laboratory results demand not only extraordinary controls but also extraordinary care in reporting. To omit such details is inexcusable.

The future of remote viewing is predictable. All over the world true believers will reproduce the experiments with positive results. Skeptical psychologists, if they bother to try, will get negative results. The general public won't hear the negative side. Too dull a story. P and T will accuse their opponents of stubbornly refusing to make a "paradigm shift" of such magnitude that the Copernican Revolution pales beside it. They may even invoke the old Catch-22 which says that skepticism itself inhibits psi.

At the moment, the two surviving SRI superstars, Swann and Geller, seem to be dimming, not because skeptics are winning (far from it!) but because people are now more intoxicated by angels, the Antichrist, and the return of Jesus. In the words of a born-again Johnny Cash song, and the text of a recent Billy Graham sermon, "Matthew 24 is knocking at the door." We can be certain, however, that new supergulls (and swans), with new bags of psi tricks under their wings, will soon be flapping out of the firmament.

Postscript

When Robert Ornstein reviewed *Mind-Reach* in the *New York Times Book Review* (March 13, 1977) he was mildly critical. The book is "pleasantly written," he said, "but slim in hard evidence." P and T "almost always go beyond evidence and claim they have proven their case when they have done nothing of the sort. In writing this book, the authors have done more harm, perhaps, to their own position and to their field of study than they have helped."

Ornstein spoke of having tried his best to replicate one of the remote-viewing experiments, with the full cooperation of P and T and the use of one of their successful subjects. No results. P and T responded with a letter in the April 10, 1977, issue, accusing Ornstein of an "inexcusable faux pas." Their letter prompted Ornstein to apologize for being too generous in his review.

Perhaps the exchange with Ornstein had something to do with P and T not sending a letter to the *NYR,* or perhaps they just considered me hopeless. However, I did see a letter they sent to Marcello Truzzi in which I was criticized only for suggesting that maybe the typing of transcripts got weaker. This is ruled out, they insisted, by the fact that all SRI typists use one-time-only ribbons.

I accept that. But when I mentioned the possibility of cues in the transcripts I wasn't guessing, because one whopping cue appears in the excerpt of a transcript published in the book. On page 65 Pat Price, rambling on about a target, says: "I'd say that is about—not half the distance they were to the marina." Price is referring to a previous test in which the marina was the target. Any judge reading this would know at once that Price could not now be describing the marina. Moreover, since the subject was taken to the site after each test, he or she would be strongly inclined to give later descriptions that would *not* apply to any previously visited site. This alone would bias the ability of the judges to match targets and transcripts.

In the transcript we are considering, Price obviously thinks he is seeing a miniature golf course, the site of a target in another experiment. He several times mentions a small Dutch-type windmill, "like the one you'd almost see in a miniature golf course." "Windmill" is mentioned seven times, and "golf course" is mentioned five times. The actual target was an arts and crafts plaza. No golf, no windmill. Yet P and T call Price's transcript "accurate in almost every detail."

The important point is that the unedited transcript contains an obvious cue that would help judges in their matching. Wilhelm's book contains a transcript of Price's description of another target. Targ, who is sitting with Price, actually says (page 217): "I've been trying to picture it in my mind and where you went yesterday out on the nature walk. . . ." This clearly tells a judge not to match the transcript with the nature walk through Baylands Nature Preserve, an earlier target.

When these cues were first pointed out, P and T replied that of course all such cues had been carefully edited out of the raw transcripts before giving them to the judges. We might still believe this had not David Marks and Richard Kammann, researching their book *The Psychology of the Psychic* (Prometheus Books, 1980), managed to obtain five Price transcripts. These transcripts, not previously published in whole or in part, had been used in an experiment for which Arthur Hastings, a friend of P and T and a true believer, had been the judge. Kammann and Marks found these transcripts (given to them by Hastings) riddled with cues about previous targets. Indeed, there were so many cues that it was easy as psi for anyone to match all five transcripts correctly to their targets without visiting any target sites. You can read about this in Chapter 3 of the Marks and Kammann book.

After Marks and Kammann published these facts in *Nature* (vol. 274, 1978, pp. 680–681), what did P and T do? They gave the entire set of nine transcripts (of which M and K had seen only five) to their friend Charles Tart, who then did what should have been done in the first place. He removed all the cues, then gave the laundered transcripts to a judge "unfamiliar with the experiment." A letter about this, signed by Tart and

Russell Targ and Harold Puthoff (photo on the cover of *Mind-Reach*, 1976).

David Marks and Richard Kammann (photo on the cover of *The Psychology of the Psychic*, 1980).

by P and T, appeared in *Nature* (vol. 284, 1980, p. 191). They do not disclose the judge's name, but they refer to the judge as "she," and we are told she had been selected after a series of tests to find a person "competent in blind matching."

The new judge matched seven of the nine transcripts correctly to their targets. P and T consider this a vindication of the original experiment! There is nary a hint of recognition that the original experiment had been flawed by cues in the transcripts which the veriest tyro in psi research would have recognized as strongly biasing the matching process. We have no way to know how completely Tart removed cues, because the edited transcripts have not been published. And how can we be sure that the new judge was completely "unfamiliar" with the experiment in view of the fact that four of the nine transcripts had already in part been published?

In my review I devoted several paragraphs to the book's system for using precognition to win at roulette. I had based these remarks on the bound galleys sent to reviewers. When the book was published I found to my surprise that the pages dealing with roulette had been entirely removed from Chapter 9. I was told that this was done at the insistence of Margaret Mead. I do not know if she objected because she thought the system wouldn't work or because she thought it unwise to publish it.

John Wilhelm tells me that lawyers for the Scientology Church tried to persuade Pocket Books to leave out his chapter called "The World of the Thetans," and one can only speculate on whether the fact that the church's tactics of harassment are now well-known had anything to do with the book's extremely poor distribution. Pocket Books says Wilhelm's book is not officially out of print, but you won't find it easy to obtain a copy.

A marvelous indication of how difficult it is to know the exact conditions under which a P and T test is made is provided by their paper on "Remote Viewing of Natural Targets" in *Quantum Physics and Parapsychology* (proceedings of an international conference held in Geneva, August 26–27, 1974), edited by Laura Oteri (1975). On page 156, under the heading "Summary of Experiments," P and T describe what they call "four experiments." In the "first experiment" H. H. (Hella Hammid) gave a description that impressively matched the target—a tiny red schoolhouse on a miniature golf course. A photograph of the "schoolhouse" is included.

The same photograph appears three years later in *Mind-Reach,* on page 95. Now we are informed that during this test the experimenter at the target site, Puthoff, was in communication with Hammid by walkie-talkie. The experiment has become a "mock experiment," not a genuine one. Writing to me in 1978 Puthoff said: "If we used walkie-talkies during experiments, it would be easy to cue subjects into a correct response—and that's as obvious to us as to anyone else! Although we have used walkie-talkies occasionally in training, we have never, never, never—not even once—used a walkie-talkie during an experiment."

Okay, but why were listeners at the conference, and later readers of the proceedings, not told that the first of the "four experiments" was not genuine but only a mock experiment using a walkie-talkie? Was it not a bit deceptive to omit this all-important fact? This has been typical of P and T reports of their "experiments." It takes years before essential information about controls, or rather lack of controls, dribbles out to those of us who weren't there.

Another aspect of remote viewing is worth mentioning. How do P and T know it is clairvoyance they are testing? From their point of view it could be: (1) telepathy from the experimenter (who may also have clairvoyant powers) at the target to the subject, (2) precognition by the subject of later visits to the targets, (3) PK influence by the subject and/or experimenters over the randomizer that selects the targets, (4) ESP by the judges (who were often persons like Duane Elgin who believe themselves to be high in psi powers) in matching targets to transcripts.

There are now many reports of attempts to replicate remote viewing by copying P and T's protocols as accurately as possible. As I predicted, true believers get positive results, and Puthoff promptly sends me their papers. Skeptics get negative results, and Puthoff never, never, never sends me their papers. The number of failed replications is now large enough to discredit the foolhardy original claim by P and T that everybody has remote-viewing ability and that replications are simple and easy. For a recent blast at P and T see Chapter 7, "The Laurel and Hardy of Psi," in James Randi's *Flim-Flam!* (Lippincott & Crowell, 1980).

31

"Learning to Use Extrasensory Perception"

My review of *Mind-Reach,* by Russell Targ and Harold Puthoff—a book on the testing of clairvoyance—appeared in the *New York Review* of March 17. Shortly thereafter the *NYR* received an interesting letter from Aaron Goldman, Sherman Stein, and Howard Weiner, three top mathematicians at the University of California at Davis. Although their letter does not deal with P and T (as Puthoff and Targ are called), it concerns the closely related work of Charles Tart, a colleague of the three mathematicians at Davis.

Tart's reputation as a parapsychologist is even higher than that of P and T. When his latest book, *Learning to Use Extrasensory Perception* (University of Chicago Press, 1976) was published last year under the imprimatur of the University of Chicago Press, it was widely hailed as a major breakthrough. (See the *New Yorker*'s "Talk of the Town," December 13, 1976.) Tart shares with P and T the conviction that ESP powers can be markedly strengthened by electronic teaching machines. The major work of P and T in this field, made possible by an $80,000 grant from NASA, was with a four-choice machine designed by Targ. I discussed this test, considered a failure by almost everybody except P and T, in my *Scientific American* column, October 1975.

Tart's work is more elaborate. He uses a ten-choice trainer, or TCT as he calls it, of his own invention. He believes it to be superior to the Targ

This review, which appeared in the *New York Review of Books,* July 14, 1977, and the letters quoted in the Postscript are reprinted with permission from the New York Review of Books. © 1977 NYREV, Inc.

machine. Indeed, his book contains severe criticism both of Targ's machine and of the protocols of P and T in their NASA experiment.

This is how Tart's TCT operates. A "sender" sits in one room in front of a console bearing a circle of playing cards from ace through ten. Beside each card is a button and a pilot light. An electronic randomizer selects a digit (zero counting as ten), then the sender pushes the button that turns on the light next to the selected card.

The "receiver" sits in front of a duplicate console in a room across the hall. A "ready" light informs him when a card has been chosen. After moving his hand around the ring of cards, searching for the "hot" one, he pushes a button to register his choice. This procedure is repeated in runs of twenty-five choices each, and twenty runs per test. Hits and misses are automatically recorded. There is no recording of the time at which any button is pressed.

As soon as the choice is made, a light beside the actual target card goes on to provide instant feedback. If the choice is a hit, a "pleasant chime" sounds inside the console. Above the console is a TV camera joined by a cable to a TV screen above the sender's console. On this screen the sender can see the hand of the receiver as he searches for the hot card. This is so the sender can concentrate more intently on where he wants the hand to stop, although Tart concedes there is no way to tell whether the receiver is getting the target information by telepathy or clairvoyance. Senders and receivers are usually college students, unpaid but sometimes given academic credit for their help.

Tart's superstar was an unnamed girl who scored so high that the results are against normal odds of more than a billion to one. She worked with unusual slowness, taking about forty-five minutes to complete each run.

Now for the letter from the three mathematicians:

To the Editors:

Readers of the review of ESP research in the March 17, 1977, issue of the *New York Review of Books* may be interested in the details of an ESP experiment conducted by Charles Tart, our colleague at U.C. Davis. Much more easily analyzed, it concerns the transmission and reception of the ten digits. . . .

[I have omitted a paragraph in which the TCT is described—M.G.]

Ten receivers in Tart's experiment had a total of 722 hits out of 5,000 trials. Statistically this is far above chance, which would be around 500 hits (since there is one chance in ten of guessing a digit). The simplicity of the experiment, together with our natural curiosity about ESP, led us to ask Tart for the raw data, which were even more spectacular than the sessions. One receiver had 124 hits in 500 trials, far above the expected fifty. This was truly astonishing.

Then, as we scanned the target digits produced by the random-number generator, we noticed that twins seemed to be sparse—that is, a digit zero following a zero, or a one followed by a one, etc. Since there is a 10 percent chance that the machine, having produced the digit X, will immediately produce X again, there ought to be some 500 such twins in the run of 5,000. Instead there were only 193. Yet, when we tested the generator, pressing the button about every three seconds, they appeared approximately 10 percent of the time. However, when we pressed the button at ninety-second intervals—which corresponds more closely to the rate in the actual ESP experiment—the generator again avoided twins as it had during the experiment.

It isn't clear how many of the 222 hits above those expected by chance can be "explained" by the peculiarities of the generator. ESP may indeed be the reason for some of these 222 hits. However, the raw data make it difficult to quantify the relative proportion of hits due to the machine's nonrandomness and those due to ESP. One way to indicate the delicacy of the analysis required is to note that there were 4,278 misses while pure chance yields some 4,500, if the generators were random.

Tart is anxious to redo the experiment, using the Rand table of random numbers, and we hope to work with him. Until the experiment is done again, we are in the position of a chemist who at the end of an experiment discovers that his test tube was dirty. Whether it was only a little contaminated or a lot doesn't matter. The experiment has to be executed with a clean test tube. Tart hopes to carry out the modified experiment before the end of the year.

I have several comments.

Tart had earlier recognized anomalies in the output of his randomizer, although he did not realize that some of them also appeared when the randomizer was operated under non-test conditions. In recent lectures he has attributed the anomalies to the ability of strongly psychic subjects to influence the randomizer by psychokinesis, the mind literally altering the operation of the device.

Tart also recognizes in his book (page 164) a clever method by which sender and receiver could have cheated, though he argues that this is unlikely. Assume that the sender, immediately after the receiver registers a target choice, operates the randomizer. Suppose it selects seven. The sender multiplies seven by, say, five to get thirty-five, then waits exactly thirty-five seconds before pressing the next target button. The receiver, watching a second hand on a wristwatch, knows the target before he starts to move his hand.

"This possibility should be eliminated in future work," Tart writes in a thumping understatement (page 104), by making "the time delay between switching off one target and selecting the next uniform and beyond the experimenter's control." Time delay techniques of secret information transmission are well known to magicians familiar with modern "mentalism." Had Tart consulted such an expert, the most glaring defect of his machine could have been remedied before he began testing.

Tart is to be commended for his willingness to provide raw data; this in marked contrast to P and T. In 1975 *Scientific American* requested permission to let a statistician inspect the raw data of their NASA experiment. The request was refused. (See the letters department of *Scientific American,* January 1976.) Tart is also to be praised for his candor in describing in his book one major defect of his experimental design.

In an article on "ESP Training" in *Psychic* magazine, March 1976, Tart wrote: "For the 10 TCT subjects, the average number of hits per run was well above chance, with odds against chance being a *million billion billions to one*" (Tart's italics). Nothing so sensational has ever been claimed before by a professional parapsychologist. The above facts make clear that until Tart repeats his tests under controlled conditions—adequate randomizing and rigid exclusion of all possible methods (there are others!) of secret coding between subject and sender—the staggering results reported in his book cannot be taken seriously even by other parapsychologists.

Postscript

Tart's comment on my criticism appeared in the *NYR,* October 13, 1977:

> Martin Gardner, writing in the *New York Review* of July 14, represents me as a severe critic of Harold Puthoff and Russell Targ's research on teaching ESP by electronic feedback machines, and while apparently esteeming me more highly than Puthoff and Targ, ends up concluding that "until Tart repeats his tests under controlled conditions—adequate randomizing and rigid exclusion of all possible methods (there are others!) of secret coding between subject and sender—the staggering results reported in his book cannot be taken seriously, even by other parapsychologists." I disagree with these three points and his conclusions.
>
> First, my review (in my *Learning to Use Extrasensory Perception,* University of Chicago Press, 1976) of Puthoff and Targ's studies was generally positive. My strongest criticism was of a *possibility* of (*not* evidence for) subject fraud on only one of their case studies, where just one subject was used, but I pointed out that their data did not indicate that a defect in the Aquarius four-choice testing machine had been utilized to inflate the scores. The Aquarius machine was not used in the mode that allowed that defect to operate in any of their other studies. Gardner also fails to report the fact that half of my own research involved using the Aquarius machine, and fifteen subjects made an overall score of 2,006 hits when 1,869 would be expected by chance. This would occur by chance alone only four in 10,000 times, and indicated good ESP results with that machine, under tightly controlled conditions.
>
> Second, I believe the letter from my mathematician colleagues, Professors Goldman, Stein, and Weiner, which was written at an intermediate

stage of our fruitful collaboration, was somewhat premature and wrongly gives the impression that my results can be readily "explained away." This is not the place for a long, technical discussion, but the basic concern is whether some of the targets could have been predicted by the subjects by using mathematical inference and knowledge of previous targets, rather than ESP. A good card player can make better-than-chance guesses at what cards are still out from keeping track of what's been played, and that is the kind of question we were investigating. A very powerful computer test of this possibility has now been completed by another colleague, Eugene Dronek of the University of California at Berkeley, and I, and we have found that such mathematical prediction cannot account for the bulk of my results: even if the subjects were trying to predict this way, there is still an enormous amount of ESP. To update the "dirty test tube" analogy, the effects of the contaminants have been estimated and found to be minor.

Third, Gardner misrepresents me in saying, "Tart also recognizes in his book (page 164) a clever method by which sender and receiver could have *cheated* . . ." (my italics). I pointed out how a sender could have unintentionally and unknowingly cued a subject; but I found no evidence that it happened: I have no evidence that my experimenters or subjects cheated. Gardner takes a position that I find morally repellent as well as scientifically invalid, namely that if a critic can think of any way a subject and/or an experimenter *could* have cheated to get apparent ESP results, *regardless of whether there is any evidence of cheating,* then the experimental results need not be taken seriously.

In science, an explanation must be capable of *dis*proof as well as proof. There is no experiment, however, in any field of science, that cannot be faked. Gardner's criticism demonstrates that he does not accept the discipline of scientific method. While he is certainly entitled to defend his personal belief system by any means he wishes, I hope the *New York Review* readers will not mistake it for science. The possibility of reliably training people to use ESP that my work raises is too important a question to be dealt with by implication and misrepresentation. The appropriate scientific response is for other researchers to carry out similar work which may confirm, disconfirm, or modify my findings.

Charles T. Tart

This was my reply:

I will comment on each of the three points:

(1) Tart's criticism of the ESP teaching machine experiments by Puthoff and Targ (pages 26–31 of Tart's book) are, in my judgment, "severe." The first subject of the pilot study was a child, the second a scientist. The child showed a mild increase in ESP ability, the scientist a remarkable increase. Tart adds: "Unfortunately, this subject, a scientist, recorded his own data, and the first subject's data were reported by his father. Since it is a general rule in parapsychological research never to allow subjects *any* opportunity to make recording errors or to cheat, these results must be

considered tentative" (p. 27, Tart's italics). Tart is politely accusing P and T of violating kindergarten canons of experimental design.

Tart then summarizes the three phases of the NASA-funded experiment that followed the pilot study. For phase I, "The total number of hits for the group as a whole was almost exactly what one would expect by chance." However, one subject scored high. (He was Duane Elgin, a self-proclaimed psychic and friend of P and T, who was then a "futurologist" at Stanford Research Institute.) Tart accepts this as genuine ESP, balanced by "significant ESP-missing" (unusually low scores) on the part of others. P and T were convinced that the overall poor results were caused by the clatter of their machine's data printer.

For phase II, P and T chucked the printer and for the first time in their experiment all scores were automatically recorded by a silent computer. As I have pointed out elsewhere, this eliminated possible sources of bias in phase I. The results showed no deviation from chance either in number of hits or learning curve slopes. In brief, the only adequately controlled phase of the experiment showed no sign of ESP.

For phase III, P and T relaxed controls, detached the computer, and went back to primitive hand-recording. Of the eight subjects, only Elgin redeemed himself. Tart summarizes the entire project as follows: "Most of their subjects showed no ESP, and of those who did, few were able to hold up in further studies."

P and T used a four-choice trainer called the Aquarius Model 100. Tart used the same machine in his early studies. However, he reports that his son discovered a way to cheat on the machine when it was in its precognitive mode (the subject guessing targets *before* they are selected), and during one of Tart's experiments the machine "broke down and began repeating one target with a very high frequency." Although Tart reports positive results with this machine, he completed his work with a ten-choice trainer of his own invention, which he clearly considers a vast improvement over the Aquarius model. It was with his own machine that Tart obtained the most sensational ESP results ever reported by a parapsychologist.

(2) I completely agree with Tart that the defect in his machine's randomizer is insufficient to account for his results of "a million billion billions to one" against chance. Nowhere have I suggested otherwise.

(3) Tart accuses me of failing to understand scientific method. Because any result can be faked, he says, there is no reason to dismiss a psi experiment merely because a loophole in the design allowed cheating. Unless one can prove that a subject cheated, Tart finds it "morally repugnant" to criticize the results.

When I read those statements I could hardly believe my eyes. Nowhere in his letter does Tart indicate an awareness of the enormous qualitative difference between testing psychics for paranormal powers and experimentation in *all other* branches of science. Blood cells, DNA molecules, gerbils, and photons don't cheat. Because of the long, sorry record (going back to ancient times) of constant cheating by self-anointed psychics, the very essence of sound experimental design in parapsychology is to close all cheating loopholes. Until they are closed, no experiment indicating sensational psi powers is worth publishing. Tart himself (in the passage quoted in my first main paragraph) takes P and T to task on just such grounds.

When subjects are objects or creatures that can't cheat, the only possibility of fraud is on the part of an experimenter. This does sometimes occur in all branches of science, usually with disastrous results. We have recently had several sad instances: the faking of mice specimens by a respected doctor at Sloane-Kettering, for example, and the scandal involving the director of J. B. Rhine's laboratory who was caught altering records. Such cases are uncommon. But cheating by self-styled psychics is not uncommon. That is why extraordinary safeguards are required in psi research that are not required in other fields.

Let me adopt Tart's technique and repeat the sentence he quotes from me, but with a different word italicized: "Tart also recognizes in his book (page 164) a clever method by which sender and receiver *could* have cheated. . . ." It was this method that I described because it is one that not many parapsychologists know. I have no idea whether it was used, or whether an isolated sender—in his enthusiasm for telepathically guiding the subject's hand to the target card—jumped up and down, thus transmitting a floor vibration to the subject across the hall who could use it, consciously or otherwise, as a cue.

Whether such methods (there are still others!) were used or not is beside the point. The point is that Tart's experimental design, because it permitted such easy ways to cheat, was incredibly poor—so poor, in fact, that it was premature for Tart to write a book about it and uncharacteristically bad judgment by the University of Chicago Press to publish it.

It would be helpful now if Tart would disclose more of his raw data. For instance, given a subject who made exceptional scores, were those scores obtained when the same person acted as sender? If so, the talented pair should be tested by better controlled replications in someone else's laboratory. Were videotapes made of any of the high scoring runs? If so, a careful study of the tapes would confirm or deny the time-delay code I described. If no videotapes were made, that is another design defect because they would have provided invaluable data. It also would be helpful if in all further testing Tart hired a knowledgeable and skeptical magician to observe the actual experiments.

Tart's reply to my note betrays a whopping misconception about the nature of the controls that are mandatory in the testing of alleged psi powers. Nevertheless, Tart's understanding of experimental design impresses me as a cut above that of most of his colleagues.

Martin Gardner

32

Seven Books on Black Holes

Black holes are hot. Although this is literally true (according to the latest theories) of some black holes, I mean they are hot as a topic. The books reviewed here are only fragments of this year's crop that deal entirely or in part with black holes. Why such obsessive interest in astronomical objects that may not even exist and that in any case cannot be fully understood without knowing general relativity theory and quantum mechanics?

Let the first paragraph of Isaac Asimov's book *The Collapsing Universe* (Walker, 1977) set the tone for what I believe is the answer.

> Since 1960 the universe has taken on a wholly new face. It has become more exciting, more mysterious, more violent, and more extreme as our knowledge concerning it has suddenly expanded. And the most exciting, most mysterious, most violent, and most extreme phenomenon of all has the simplest, plainest, calmest, and mildest name—nothing more than a "black hole."

Black. Black is beautiful, black is ominous, black is awesome, black is apocalyptic, black is blank. "A hole is nothing," Asimov continues, "and if it is black, we can't even see it. Ought we to get excited over an invisible nothing?"

Nothing. Why does anything exist? Why not just nothing? This is the superultimate metaphysical question. Obviously no one can answer it,

This review, which appeared in the *New York Review of Books,* September 29, 1977, and the letters quoted in the Postscript are reprinted with permission from the New York Review of Books. © 1977 NYREV, Inc.

yet there are times (for some people) when the question can overwhelm the soul with such power and anguish as to induce nausea. Indeed, that is what Sartre's great novel *Nausea* is all about.

Suddenly we are being told that if a star is sufficiently massive it eventually will undergo a runaway collapse that ends with the star's matter crushed completely out of existence. Not only that, but our entire universe may slowly stop expanding, go into a contracting phase, and finally disappear into a black hole, like an acrobatic elephant jumping into its anus. There is speculation (not taken seriously by the experts) that every black hole is joined to a "white hole"—a hole that gushes energy instead of absorbing it. The two holes are supposedly connected by an "Einstein-Rosen bridge" or "wormhole." When a huge sun collapses into a black hole, so goes the conjecture, a companion white hole instantly appears at some other spot in space-time. This could explain the incredible outpouring of energy from the quasars, those mysterious objects, apparently far beyond our galaxy, that nobody yet understands. Was the big bang which created our universe the white hole that exploded into existence after a previous universe collapsed into its black hole?

It is easy to understand why the religiously inclined are excited by such wild, speculative cosmology. The heavens declare the glory of God and the firmament showeth his handiwork. Nor is it hard to understand why those who are into Eastern philosophy, pseudoeastern cults, parapsychology, and unorthodox science are also fascinated. If the universe can be *that* crazy, so goes the argument, then why be disturbed when the Maharishi announces, as he recently did, that transcendental meditation can enable one to levitate and become invisible? Black holes are the latest symbols of unfathomable mystery. Public interest in them is, I am persuaded, no indication of interest in science, but rather a peculiar by-product of the specter of the supernatural that is now haunting North America.

For the reader with no understanding of relativity and quantum theory—that is, the average reader—Asimov's book is the best of the lot. The old maestro writes with his unfailing clarity, humor, informality, and enthusiasm. Like all top science-fiction writers he knows exactly where to draw the line between serious science and fantasy. Periodically he reminds his readers that there is as yet no clear observational evidence that black holes exist and that "almost anything some astronomers suggest about a black hole is denied by other astronomers."

Cautiously, step by step, Asimov sketches the necessary background for understanding a black hole's properties. He begins with gravity, that gentle, all-pervasive, poorly understood force that holds together the matter of galaxies, stars, and planets. At the centers of planet-size bodies the pressure of gravity is insufficient to overcome the opposing electromagnetic force that binds the molecules of the matter at the core, and the matter remains intact. However, if the body is large enough (about the size of Jupiter) the pressure of gravity becomes so strong it triggers a hydrogen fusion reaction. The body becomes a sun.

There are three ways a sun can die. If a star is close to the size of our sun it will exhaust its hydrogen fuel, expand to a red giant, then slowly contract to a white dwarf. Eventually it will cool to a black dwarf, a permanently embalmed corpse that never changes unless it happens to be eaten by a black hole.

If a star is moderately greater in mass than our sun, its fate is more interesting. It is likely to explode into a supernova; then part of its mass instantly shrinks to a size smaller than the earth. So great is the density of this body that its gravitational force overcomes the opposing electromagnetic force and the structure of the star's matter disintegrates. It becomes a fast-spinning neutron star.

Most astronomers are convinced that pulsars are neutron stars. These are small stellar objects inside our galaxy that send out absolutely regular beeps of radio waves, sometimes beeps of visible light. There are probably millions of them in the Milky Way that are within the range of today's radio telescopes.

If a star is much more massive than our sun, it is expected to expire in a manner so bizarre that its fate is still shrouded in mystery. After it completes its catastrophic implosion, not even neutrons can withstand the enormous gravitational compression. All particles are totally destroyed, and the laws of physics no longer have meaning. The star has entered a black hole.

Black holes were crudely anticipated in 1798 by the French mathematician Pierre Simon de Laplace. His predecessor Isaac Newton believed that light consists of particles that are affected by gravity. If a star is large enough, Laplace pointed out, its gravitational force will prevent all light from escaping from it. This is not strictly true. In Newtonian physics the speed of light approaching a star, no matter how massive, would be so greatly accelerated that it could bounce off a reflecting surface and escape.

In relativity theory, light also consists of particles (photons) that are influenced by gravity, but their speed is a constant that cannot be exceeded. A few months after Einstein published his general relativity theory a German astronomer, Karl Schwarzschild, made exact calculations of what is now called the "Schwarzschild radius." This is the radius of a body, given its mass, below which gravity is strong enough to prohibit light, matter, or any kind of signal from escaping. It is the critical radius below which matter becomes an invisible black hole. For a mass equal to our sun's, the radius is a few kilometers. For a mass equal to the earth's, it is the radius of a large pea.

In 1939 J. Robert Oppenheimer and his student Hartland Snyder made some surprising calculations. Assuming the truth of relativity, there are no laws to prevent the gravitational collapse of a sufficiently massive sun from compressing the sun's matter within the Schwarzschild radius and forming a black hole. Moreover, the calculations lead to something even more mind boggling. At the core of every black hole there has to be a space-time "singularity"—a term mathematicians use for a point at which something catastrophic happens to the solution of an equation. In this case, calculations show that space-time curvature becomes infinite, which is to say that it becomes a single point. At that point, gravitational force and density (mass per unit volume) also become infinite.

If a space-time singularity actually occurs and can be observed, it is called a "naked singularity." So far, no one has seen a naked singularity. Perhaps its equations tell only part of the story and there are forces not yet understood which prevent singularities from existing. Roger Penrose, a brilliant theoretical physicist at Oxford University and a chief architect of black holes, believes that space-time singularities *can* occur, but a "cosmic censor" prevents them from becoming naked. It conceals them, so to speak, inside an "event horizon" that prevents them from interacting in any way with the universe.

For twenty years the calculations of Oppenheimer and Snyder were considered no more than eccentric exercises for graduate students. Then in 1962 the quasars were discovered, and five years later the pulsars. Suddenly the astrophysicists realized that maybe they were seeing objects in the final stages of just the sort of catastrophic collapse that had been worked out on paper. At first it was hoped that if the collapsing mass of a big star were a bit lopsided, the singularity could be avoided. But Penrose proved otherwise. The singularity is unavoidable. Regardless of the size, shape, or chemical constitution of a sun, if it is massive enough to collapse into a black hole, it will have that awful singularity at its center. As for the hole itself, all the structural peculiarities of the sun that formed it will be obliterated. "Black holes have no hair," so goes a theorem. It means that all black holes, aside from mass, spin, and electric charge, are identical.

It is possible, as Philip Morrison and other cosmologists have emphasized, that laws not yet known may prevent the formation of black holes. True, there are some spots in the sky where astronomers think they see something going on that can be explained only by a black hole—notably the strong X-ray radiation coming from the vicinity of a giant star in the constellation of Cygnus (the Swan)—but what they see may have conventional explanations. There is no hard evidence, though the prevailing opinion is that black holes do exist. Some astronomers suspect that a giant black hole squats at the center of every galaxy, spinning silently while it slowly gobbles up nearby suns. As of today, however, black holes are theoretical constructions supported mainly by the fact that relativity theory requires them, by the rule that anything not excluded by theory probably exists, and by observed phenomena in the heavens that cannot be explained in any better way. Either black holes are real, it is said, or there are holes in relativity.

There is, of course, nothing wrong in building theoretical models for structures before they are observed. Sir Arthur Stanley Eddington once remarked, only half in jest, "You cannot believe in astronomical observations before they are confirmed by theory." Eddington, by the way, a few years before Oppenheimer's calculations, came remarkably close to constructing a model of a black hole.

"The star," Eddington wrote, "apparently has to go on radiating and radiating and contracting and contracting until, I suppose, it gets down to a few kilometers radius, when gravity becomes strong enough to hold the radiation, and the star can at last find peace." So far, so prophetic! But Eddington went on: "I felt driven to the conclusion that this was almost a *reductio ad absurdum* of the relativistic degeneracy formula. Various accidents may intervene to save the star, but I want more protection than that. I think that there should be a law of nature to prevent the star from behaving in this absurd way."

It is too early to know whether Eddington's conclusion is right or wrong. In a few years astronomical evidence for black holes may be overwhelming. Or it could go the other way. Today black holes are the fashionable playthings of clever astrophysicists. Tomorrow their models may collapse to take their places alongside phlogiston and the epicycles of Ptolemy.

The most sensational of recent conjectures is that our entire expanding universe is destined to enter a black hole. If there is enough matter in the universe (much of it could be hidden inside black holes), gravity will halt the expansion and the universe will start going the other way. Cosmologists can think of nothing to prevent this collapse from plunging the cosmos into a black hole. As for what happens next, who knows?

Readers who want to go more deeply into the structure of black holes will find Robert Wald's book *Space, Time, and Gravity* (University of Chicago Press, 1977) a worthy purchase. He is a physicist at the University of Chicago's Fermi Institute, and his book is based on a series of lectures he gave at the university in 1976. It covers the same ground as Asimov's book, but with more technical information. The last chapter is particularly good in summarizing the recent discoveries of the young Cambridge mathematical physicist Stephen Hawking.

Hawking's combination of courage, optimism, and intellectual virtuosity is already legendary. For years he has been almost totally paralyzed by a progressive nerve and muscle disease. Although he can move himself about in a motorized wheelchair, he cannot write, and he speaks with enormous difficulty. But his mind still operates with crystal clarity, and his calculations are still staggering his colleagues.

Hawking's major discovery is that black holes are not black. Quantum theory, it turns out, implies that in the powerful gravitational field surrounding a black hole there is constant creation of particles (of every kind) and their antiparticles. Some of these particles fall into the hole, others escape as radiation. There is thus a constant leakage of energy, and a flux around the hole that could be observed.

If black holes are large, this loss of energy is slow and negligible. Hawking believes, however, that the big bang may have been chaotic enough to have fabricated billions upon billions of micro black holes, each smaller than a proton, but containing a mass of a few hundred million tons. These "primeval" miniholes would now be in their final stages of evaporation. They would get hotter and hotter, smaller and smaller, and finally explode in a tremendous burst of particles and gamma rays.

Nigel Calder's big, handsome *The Key to the Universe* (Viking, 1977) devotes only two chapters to black holes, but they are excellent nontechnical summaries, and the other chapters are a splendid introduction to the latest theories of matter. Calder is one of the most reliable of British science writers. His book, based on a popular BBC television show which he wrote and presented last January, is abundantly illustrated with diagrams and photographs, including pictures of famous physicists whose faces the public seldom sees.

Calder is unusually skillful in explaining quark theory and why it is rapidly outrunning its nearest rival, the "bootstrap" theory. The bootstrap hypothesis is the "democratic" view that none of the particles that make up matter is more fundamental than any other. Each is simply an interaction of a set of other particles. The entire family thus supports itself in mid-air like a man tugging on his bootstraps, or a transcendental meditator in the lotus position, suspended a few feet above the floor.

Quark theory is the aristocratic view that particles are combinations of more elementary units which Murray Gell-Mann named quarks after the line in *Finnegans Wake*, "Three quarks for Muster Mark!" At first only three kinds of quarks were believed necessary: up, down, and strange, together with their antiparticles. The three kinds are called "flavors." There are now reasons for thinking there is a fourth flavor, "charm." Each flavor comes in three "colors." In the U.S. the colors naturally are red, white, and blue. (Calder's plates use red, blue, and green, with turquoise mauve, and yellow for the anticolors.) This makes twelve quarks in all, with their twelve antiquarks.

Color and charm are, of course, whimsical terms unrelated to their usual meaning, although the mixing of quark colors does obligingly correspond (as Calder shows) to the mixing of actual colors. Some theorists think there are still other quark properties such as truth, beauty, and goodness. Abdus Salam, the noted Pakistani physicist, is now promoting a "quark liberation movement" that regards quarks as made of "pre-quarks" or "preons." In Peking a group of young physicists have a similar view involving "stratons" that form an infinite nest like a set of Chinese boxes.

The essences of these debates are skillfully outlined in Calder's book. He even leads you to the brink of the new, exciting "gauge theories" that may someday unify the strong, weak, and electromagnetic forces—perhaps even gravity—in one fundamental theory.

P. C. W. Davies's *Space and Time in the Modern Universe* (Cambridge University Press, 1977), although it too contains an excellent account of black holes, is mainly a summary by a British physicist of modern views about space and time. Davies has long been troubled by why events in our universe go only one way in time. His earlier book, *The Physics of Time Asymmetry,* was fairly technical. This volume covers the same ground, but on a level more readily understood by laymen.

Time has at least five different "arrows":

1. The arrow of psychological time—our consciousness of the flow of events from past to future.

2. The arrow of certain weak interactions involving K mesons. All other particle interactions are "time reversible" in the sense that, if you take a motion picture of them and run it backward, you see nothing to indicate the film has been reversed. The K-meson events, inexplicably, violate this reversibility.

3. The arrow of entropy—the movement of macrosystems such as galaxies toward increasing disorder (comparable to the destruction of order in a deck of cards by random shuffling).

4. The arrow of radiation from a center, such as the expanding concentric circles produced by a stone dropped into a pond, or the radiating light of a sun.

5. The monstrous arrow of the expanding universe.

How these five arrows are related to one another, and whether universes can exist with one or more (perhaps all) arrows pointing the opposite way from those of our own universe, is a singular story, nowhere better told than in Davies's book.

Fred Hoyle's *Ten Faces of the Universe* (W. H. Freeman, 1977) is the latest of his seemingly endless popular surveys of modern astronomy, all lavishly illustrated and entertainingly written. Hoyle likes to beat loudest on the drums of his own inventions, but it doesn't matter because his speculations are never boring. It must have been a tragic experience for him to watch his beloved steady-state theory of the universe go down the drain as the big-bang theory became accepted, but this seems not to have diminished his mental energy or his fondness for funky theories. Hoyle's book has less to say about black holes than the others, but this is because it cuts a broader swath. The book includes chapters on earth geology, on biology, and on the importance of population control. (He sees the unchecked growth of population as the greatest of all threats to humanity.)

Our two remaining books, entirely about black holes and related matters, plunge into unrestrained fantasy. Adrian Berry, science writer for a

London newspaper, makes only a feeble attempt to separate fact from reasonable conjecture, or reasonable conjecture from eccentric conjecture. His *The Iron Sun* (Dutton, 1977) is best read as you would an Asimov science-fiction novel. Indeed, some of Asimov's novels anticipate much of what Berry has to say.

Berry is concerned mostly with the conjecture that every black hole is joined by a wormhole to a white hole in some other part of the cosmos, or to a white hole in a completely different cosmos. Perhaps the "other" world is made of antimatter, like the Antiterra of Nabokov's novel *Ada*. Matter pours into our black holes to emerge as antimatter in the other world's white holes, while its antimatter pours into its black holes to emerge from our white holes.

Some recent calculations have suggested that a spaceship just might be able to rocket into a black hole and avoid hitting the dreadful singularity. Berry imagines a future in which spaceships use black and white holes as entrances and exits for instantaneous travel across vast distances. When this becomes possible, he writes, mankind will be able to roam and colonize the entire universe. Science-fiction heroes have been doing this for decades, but Berry dresses it up in the latest jargon, and his book is fun to read if you don't take it seriously.

John Gribbin's book on white holes carries this kind of fantasy to still greater heights. Indeed, his book is almost as funny as John G. Taylor's *Black Holes*, published in 1973. Taylor is the mathematical physicist at the University of London whose latest book, *Superminds*, is about British children who Taylor is convinced can bend spoons by paranormal powers better than Uri Geller's. The psi force is probably electromagnetic, Taylor argues. His black hole book is less preposterous, but it does use the black hole as a jump-into point for occult speculations. (For Taylor's later about-face, see Chapter 16.)

Gribbin, who has a doctorate in astrophysics, is the co-author of an earlier book of quasi-science, *The Jupiter Effect*. This great work explains why "there can be little doubt" that in 1982 Los Angeles will be the site of "the most massive earthquake experienced during this century." In 1982 all nine planets will be on the same side of the sun. Jupiter's pull on the sun will thus be augmented by the other planets. This will cause unusual sunspot activity which will agitate the earth's atmosphere. This in turn will agitate the earth's crust, especially along the San Andreas fault. "There can be little doubt" is the phrase that should have forewarned the good Dr. Asimov before he wrote his introduction to this book.

The most absurd passage in Gribbin's new book, *White Holes* (Delacorte, 1977), the one on white holes, speculates on how tachyons may explain psychic spoon-bending. Tachyons are conjectured particles that go faster than light. There is not the slighest evidence they exist, but if they do they would, for certain observers, move backward in time. "Perhaps," writes Gribbin,

> the spectacular production of bent spoons produces the wave of astonishment from the audience, releasing a flood of tachyons which travel backward in time to cause the spoons to bend just before they are produced to cause the surprise. If such a process could be triggered deliberately, it would explain telepathic phenomena as the direct tachyonic communication between minds, but something as physical as spoon bending seems to require the pooled effort of many minds—except, according to John Taylor, in the case of children. This should be no surprise in the light of the above; children have more vivid imaginations than most adults, with more powerful emotions presumably releasing stronger tachyonic vibrations. Perhaps this tachyonic link even provides a clue to such mysteries as poltergeists!

Black holes and bent spoons. The healthy side of the black-hole craze is that it reminds us of how little science knows, and how vast is the realm about which science knows nothing. The sick side of the black-hole boom is the appropriation of astrophysical mysteries to shore up the doctrines of pseudoscientific cults, or the shabby performances of psychic rip-off artists.

Penrose is now doing research and publishing papers on a bizarre mathematical entity he has invented, called a "twistor," that he hopes will clear up some thorny problems about the linking of gravity and quantum theory. A twistor is a type of "spinor," a mathematical operator that calculates what happens when rotations are combined. Penrose's twistors are sort of halfway between particles and pure geometry. I wouldn't be surprised to learn that even now some hack journalist is working on an article for *Reader's Digest* titled "Twistors: Cosmic Carriers of Psychic Energy?" With a little help from the media, twistors could become hotter than black holes.

Postscript

John Gribbin, in a letter published in *NYR* (December 8, 1977), insisted that his wild science speculations should all be taken as jokes:

> There is a great danger in asking one person—however brilliant—to review seven books simultaneously. It is possible, perhaps even likely, that not *all* of the books will be read with quite the same loving care that would be lavished if the reviewer were dealing with them individually. This seems to have happened with Martin Gardner's recent review of a bumper bundle including my own *White Holes,* and I would be grateful for the opportunity to comment on his rather strange remarks.

It is, of course, the reviewer's privilege to choose to comment not on my current book but on a book published three years ago on a totally unrelated topic, *The Jupiter Effect*. I am pleased to learn that Gardner has at least read one of my publications. But his comments on *White Holes* are ludicrous.

Even from the passage on spoon bending quoted, any intelligent reader — such as a typical reader of your journal — must surely realize that my comments on tachyons and spoon bending are presented tongue in cheek. It is, Mr. Gardner, a *joke*. May I quote the sentence which immediately follows the paragraph Gardner chose to quote: "So much for science fiction. . . ." Even the unintelligent should realize from these words, had they been read at all, my opinion of the bent-spoon cult.

If Gardner has any serious comments on the book, I would like to hear them. For now, in view of the incompetence of his so-called "review" I feel that the least you could do is ask someone — *anyone* — actually to read the book and then to provide a genuine review which does not present as my own beliefs ideas which I have discussed only in order to hold them up to ridicule myself.

John Gribbin

Puzzled by this unexpected ploy, I responded as follows:

Mr. Gribbin quotes only the first half of the sentence which he says proves that the passage I quoted was intended as a joke. The complete sentence is: "So much for science fiction, for the time being at least."

Now this sentence is not sufficient to establish that the theory Gribbin presented — tachyons as an explanation for psychic spoon bending — was intended as a joke. The reason is simple. Eighty percent of the theories discussed in *White Holes* are "science fiction, for the time being at least." The notion that some day we may be able to rocket around the universe by going in and out of black and white holes is science fiction at its wildest. Indeed, the very concept of a white hole is "science fiction, for the time being at least."

Does Gribbin consider tachyons a joke? In his book's introduction Gribbin writes: "The concept of tachyons — those faster-than-light particles — has not yet gained general acceptance. . . . How long will it take, I wonder, before this new imaginative leap is used by sober engineers to design communicators with which we can transmit and receive messages that traverse the Galaxy faster than the speed of light?"

To send a message faster than light means sending it back in time, and this leads directly to logical contradictions. If *A* sends a tachyonic message to *B*, in another galaxy, and *B* replies tachyonically, then *A* gets the reply before he sent the message. Gribbin nowhere mentions this well-known paradox, which renders tachyons (if they exist) useless for communication, but perhaps he knew about it all along and intended his remarks on tachyons to be taken as tongue in cheek.

It is good to know that Gribbin regards psychic spoon-bending as a joke and that his several pages about it (with many references to his colleague

John Taylor who wrote an extremely serious book on the "Geller effect") are there only to hold spoon bending "up to ridicule." But now I am puzzled about the rest of the book. Can it be that Gribbin also intended to hold white holes up to ridicule and that his entire book is a joke?

Martin Gardner

My guess that maybe Gribbin's entire book is a joke was a good one. In two later personal letters Gribbin clarified his position. "I regard all abstract theoretical science as a joke." He adds that he considers tachyons especially funny.

I doubt if many admirers of his later book, *Timewarps* (Delacorte Press/Eleanor Friede, 1979), take this book as another joke. As before, nothing in the text suggests that Gribbin has his tongue in cheek. He defends reincarnation throughout, citing the Bridey Murphy case as good evidence, without giving any hint about how Bridey was exposed. He praises the books of Michel Gauquelin, the Frenchman who has his own peculiar brand of astrology (correlating professions with planetary positions at birth), praises books by occult journalist Lyall Watson, speaks highly of Jack Sarfatti's theories (strongly recommending the crazy book Sarfatti co-authored, *Space-Time and Beyond*), regards the reality of ESP and precognition as "proven" (page 140), finds the evidence for dream telepathy "convincing" (page 142), and so on. Of tachyons he writes that the "balance of evidence now is in their favor" (page 109), which surely he must know is totally untrue. The book is filled with spaceships plunging through black holes into other universes, some of the universes running backward in time.

The history of his *Jupiter Effect*, written in collaboration with Stephen Plagemann, is a classic example of what little effort editors make to check eccentric-science manuscripts with experts. The book contains no picture showing the Grand Alignment of the planets in 1982. As a result, readers took "alignment" to mean that the nine planets would be in an approximate straight line. Indeed, when the Library of Science took this book, their gee-whiz brochure stated that for the first time in 179 years "all the planets will be perfectly aligned on the same side of the sun."

What the authors really meant was that the planets would be within the same semicircle. They come nowhere close to being in line. I was surprised to get another letter from Gribbin in 1978, saying he and Plageman had discovered that the planets bunch up even better in December 1980; this was now their preferred "trigger date," with November 1982 their second choice. He said I should feel free to "broadcast this forecast generally."

When the book first appeared, no reputable geophysicist or astronomer was impressed. *Time* (October 7, 1974) quoted experts as making such remarks as "astrology in disguise" and "pure fantasy." A check of

179 years before 1982 showed no unusual seismic activity in any quake-prone areas of the earth. Astronomer George Abell pointed out that Jupiter and Saturn are so large that their combined mass is twelve times that of all the other planets together, yet their frequent lineups correlate with no earthquakes or solar activity. For strong critiques of the book see two papers by Jean Meeus: "Comments on the Jupiter Effect" (*Icarus*, vol. 26, 1975, pp. 257–268) and "Planets, Sunspots and Earthquakes" (*Mercury*, July–August 1979, pp. 72–74); "The Jupiter Effect," by D. Anderson (*American Scientist*, November–December 1974, p. 72); and "The Great Earthquake Hoax," by E. Upton (*Griffith Observer*, January 1975).

I found it surprising that, when the Book-of-the-Month Club rejected it after being told by experts that it was worthless, the book was picked up by a *science* book club! I once asked an editor at Random House, who had bought the book for paperback reprint, if she had checked on the book with any astronomer. She looked surprised. "No, why should I? Gribbin has a doctorate in astrophysics."

The climax to this Jovian comedy was Gribbin's article on "Jupiter's Noneffect," in *Omni*, June 1980. Gribbin totally rejects the effect. "The book has now been proved wrong; the whole basis of the 1982 prediction is gone. . . . There is no reason now to expect any unusual seismic disturbance in 1982 from the causes given in the book. This does not, of course, rule out the possibility of big earthquakes then. But if you want an astrological prediction, I'm afraid you are going to have to ask someone else."

What changed his mind? The sun. In 1979 the sun's activity rapidly increased and is expected to pass its peak by the end of 1980. "Plagemann and I definitely got the year wrong. . . . There is every prospect that 1982 will be quieter in seismic terms than 1979 and 1980. . . . If Los Angeles is still standing by the end of the year, the rest of our forecast will have been invalidated."

Gribbin even allows that his critics were justified. "I am older now and, I hope, wiser." He is sorry that "half-baked cults" and "weirdos" latched onto his ideas. However, he still believes that earthquakes do correlate with solar activity, and he closes his apology with: "Mind you, as long as the sun continues to be active this year, I'm keeping my fingers crossed for the sake of L.A."

I predict that if L.A. is destroyed before the end of 1982 Gribbin will find a clever way to revive the Jupiter effect. And if it isn't, well, didn't he tell us in 1980 that his theory had been discredited?

33

"Close Encounters of the Third Kind"

Close Encounters of the Third Kind opens with a bang. At first the titles flash on and off in eerie silence, then a faint sound slowly swells in volume until it explodes. A symbol of the explosion that created the universe? The producers' hope that the movie will blow everybody's mind?

It is too early to know whether young (age thirty) Steven Spielberg, the director who gave us *Jaws,* has done it again, this time without a bare nipple or a spurt of blood. The film's dazzling photography, high decibel score, and tolerable acting make it hard to see how bad the film really is, but of course that is the secret of blockbusting. Douglas Trumbull, who created the special effects for *2001: A Space Odyssey,* is indeed a genius, and his contributions to *Close Encounters* are everything the film's publicity says. Alas, beneath the visual hanky-panky stretches a thin, hackneyed plot that was done to death in the SF magazines and third-rate films of the fifties.

This is easier to comprehend if you read Spielberg's ghost-written version, *Close Encounters,* issued by Dell paperbacks in 1977 as a movie tie-in. Here on the stark pages, uncontaminated by clanking sounds and flashing colors, you can savor the film's dull story, cardboard characters, and dreary dialogue in all their pure, clean, adolescent banality. Both novel and movie, however, have one thing going for them that could make the film as whopping a success as *Star Wars.* More than any other

This review, which appeared in the *New York Review of Books,* January 26, 1978, and the letters quoted in the Postscript are reprinted with permission of the New York Review of Books. © 1978 NYREV, Inc.

SF novel or movie, they reflect the extent to which ufology has become a pop religion.

Millions of Americans, disenchanted with science and politics, are longing for apocalypse—for a mystical explosion that will instantly solve the world's problems and start a new age of love. For Protestants who haven't left, or who are able to return to, evangelical Christianity, expectation of the Second Coming is rapidly rising. Billy Graham more and more thumps on the theme of a hopelessly corrupt world, firmly in Satan's grip, but any day now—surely soon!—the Lord will return. Eccentric cults based on Parousian nearness are flourishing as seldom before. Shabby books like Hal Lindsey's *Late Great Planet Earth* sell by the millions.

For those who cannot believe in the Second Coming, or the Messianic hopes of orthodox Judaism, there are the UFOs! If the earth is being visited by extraterrestrials, if the sky (as an Indian sadhu puts it in *Close Encounters*) is singing to us, surely the aliens must be friendly or by now we would have learned otherwise. It is this childish possibility that has kept the flying saucers aloft for thirty years. Thirty years! Exactly the age of Mr. Spielberg.

Strange things have, of course, always been happening in the heavens, but the first flying saucer "flap" had a precise beginning. It was June 24, 1947. Kenneth Arnold, flying his private plane near Mt. Rainier, saw nine disklike objects flipping through the firmament. A wire service man called them "saucers," flurries of new sightings followed, and ufology arrived to stay.

The press and radio responded quickly to the growing public interest in UFOs and, as always, the sensational books and magazine articles boosted the mania even higher. A few government and military officials at first took the saucers seriously, but after twenty years of investigation the Air Force finally decided that nothing extraordinary was going on overhead. To settle the matter, a distinguished physicist, Edward U. Condon, was handed half a million dollars by the Air Force to produce the definitive "Condon report"—a 1,000-page document that can be summarized in one sentence. There are no UFOs that can't be explained as hoaxes, hallucinations, or honest misidentifications of such natural objects as meteors, Venus, huge balloons, conventional aircraft, reentering satellites, and atmospheric illusions.

Of course the Condon report, when it came out in 1968, no more settled the matter than the Warren Commission report settled the question of who killed the president. Indeed, even before the Condon report was published a leading occult journalist, John G. Fuller, blasted it with an article in *Look*: "The Flying Saucer Fiasco."

Obviously there is no way that the Air Force or anybody else can prove that alien spacecraft are not visiting us. Is there a tooth fairy? No amount of cases in which a grownup is caught pushing a quarter under a child's pillow will add up to irrefutable negative evidence. Always there is a small residue of cases in which grownups are not caught, and the morning appearance of money remains mysterious. No matter how many sightings of UFOs are shown to have natural explanations, there is always a residue—how could it be otherwise?—of cases for which information is insufficient for judgment.

The cast of mind of true believers in alien UFOs is remarkably similar to that of true believers in spiritualism when it was in its heyday. It mattered not a spirit rap how many mediums were caught in fraud. Every time this happened, Sir Arthur Conan Doyle, who believed in the reality of fairies as well as of ghosts, would sigh and point out, as if talking to a child, that some mediums do indeed cheat, but not all of them and not all the time. Always that residue of the unexplained.

Dr. J. Allen Hynek, professor of astronomy at Northwestern University, is the Conan Doyle of ufology. He started out as a debunker, but now is firmly persuaded that something paranormal—he doesn't know just what—is behind the UFO flaps. In his latest book, *The Hynek UFO Report,* issued by Dell in 1977 as a companion to Spielberg's "novel," he writes:

> Today I would not spend one additional moment on the subject of UFOs if I didn't seriously feel that the UFO phenomenon is real and that efforts to investigate and understand it, and eventually to solve it, could have a profound effect—perhaps even be the springboard to a revolution in man's view of himself and his place in the universe.

The title of Spielberg's movie is from Hynek's 1972 book, *The UFO Experience.* Close encounters of the first kind are mere sightings. The second are physical interactions. The third are meetings with the aliens. Spielberg, a long-time UFO enthusiast, hired Hynek to be his technical consultant. Sure enough, in the movie's climactic scene when the great encounter of the third kind occurs, there is Dr. Hynek himself, standing among the observers, puffing contemplatively on a pipe and looking very unsurprised.

The Hynek UFO Report contains nothing of substance that Hynek has not said many times before. Four-fifths of all UFO reports, he cheerfully admits, are easily explained. But that damnable residue! The government is attacked once more for "suppressing" data. The Condon report is again branded a huge fraud whose "cold and clammy" hand was lifted by the last big UFO flap in the fall of 1973 when four planets were exceptionally bright, and there had been a widely publicized claim by two fishermen in Pascagoula, Mississippi, that they had been kidnapped by a flying saucer. UFO debunking books by top scientists and writers are dismissed as flimsy efforts of the "establishment" to sweep truth under the rug. Hynek likens his detractors to those who refused to look through

Galileo's telescope for fear of seeing something that might damage their "belief systems."

The best insight into Hynek's own belief system can be gained from an interview published in the June 1976 issue of *Fate,* a tawdry occult pulp magazine that thirty years ago was the first to publish articles about UFOs as extraterrestrial objects. Hynek once favored the "nuts and bolts" theory that UFOs are physical things, but now, he tells us, he inclines to the view (proposed by Jung) that they are psychic projections. "Perhaps an advanced civilization understands the interaction between mind and matter. . . . Perhaps it is a naïve notion that you've got to build something physical, blast it off with sound and fury to cross vast distances and finally land here. . . . There are other planes of existence— the astral plane, the etheric plane and so forth."

"I believe," he goes on, "the world is in a psychic revolution that most of us are not aware of. And least aware are the *Establishment* scientists. . . . The new puzzle pieces are being given to us by the whole parapsychological scene—ESP, telepathy, the Uri Geller phenomena, psychic healing and particularly psychic surgery."

But now Hynek is troubled by a seeming contradiction. If UFOs are psychic constructs, how come they leave physical traces? "UFOs break tree branches, appear on radar and are photographed. Maybe they're an example of the Uri Geller-type phenomena in which physical effects occur apparently without physical causes. . . ."

Do the aliens come from beyond Pluto, or from "parallel or interlocking worlds"? Hynek wishes he knew. (The fairies, Conan Doyle believed, live in an interlocking world of "vibrations" different from ours. See his book, *The Coming of the Fairies,* with its splendid photos of the winged creatures, far more convincing than the blurry, easily faked photos of ufology.)

"I ran across two contactee cases just recently," Hynek told *Fate,*

in which the witnesses said they were impelled to do something; they were compelled to sleepwalk, to leave their beds and to go where the spacecraft was waiting. There they saw the creatures. They were without will of their own and suffered very bad effects from the experience—nausea, headaches, etc. The modern psychiatrist might label these persons "disturbed." Sure they're disturbed. But why?

Hynek's remarks outline the central plot of *Close Encounters.* Roy Neary ("near" the Great Truth?), played by Richard Dreyfuss, who also starred in *Jaws,* is a power company lineman in Muncie, Indiana—the "Middletown" chosen by the Lynds for their classic sociological study of ordinary Americans. (No doubt the aliens read the book in kindergarten.) When Roy is sent to investigate a mysterious blackout, he has a dramatic encounter of the second kind. Back home, he becomes increasingly haunted by a mountain shape. He sees it first in a glob of shaving cream, and at dinner tries to build it with mashed potatoes while a son's eyes fill with tears. He thinks poor dad is losing his marbles. A few days later Neary is yanking up shrubs to garnish a large model of a mountain he has constructed in his hobby room. His distraught wife, a prototype of the stubborn UFO skeptic, packs the children in a car and leaves.

As luck would have it, Roy sees on a TV newscast the very mountain he has modeled. It is Devil's Tower, a steep-sided mesa in Wyoming. There has been, it seems, a derailment, and the area is being evacuated because a nerve gas has contaminated the region. Roy feels compelled to go there.

A young widow, Jillian Guiler, lives not far from Roy with her four-year-old son Barry. When a UFO passes over her roof at night, all the electrical toys and devices in the house turn on and go haywire. (This, by the way, is something new in ufology. It will be interesting to see if similar events are reported in the 1978 UFO flap expected as a consequence of the movie.) Barry, enjoying the poltergeist fun, runs out of the house. Jill finally catches him, but not before the two are almost killed on the highway by Neary, who is chasing a chain of UFOs around a dangerous curve.

A few nights later, when the UFO returns, the force inside Jill's house is even scarier. She tries to bolt the doors and windows, but the force drags Barry through the dog's kitchen entrance. This time, for reasons never made clear, the aliens kidnap him. Now Jill is obsessed by the mountain shape. She, too, seeing it on the news, can't avoid the trek to Wyoming. Near Devil's Tower, Jill and Roy meet again.

The chemical derailment is only a cover for Project Mayflower. The aliens have made computer contact with an international group of ufologists, headed by a handsome expert acted by the French movie director François Truffaut. Spielberg had in mind Jacques Vallee, a French-born ufologist who collaborated with Hynek on their 1975 UFO book, *The Edge of Reality*.

What the Devil is going on? Well, the aliens want a rendezvous on Devil's Tower. Technicians have blasted out a clearing on the mountain and surrounded it with floodlights, computers, television cameras, portable toilets, and so on. A Moog synthesizer is connected to a big display screen on which each tone lights a differently colored rectangle.

Project personnel try to hustle Roy and Jill out of the area, along with a small group of "nobodies" who also have been inexplicably drawn there, but the pair escape. After strenuous exertions they finally make it to the clearing just in time to hear the loudspeakers boom: "Take your positions please. This is not a drill."

To prove how friendly they are, the aliens put on a stupendous aerial show. For an opener they form stars in the dark sky that duplicate the

Big Dipper. Then their small craft, seemingly made entirely of colored lights, swoop here and there, flying through one another and through the mesa just like Jonathan Livingston Seagull.

The mother ship, a monstrous wheel of light, slowly settles over the clearing and hangs there like a mammoth Victorian chandelier. Swing low sweet chariot. In the novel it generates a negative gravity field that makes everybody feel 40 percent lighter. The mother ship is Spielberg's beatific vision, his poor replica of Dante's vision of the Godhead in the last canto of *The Divine Comedy.* Some observers actually fall on their knees in awe.

On the Moog synthesizer a musician plays a corny five-note theme that the aliens have taught the earthlings as a kind of password. The mother ship breaks into deep organ tones. A computer gets into the act, and there is an idiotic jam session that Spielberg describes as "very strange music—at one moment melodic and the next atonal, sometimes jazzy, then a little country western. . . ."

Pauline Kael, in the *New Yorker,* calls this "one of the peerless moments in movie history—spiritually reassuring, magical, and funny at the same time." Apparently Ms. Kael sat bug-eyed through the film, finding it an innocent fantasy of such "immense charm" that she could only liken it to *The Wizard of Oz.* "It's trying to teach us something," says a technician during the film's peerless moment. "It's the first day of school, fellas!"

Like Dante's, Roy's desire and will are now rolling with the divine wheel of cosmic love. His wife and children? Who cares. Truffaut, sensing Roy's desire, recruits him on the spot to join a team of a dozen astronauts (the twelve disciples?) who are waiting in helmets and red jumpsuits to go aboard.

The ship disgorges a group of dazed U.S. Navy men. Surprise! They are the crew of the famous lost patrol of Flight 19, a squadron of five Avenger torpedo bombers that vanished into the Bermuda Triangle in 1945.

Someone says: "Lieutenant, welcome home. This way to debriefing." It would be hard to top that in bathos, but Spielberg does it. "They haven't even aged!" a civilian shouts. "Einstein *was* right!" To which a team leader responds: "Einstein was probably one of them."

And now, toddling out of mother ship, still enjoying the fun and games, is little Barry. Jill rushes forward, accompanied by handclaps in the theater.

Tall creatures start to emerge. They are hard to see, silhouetted against a blinding white light, but we can make out enormous heads, long necks, and pipe-stem arms and legs of great flexibility. They are followed by their children—twittering, lovable little things who rush around touching everybody, "feeling human groins, human faces, human backsides." It's a group grope at Esalen. "If the human didn't like it, they moved on to someone who did . . . an orgy of touching, palpating, feeling, stroking."

The thirteen red-clad astronauts (for Roy is now among them) march solemnly into the mother ship. Presumably they will be brought back later, from wherever they are going, stuffed with transcendent wisdom. The Age of Aquarius has dawned. Jill watches through happy tears, snapping photos with her Instamatic, and too freaked out to guess that when Roy returns she'll be an old lady and he'll still be thirty-two.

At last—a close-up of an alien. It has a big balloon of a face, with enormous Kewpie-doll eyes. Responding to Truffaut's noble, transfigured countenance, the face manages a feeble, crooked smile before he, she, or it returns to the mother ship. The picture ends, not with a bang but a simper.

Before the brave astronauts go aboard there is a crude church service during which a priest intones: "God has given you his angels' charge over you." Could it be that these friendly humanoids are the angels of the Bible? Billy Graham and Father Andrew Greeley won't buy it, but millions of Laodicean Protestants will have no trouble stashing the notion into their brains alongside demons and other moldering vestiges of Christian mythology.

It is this pretentious, quasi-religious, Nirvana-like finale that may well keep Spielberg's ridiculous script from sending Columbia Pictures stock into a tailspin. Or maybe not. Maybe enough ordinary souls out there, even in Muncie, are capable of smelling the spiritual fakery of it all. For it is not God who comes to rescue humanity. It is just another race of humanoids.

"It turns me on," Spielberg told *Newsweek,* "to think that when we die we don't go to heaven but to space, to Alpha Centauri, and there we're given a laser blaster and an air-cushion car." Does this not say it all? Gee Whiz, fellas! Jesus (a superhumanoid from another galaxy?) once prayed (Luke 10:21): "I thank thee, O Father, Lord of heaven and earth, that thou hast hid these things from the wise and prudent, and hast revealed them unto babes." This is why the aliens are so interested in Barry and simple nobodies like Neary. This is why the wise scientists won't look through Hynek's psychic telescope.

In the original film version of *Close Encounters* a song from Roy's childhood floats into his head just before he boards the celestial chariot. You won't believe it, but the song is from Walt Disney's *Pinocchio,* and its stanzas still grace the last pages of the novel.

> *When you wish upon a star,*
> *Makes no difference who you are,*
> *Anything your heart desires will*
> *come . . . to . . . you.*

As Roy, eyes shining, tramps like a Cub Scout into the Great Mystery, another stanza jogs through his mind:

> *Like a bolt out of the blue,*
> *Fate steps in and sees you through.*
> *When you wish upon a star your*
> *dream . . . comes . . . true.*

After this scene provoked derisive snorts at a Texas preview, Spielberg had enough sense to recognize that its effect was about the same as having Roy burst into the lyrics of "On the Good Ship Lollipop." I'll bet a dime that even Dr. Hynek was happy to see Pinocchio go.

What remains is not much better. It is fashionable now to describe Spielberg as a terribly gifted but innocent prodigy, bug-eyed with wonder and lost in the Ozzy worlds of modern technology and the silver screen. It will be interesting, concluded *Newsweek,* to watch him grow up. Yes. And the more he grows the less likely he'll make another blockbuster.

Postscript

In its March 23, 1978, issue the *NYR* printed the following letter from Budd Hopkins:

Masters of rhetorical overkill must know not only where to place the howitzers but more importantly, how to shackle one's enemies to highly visible straw men. In his January 26 piece entitled "The Third Coming," Martin Gardner gave a seamless demonstration of these skills. He didn't like the film *Close Encounters of the Third Kind* and used it to attack a recent book, *The Hynek UFO Report,* one of the three items listed as being under review in his article. The author, Dr. J. Allen Hynek, is introduced simply as "professor of astronomy at Northwestern University. . . . He started out as a debunker, but now is firmly persuaded that something paranormal— he doesn't know just what—is behind the UFO flaps." Then Gardner fires his howitzers, an informal and unimpressive magazine interview with Hynek, published elsewhere, being the target for his attack. Gardner also presents as evidence for the foolishness of the UFO phenomenon this information: ". . . after twenty years of investigation the Air Force finally decided that nothing extraordinary was going on overhead."

The innocent reader might have liked to know that for those twenty years this very same Prof. Hynek was the Air Force's scientific consultant on UFOs, that he spent a number of those years as the leading official debunker—remember "swamp gas," one of his less inspired explanations?— and that he takes the UFO phenomenon very seriously now precisely because of the persuasive evidence accumulated by this twenty-year Air Force investigation.

Another thing Gardner could have told us about the book he was allegedly reviewing was its subject. *The Hynek UFO Report* is a study and statistical analysis of the Air Force's own files on UFOs, a review of the 13,134 reports collected during the twenty years Hynek served as consultant. As such, the book is a narrowly focussed and important document, written by a participant.

Concealing the subject matter of the book under review and neglecting to mention its author's unique credentials and vantage-point are bad enough, but Gardner's use of a movie he didn't like to attack something as complex and multi-levelled as the UFO phenomenon is absurd. It's a little like using the banality of Stanley Kramer's *The Defiant Ones* to discredit the southern civil rights movement, or marshaling Méliès films and H. G. Wells's fictions to belabor NASA's more complicated projects.

Misinformation and rhetorical sleight of hand abound in Gardner's review: "the Condon report . . . can be summarized in one sentence. There are no UFOs that can't be explained as hoaxes, hallucinations, or honest misidentifications. . . ." The fact is that over 25 percent of the UFO reports studied by the Condon scientists remained unidentified. The case conclusions studied one by one are particularly revealing, as in this example: "although conventional or natural explanations cannot be ruled out, the probability of such seems low in this case, and the probability that at least one genuine UFO was involved appears tò be fairly high." Again, "It does appear that this sighting defies explanation by conventional means." And a marvelously sophistic "solution": "This unusual sighting should therefore be assigned to the category of some almost certainly natural phenomenon which is so rare that it apparently has never been reported before or since"— in other words, I don't like the implications of this report so I'll invent a "natural" miracle in place of a disturbing, possibly "unnatural" one.

Gardner's article also implies a consensus among scientists that the UFO phenomenon is beneath contempt, but again this consensus is a rhetorical invention. Recently the 2,611 members of the American Astronomical Society were sent questionnaires about the UFO issue. More than half responded; 53 percent of these said the UFO phenomenon "probably" or "certainly" deserves scientific study (17 percent said "probably not" and a mere 3 percent said "certainly not").

Gardner's assault involved an attempt to tie the UFO phenomenon to everything from Christian theology to the tooth fairy, but the problem is that Astronauts Slayton, Cooper, and McDivitt reported UFO sightings, not the tooth fairy. President Carter so far as I know has never claimed to see an angel, though he did, while he was governor of Georgia, file a UFO report. No full-scale scientific inquiry, to my knowledge, has ever been launched to investigate elves or ghosts, nor would the hundreds of astronomers mentioned above suggest such an investigation "probably" or "certainly" should be undertaken. Mr. Gardner finds it more comfortable to view Spielberg's film as if it, rather than the Air Force data Hynek presents, constitutes the UFO problem. He should have skipped the movie and read the book.

Budd Hopkins

To this I replied:

Budd Hopkins's most glaring bit of verbal obfuscation is one that I tried to clear up in my review. It is one thing to say that an object in the sky is "unidentified," and quite another to call it an extraterrestrial spacecraft. UFO buffs are forever pointing out that astronauts have reported UFOs. This sounds impressive until you realize that it means nothing more than that they reported seeing something they couldn't identify. In this literal sense, almost everybody, including Jimmy Carter, have seen UFOs. I myself saw an awesome one at the age of about ten when I was lying awake one summer night in Oklahoma and gazing out a window. My guess now is that I saw a fireball (it split into two parts, each of which continued through the sky), but how can I be sure?

Hopkins's blurring of the distinction between "unidentified" and "extraterrestrial spacecraft" is behind his objection to my summary of the Condon report. I did not say that this report said that all data had been explained. I said, correctly, that the report concluded that all data *could* be explained. I went to considerable pains to give reasons why it is the nature of the case that there be many sightings for which information is insufficient to pinpoint an explanation. It no more follows that these UFOs are alien spacecraft than it follows from unidentified noise in a radio telescope that aliens are trying to signal us.

My assertion that it is a consensus among scientists that UFOs are not alien spacecraft is not in the least belied by the astronomers who replied to the questionnaire cited by Hopkins. On this survey see the criticisms of Philip Klass and John Robinson in the Fall/Winter 1977 issue of the *Zetetic,* with replies by P. A. Sturrock, the ufologist who made the survey. Among those who troubled to respond to the multiple-choice quiz, 23 percent thought the "UFO problem certainly deserves scientific study," 30 percent thought it "probably" does, 27 percent checked "possibly," 17 percent checked "probably does not," and 3 percent checked "certainly does not."

Had I taken this quiz I would unhesitatingly have checked "certainly deserves." No one doubts that there is a "UFO problem." I believe it deserves serious study by psychologists and sociologists, as do all such long-lasting belief manias. Hopkins does not add that another part of the same questionnaire offered astronomers a choice of eight explanations for UFOs. Ninety percent thought they had "prosaic/terrestrial" explanations, 7 percent checked "a cause which respondent cannot specify," and 3 percent checked the extraterrestrial technology hypothesis. Note the exquisite balance. 3 percent were true believers like Dr. Hynek, and 3 percent thought it was a waste of time to investigate UFO reports.

Hopkins's use of this survey to give the impression that 53 percent of American astronomers share Dr. Hynek's wild views is a sterling instance of how the results of the survey have been distorted by believers, and about which Sturrock himself complains in his replies to his critics.

Martin Gardner

While I type, *Close Encounters* is back in the theaters. Spielberg has chopped out fifteen minutes and added twenty minutes of old out-takes and new footage, mainly to provide scenes inside the mother ship and suggest a sequel. I haven't seen the revised version.

I did see Hynek, along with Edgar Mitchell, Betty Hill, George Barski, and Robert Jastrow, as a guest on the Stanley Seigel TV show, December 30, 1977. Mitchell said he thought it "probable" that UFOs were extraterrestrial, that he was a "religious" man, and that the UFOs "might" be here to aid humanity. Hynek was even more cautious. He allowed that *Close Encounters* was overdramatic, but he praised Spielberg's knowledge of ufology and said that everything in the movie was based on actual reports.

Betty Hill, whose close encounter (along with her now-deceased husband, Barney) was the basis of a book by John G. Fuller, retold her wild story for the umpteenth time. She displayed a modeled head of one of the creatures who had "dragged her aboard" the UFO, and she said that the kewpie-doll head in *Close Encounters* had been based on it. One of the aliens spoke to her in English, and she has a torn dress to prove she struggled with them. Both Mitchell and Hynek looked uncomfortable while she rattled on, though neither of them uttered a single critical remark.

Barski, an elderly gentlemen who owns a liquor store, told the story of his encounter of the second kind. All he did was watch from his car while ten or more creatures from a cylindrical UFO scooped up samples of earth. They were three feet high and looked, he said, like children in jumpsuits.

Jastrow pointed out that if there are intelligent beings elsewhere in the universe they are likely to be billions of years ahead of us and therefore extremely unlikely to resemble human children in jumpsuits. The trouble with all contactee stories, he said sensibly, is that the creatures never talk or behave in a way commensurate with advanced beings. As I would put it, they talk and act exactly like the notions that simpleminded people have of creatures from other worlds.

The friendship between Hynek and Vallee suddenly cooled in 1979, but let me first back up and summarize Vallee's strange career. His first two books of ufology, *Anatomy of a Phenomenon* (Regnery, 1965) and *Challenge to Science* (Regnery, 1966) leaned toward the then-prevailing view among UFO buffs that UFOs are alien spacecraft. The second book, which Vallee wrote with his wife, Janine, has an introduction by Hynek.

Vallee became a U.S. citizen in 1967 and now lives near San Francisco, where he heads his Infomedia Corporation. He has a French master's degree in astrophysics and a doctorate in computer science from Northwestern University.

In 1969 Regnery brought out Vallee's third book, *Passport to Magonia*. It marked his first big turn in mental space. UFOs, he argued, are probably *not* nuts-and-bolts spacecraft. More likely they are paranormal phenomena, as Jung suggested. Vallee explicitly likened them to the fairies that everybody seemed to see in Conan Doyle's day.

Vallee's next book, *The Invisible College* (Dutton, 1975), introduced his concept of a "control system." UFOs are myths created by unknown paranormal forces. The book's title refers to an underground network of people, organized by Vallee, who are studying UFOs seriously. The same paranormal hypothesis dominates *The Edge of Reality* (Regnery, 1975), which Vallee wrote in collaboration with Hynek. "There is a physical object," Vallee said in an interview published in *Fate* (February 1978). "It may be a flying saucer or it may be a projection or it may be something entirely different." Whatever it is, it "has the ability to create a distortion of the sense of reality or to substitute artificial sensations for the real ones. . . . A strange kind of deception may be involved."

Deception! This bold thought reached full flower in Vallee's latest book, *Messengers of Deception* (And/Or Press, 1979). In this stupid, paranoid work Vallee puts forth the hypothesis that UFOs are the product of deliberate human deception by high government officials, possibly a collaborated effort by the major governments of the world similar to deceptions they used against Hitler in the Second World War. The purpose of the plot is to spread irrationalism around the world, an irrationalism that could topple humanity and lead to another Dark Age. UFOs are not from outer space. They are created right here on earth by an advanced "psychotronic technology." They are real, all right, and can do all sorts of terrible physical things, such as mutilate cattle. Exactly what is a UFO? "I don't know what it is," Vallee told Christopher Evans in an *Omni* interview (January 1980). "It seems to be a lot of electromagnetic energy in the form of microwaves, in a small space, and an intense, colored 'light.'"

Hynek does not buy this earth-based conspiracy theory. He is particularly incensed by Vallee's suggestion that agents of the Great Plot have infiltrated numerous UFO organizations and cults, including Hynek's own Center for UFO Studies! That was a bit much. Hynek blasted the plot theory in his article "Messengers of Deception, Or Who's Manipulating Whom?" in *Second Look* (May 1979).

Hynek says he doesn't know what UFOs are either. In his lecture entitled "What I Really Believe About UFOs" (*Proceedings of the First International UFO Congress*, compiled and edited by Curtis G. Fuller, Warner paperback, 1980), Hynek said essentially the following. There is strong evidence that UFOs are nuts-and-bolts spacecraft controlled by ETI (extraterrestrial intelligence) and equally strong evidence they are psychic phenomena controlled by EDI (extradimensional intelligence in some parallel reality). Hynek proposes a third possibility: They are both physical and psychic, both material and mental.

Hynek says he doesn't "support" any of these theories. They were hotly debated at the 1977 congress at which Hynek spoke, a congress sponsored by that great "scientific journal," *Fate* magazine. The proceedings, unless you are a confirmed UFO believer, are funnier than a book by

Velikovsky. Hynek didn't want to offend anybody listening to him, but he left no doubt about his belief that UFOs present a deep mystery for science and that we are in a position similar to that of Galileo trying to understand sunspots. We are on the threshold of a major scientific breakthrough, although Hynek doesn't know just what it will be. People in future centuries, he says, will look back on us and say, "They were really dumb in those days. They didn't even know what UFOs were." Doyle felt exactly the same way about his ghosts and fairies, but instead of people looking back now and saying how dumb the skeptics were—they didn't even understand fairies and ectoplasm—they look back and marvel at how dumb Doyle was.

As for dumb skeptics like Carl Sagan, Phil Klass, and myself, who see the UFO mania as nothing more than a social-psychological phenomenon, we are what Vallee calls the "useful idiots" who are being manipulated by whatever diabolical government and military forces are behind the great UFO deception. Does Vallee really believe all this? I wish I could say no and call him an interesting charlatan, but I fear the answer is yes.[1]

I am not so sure about Charles Berlitz. He made his first pile of dough with *The Bermuda Triangle* and is now trying again with *The Roswell Incident* (Grosset & Dunlap, 1980), in which he and his co-author tell us all about the flying saucer that crashed near Roswell, New Mexico, in 1947. The Air Force keeps saying it was just a weather balloon and its instruments, but Berlitz knows better. It was a craft from outer space, and the CIA is hiding the wreckage and the bodies of the extraterrestrials in a secret warehouse in Virginia. Yes, Virginia, the aliens are right there, and we "sit at the verge of the greatest news story of the twentieth century. . . ."

Said *Discover* (in its October 1980 review of this latest slice of ufological baloney): "Anyone who will believe that such a secret could be kept through six different administrations in a loose-lipped city like Washington deserves Charles Berlitz."

NOTE

1. A qualified yes. In his preface to the Bantam paperback edition of *Messengers of Deception*, Vallee reveals that Major Murphy, so often quoted in the book, is a fictional character. "I invented him because I felt that there was a need to bring out a controversial and novel perspective on the entire problem of UFOs."

34

"Four Arguments for the Elimination of Television"

Several years ago I read about a man who was so annoyed by the drivel on his television set that he blasted the screen with a shotgun. The drivel has since grown worse—bad drivel drives out good drivel—and now along comes *Four Arguments for the Elimination of Television* (Morrow, 1978), by an ex-advertising executive with the unlikely name of Jerry Mander, that tells us exactly what to do. Destroying a single television set won't solve the problem. We must, says Mander, exterminate television altogether.

He's not joking. The nature of television, Mander argues, is so insidious that it is "not reformable." He likens it to military weaponry. You can't rehabilitate a bomb. "Television . . . must be gotten rid of totally." Otherwise society will move straight into the nightmares of *Brave New World* and *1984*.

In Mander's mind the United States is already a dictatorship of sorts. Big Brother is Big Business. Only Big Business can afford network television time. This enables Big Business to control programming and use television for reducing viewers to robots. Unable to distinguish reality from television fantasy, the homogenized, mesmerized public is brainwashed into buying endless products that nobody needs. The rich get richer while the bamboozled consumers, sitting trancelike in front of their television screens, steadily deteriorate mentally, emotionally, and physically.

If Mander had written only his first chapter about the baleful effects of television, then used the rest of his book to show why television can't

Reprinted with permission from the June 1978 issue of *American Film Magazine.* © 1978 The American Film Institute, J. F. Kennedy Center, Washington, DC 20566.

be reformed, to explain how to get rid of it utterly, and to explore the implications of such a radical step, his book might have been worth reading. Instead, Mander meanders on and on, detailing evils we already know and mixing them up with dubious evils we don't know.

We all know about television's monstrous advertising power. We all know how television corrupts democracy by concealing the true character of political leaders behind carefully contrived television images. We all know that every year more people sit for longer periods with glazed eyes glued to the boob tube. Who can doubt that television-addicted children would be healthier romping outside in the fresh air and sunshine? Unfortunately, Mander is not content with the obvious—he has to hammer home his points with pseudoscience.

Consider John Ott, a former banker who published a book in 1973 called *Health and Light.* According to Ott, all artificial lights are harmful, but some are more harmful than others. Fluorescent light in particular. If pink, it can cause cancer in rats. Color television sends out less dangerous visible light than black-and-white because the light covers a wider spectrum, but it harms us more because it also shoots out X-rays. If you check the section on Ott in that great scientific treatise *The Secret Life of Plants,* [1] you'll get more details on why Ott thinks "mal-illumination" is as dangerous as malnutrition. His "research institute" at Sarasota has shown that color television makes children hyperactive, withers bean plants, and reduces the litters of rats living fifteen feet away. Mander takes all of this seriously. He thinks it entirely possible that color television can cause human cancer.

That isn't all. Drawing on Anne Kent Rush's book *Moon, Moon,* he explains how wavelengths of light correlate with the "resonances" of food. Unless the light you are getting has the right wavelengths, the iron or calcium in your food is no good. Food colors are also important. For lung ailments you should eat white foods like turnips. For heart disorders you must eat red food like beets. Intestines need pink, the spleen needs green. Mander buys it all. He suspects that the phosphorescent light from color television, "projected at 25,000 volts directly into human eyes and from there to the endocrine system," is doing awful things to us.

Not only does television flood our bodies with horrendous vibrations, it also saturates our minds with terrible images—images that stress war over peace, violence over nonviolence, charisma over wisdom, sex over love, the bizarre over the normal, and so on. To bolster his views that such images can damage one's health ("viewing 'Kojak' means absorbing his character"), Mander cites the "autogenic therapy" of J. H. Shultz. This European healing fad takes people through imaginary tours of their insides, "visually discovering their organs . . . and picturing them as functional and healthy." Carl Simonton is using similar "image therapy" to cure cancer. [2] Good images heal. Bad images make us sick.

Mander's greatest heights of absurdity are reached when he belabors the television screen for being an artificial substitute for the real world. For starters, the screen is flat. Mander also reminds us, so help me, that it fails to supply taste, smell, or touch. (You'd think he would point out how much better color television is than black-and-white movies, but his book is devoid of such subtleties.)

What's worse, television images keep changing abruptly to hold our attention. The real world doesn't behave like that. Even motion on the screen, he explains, is fake—an illusion produced by little dots flashing on and off as the screen is scanned by the electron beam. It never occurs to Mander that staring at a printed page, which I was forced to do for several hours in order to get through his vacuous and irrelevant book, is to stare at something even less like the real world than television.

Nearing the end of Mander's diatribe, I kept wondering when he would get around to revealing how society can abolish television without abolishing technology. It turns out that he wants to abolish technology, too. Well, not quite—just all of it past a certain line. And where is the line? It is the line beyond which the operations of technology are "too complex for the majority of people to understand."

Mander realizes that this cuts a wide swath. He wants to get rid of all nuclear power, satellite communication, microwave technology, laser technology, control computer banks, genetic engineering, and a "thousand other processes." Surely we would have to abandon computers, cars, airplanes, ships, telephones, radios, and motion pictures. Washing machines would have to go (who knows how their timers work?), even modern flush toilets. Does Mander understand his typewriter well enough to repair it? Did he write his book with a quill pen?

Mander is fully aware that advanced technology can't be abolished without overthrowing capitalism. Such a revolution, he allows with cheerful Mander candor, would be easier than reforming television. We might as well try, as he unfelicitously phrases it, "for the hole in one."

After we get rid of capitalism, what goes in its place? Not socialism, because Mander assures us that state control of television, as in democratic Sweden or the totalitarian Soviet Union, is just as bad if not worse than control by Big Brother Business. Alas, he ends his book with no more of a hint about the nature of the new order he so passionately desires than one could have got from any of the leaders of the old New Left. We can only surmise. Presumably it is some form of agrarian anarchism— everybody down on the farm, living the simple life.

A native New Yorker and a graduate of the Wharton School of Finance and Commerce, Mander spent fifteen hollow years as a public relations and advertising man before he discovered yoga, mandalas, the ecology movement, and how much he hated his work. His San Francisco advertising agency was responsible for *Scientific American's* great paper airplane contest, and Mander's only previous book is on how to fold such toys. At least the man who shot his television destroyed one set. The British Luddites managed to smash hundreds of their hated machines. On the back cover of Mander's book, the editor of *Film Quarterly* likens him to David with his slingshot. It is a naive metaphor. The book is more like a paper bomber folded from a sheet of Kleenex and tossed against the wind in the general direction of Madison Avenue.

NOTES

1. *The Secret Life of Plants,* by Peter Tompkins and Christopher Bird (Harper and Row, 1972), was surely one of the funniest psi books of the seventies. Portions of it actually ran in *Harper's Magazine* (November 1972). No wonder *Harper's* later went broke.

Peter Tompkins had earlier written *Secrets of the Great Pyramid* (Harper and Row, 1971), almost as funny as his book on plants. Chapter 21 of this work tells you how little models of the Great Pyramid will keep food fresh and razor blades sharp. Could it be, Tompkins wonders, because the shape of the pyramid concentrates Wilhelm Reich's orgone energy? If you're curious about the history of these crazy claims, see Chapter 17 of my *Incredible Dr. Matrix* (Scribner's, 1976). On the earlier literature about pyramidiocy see Chapter 15 of my *Fads and Fallacies.*

2. O. Carl Simonton is medical director of the Cancer Counseling and Research Center in Fort Worth. He and his wife co-authored *Getting Well Again*, published in 1978 by J. P. Tarcher, Inc. (Tarcher is the husband of Shari Lewis, one of the greatest and certainly the prettiest ventriloquists of all time. He is currently turning out a dozen or so books a year that deal with offbeat science and health.)

The notion that cancer is essentially psychosomatic has been around for many decades and is frequently intertwined with beliefs in ESP and PK. Lawrence LeShan, for example, whose book *You Can Fight for Your Life: Emotional Factors in the Causation of Cancer* was published by M. Evans in 1977, has written several books on parapsychology and psychic healing. Simonton's work is not in this psychic area of "holistic health," but he firmly believes that cancer is primarily caused by emotional stress and that it is best combatted by image therapy.

According to Maggie Scarf's article "Images that Heal" (*Psychology Today*, September 1980), the Simonton technique is roughly as follows. One imagines the healthy white blood cells transformed into "voracious sharks" that attack and kill the "frightened, grayish little fish," which are the malignant cells. The dead fish are then flushed out with body fluids, and the sharks revert back to peaceful white cells.

Simonton himself had a skin cancer on the side of his nose that was surgically removed when he was 16. Later he developed another skin cancer on his face. He told Scarf that he had made it "go away" by practicing his imaging method.

There have been many studies of the relation of cancer to earlier emotional stress, but they are contradictory and inconclusive. Scarf does a good job in summarizing both sides of the controversy. Most cancer experts do not look favorably on Simonton's work, and some even call it a "cruel hoax" in the same category as Krebiozen and Laetrile. For another recent defense of "mental image" therapy see *Imagery and Cancer,* by two Dallas psychologists, Jeanne Achterberg and G. Frank Lawlis (Institute for Personality and Ability Testing, 1978).

35

Four Books on Catastrophe Theory

If white and black blend, soften, and unite
A thousand ways, is there no black and white?
Pope, An Essay on Man

"All things," said Charles Peirce, "swim in continua." At what wave length does blue become green? When does a child become a grown-up? Are viruses alive? Do cows think? It is also obvious that there are discrete "things" that swim in these spectrums, and sometimes jump from one part of a spectrum to another. Day fades into night, but a flicked switch produces instant darkness. One can imagine a hippopotamus changing by imperceptible degrees into a violet, but, as Charles Fort once asked, who would send a lady a bouquet of hippopotami?

The abstract world of pure mathematics displays the same crazy mixture of continua and discreteness. The counting numbers rise by jumps, but the real numbers form a continuum so dense that it is meaningless to ask what real number comes next after any counting number. Between 2 and 2.000 . . . 1, where the dots represent, say, a billion zeros, there is an uncountable infinity of other real numbers. Continuous functions often graph as curves with well-defined maxima and minima, and with singularities that can be sharper than spearheads.

The continuities and discontinuities of mathematics, for reasons that trouble philosophers, fit the real world with incredible accuracy. Add

Reprinted from the *New York Review of Books,* June 15, 1978, with permission from the New York Review of Books. © 1978 NYREV, Inc.

two cats to two cats. Lo and behold, you get four cats. Apply calculus to the smooth motions of the earth, sun, and moon, and the abrupt start of an eclipse can be predicted with fantastic accuracy.

It is a naïve error to suppose that calculus is concerned only with smoothness. Like the real world, it too is riddled with abruptness. Toss a ball from here to there. At the top of its trajectory it enters a singularity where it makes an abrupt change from one type of behavior (rising) to another (falling). In the language of calculus, the derivative that measures its rate of vertical change goes to zero. More complicated events may involve many variables, each changing smoothly, yet the system may reach critical points at which it suddenly flips from one state to another. For centuries mathematicians applied calculus to such singularities, but it was not until the mid-sixties that René Thom, a distinguished French topologist at the Institut des Hautes Études Scientifiques near Paris, hit on a startling new insight. Thom called his discontinuities "catastrophes." His work in this field, and the work of his followers, quickly became known as "catastrophe theory," or CT for short.

Thom's fundamental discovery was that under certain precisely defined conditions there are just seven types of elementary catastrophes. Each involves no more than four variables and can be modeled in what physicists call a "phase space" (in CT it is a "behavioral space") of two through six dimensions. In these abstract spaces the change of a system is diagrammed by the path of a single point that moves over a smooth "behavior surface." The catastrophe occurs when the point is forced by the structure to jump from one sheet of the surface to another.

Thom's seven types are defined topologically, which means that the actual magnitudes of each variable are irrelevant. Topology is a branch of geometry concerned only with properties that remain the same regardless of how a figure is deformed in a continuous way. Think of the figure as drawn on a rubber sheet. If you stretch, shrink, or bend the sheet, the figure's topological properties are unaltered; for example, if points on a line are labeled 1,2,3,4, . . . , they cannot lose this ordering. There is no way to deform the sheet so that 3 is no longer between 2 and 4. The seven elementary catastrophes are like scaleless maps drawn on rubber surfaces. In their own way they are as beautiful as the five Platonic solids.

The tossed ball illustrates the simplest of the seven. Thom called it a "fold" catastrophe because it can be graphed on an ordinary sheet of paper creased along a line that goes through the curve's singularity. The ball's behavior is shown by a point that moves along this curve. When it crosses the crease it jumps from going up to going down.

The next simpler catastrophe, the "cusp," is less trivial. Its model is three-dimensional: a two-dimensional surface pleated in three-dimensional space so that when it is projected on a plane the pleat maps as a cusp-shaped curve. (A familiar example of a cusp is the bright line, shaped like the top of a heart, that you sometimes see on the surface of a cup of coffee under strong slanting light. In the drawing below, the cusp is shown at the bottom of the diagram.) The "swallowtail" catastrophe is modeled in four dimensions. The "butterfly," "hyperbolic umbilic," and "elliptic umbilic" catastrophes are five-dimensional. The "parabolic umbilic" requires six dimensions. There are infinitely many catastrophes in higher spaces, but all with four or fewer variables are topologically equivalent to one of the seven.

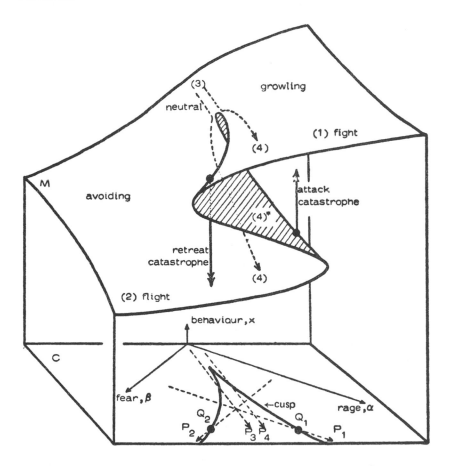

Zeeman's cusp-catastrophe model of a provoked dog. The pleat in behavior surface M projects as a cusp on control surface C where the horizontal axis represents rage versus fear, and the vertical axis represents change from neutral behavior to attack or flight. As the dog is provoked, its behavior is modeled by a point that ranges over the behavior surface. Its path projects as a trajectory on the control surface below.

The point starts at 3. As the dog is increasingly provoked, the point is most likely to enter the fight sheet on the right or the flight sheet on the left. But under certain circumstances, when fear and rage tend to be equal, the point crosses an edge of the pleat where it is forced to jump abruptly either down to flight or up to fight. On the lower surface, these trajectories plot as paths P1 and P2. Points Q1 and Q2, where a trajectory intersects the cusp, are the catastrophe points at which the dog's behavior suddenly alters.

Thom discusses his work and its biological implications in a remarkable book, *Structural Stability and Morphogenesis,* published in French (by W. A. Benjamin, Reading, Mass.) in 1972 with a foreword by the late C. H. Waddington, a British geneticist who was the first top scientist to greet CT with enthusiasm. (An English translation by D. H. Fowler was published by Benjamin in 1975.) E. Christopher Zeeman, who heads the Mathematics Institute at the University of Warwick, was another British friend of Thom who became an enthusiast. Slightly younger than Thom, he too had specialized in topology; his doctorate was in knot theory. In a few years, after much writing and lecturing on CT, Zeeman became the theory's foremost fugleman. In 1977 Addison-Wesley published his *Catastrophe Theory: Selected Papers 1972–1977.*

And now a funny thing happened in the history of modern mathematics. The event can be modeled—Zeeman himself in a playful mood so modeled it—as a catastrophe. What normally would have been a minor technical discovery in topology abruptly flowered into a crusading cult.

It all started in England in 1975 with a burst of publicity. The BBC's television program *Horizon* hailed CT as a big scientific breakthrough. The *New Scientist,* a popular weekly, featured the theory on a cover that Alexander Woodcock and Monte Davis describe in their book, *Catastrophe Theory* (Dutton, 1978), as resembling an advertisement for a Hollywood disaster film. In the United States the theory got its first big boost in *Newsweek* with a two-page article by Charles Panati, a journalist of the paranormal who gave us that great scientific anthology, *The Geller Papers.* Panati stressed the application of CT to human behavior, quoted Zeeman's ringing pronouncement "Catastrophe theory is a major step toward making the inexact sciences exact," and declared that CT was the most important development in mathematics since calculus. In April 1976 Zeeman's *Scientific American* article on CT added more prestige to the movement.

As I see it, three leading variables underlay this extraordinary surge of interest. "In the beginning," as Tim Poston and Ian Stewart put it in their book, *Catastrophe Theory and Its Applications* (Fearon-Pitman, 1978), "was Thom." His book is a provocative blend of mathematics, wide-ranging science, misty metaphysics, impenetrable speculation, and purple propaganda. Although modest in his expectations of immediate applications, Thom offered CT as a new "language" for science, a new "paradigm" that eventually would have revolutionary consequences.

The second factor in CT's abrupt rise was the charismatic, witty personality of Zeeman. He delighted his students by showing them his "catastrophe machine"—a rotating cardboard disk with two attached rubber bands—that demonstrates the behavior of a cusp. In 1975 he precipitated the first major battle over the theory by applying it to a 1972 riot in England's Gartree prison. His enthusiasm was as catching as prisoner discontent. Soon a small band of followers, some of them his students, were beating the catastrophe drums with the zeal of a Salvation Army street band.

The third factor was the interplay between CT's terminology and the temper of the times. The word "catastrophe" resonates with apocalyptic hopes and fears. One thinks of atom bombs, cosmological big bangs, black holes, political terrorism, airplane crashes, earthquakes, floods, fires, revolutions, third encounters, and the Second Coming. Catastrophe theory! What PR expert could have devised a better name? Had Thom called it "discontinuity theory" it is likely that only mathematicians would have learned of it.

Because any abrupt change that springs from a confluence of smoothly changing variables can be described by a catastrophe model, it follows that CT can invade any branch of science. In optics it has been most successfully applied to caustics—curves produced by reflection or refraction such as the coffee-cup cusp. A rainbow is a color caustic modeled by the trivial fold catastrophe. The rippling white lines you see at the bottom of swimming pools are caustics. One of the early triumphs of CT was a proof that these fluctuating lines are not like mudcracks but are elongated triangles with cusped corners.

CT applies to any kind of sudden bending, such as the buckling of girders when a bridge collapses. A familiar novelty called a jumping disk is presented by Woodcock and Davis as an example of a cusp catastrophe. The slightly buckled metal disk is warmed, then pressure of the thumb forces the buckle to go the other way. Placed on the floor, the disk is stable until cooling causes the buckle to flip back again and send the disk several meters into the air. Magicians have a similar trick with a tennis ball. Make a tiny air hole so that pressure with the thumb will produce a small dimple in the ball's surface. Set the ball on a slight incline, balanced on the invisible concavity. As air slowly reenters the ball, the dimple disappears and the ball suddenly rolls down.

Any phase transition or simple threshold phenomenon in physics lends itself to CT: the abrupt freezing or boiling of water, the sudden shock wave produced by the crack of a whip or a plane breaking the sound barrier, and so on. An earthquake is an obvious example. Biological phenomena to which Zeeman and others have tried to apply CT are all over the place: the alteration of protein when you boil an egg, the sudden division of a cell, the firing of a nerve impulse, the beat of a heart, mutations, insect plagues, the rapid evolution of a new species or the vanishing of an old one, the differentiation of embryonic cells, the orgasm.

Animal behavior swarms with catastrophes. Zeeman's favorite example is the cusp behavior of a slowly provoked dog. The conflicting drives of rage and fear do not cancel to produce neutral behavior. According to Zeeman and the cusp model, the dog's behavior diverges in one of two directions that lead either to abrupt attack or to sudden flight. (See box on page 367.)

Catastrophists are busy investigating psychological phenomena of the sort stressed by the Gestalt school: reversals of perspective in optical illusions, explosions of anger, bursts of tears, quick decisions to get married or divorced, nervous breakdowns, the "aha" reactions in creative thinking. When Archimedes ran naked down the street shouting "Eureka!" he was celebrating a cerebral catastrophe. Zeeman has applied the butterfly to the psychosis called *anorexia nervosa* (compulsive fasting). John Allen Paulos, at Temple University, is writing a book about CT and humor—the slow build of a joke to a punch line that triggers the explosive guffaw. Going to sleep and waking up are abrupt transitions with cusp models. I cannot resist a juicy quote that I found in William James's *Principles of Psychology* on the catastrophic decision to get up on a cold morning:

> We know what it is to get out of bed on a freezing morning in a room without a fire, and how the very vital principle within us protests against the ordeal. Probably most persons have lain on certain mornings for an hour at a time unable to brace themselves to the resolve. We think how late we shall be, how the duties of the day will suffer; we say, "I *must* get up, this is ignominious," etc.; but still the warm couch feels too delicious, the cold outside too cruel, and resolution faints away and postpones itself again and again just as it seemed on the verge of bursting the resistance and passing over into the decisive act. Now how do we *ever* get up under such circumstances? If I may generalize from my own experience, we more often than not get up without any struggle or decision at all. We suddenly find that we *have* got up. A fortunate lapse of consciousness occurs; we forget both the warmth and the cold; we fall into some revery connected with the day's life, in the course of which the idea flashes across us, "Hollo! I must lie here no longer"—an idea which at that lucky instant awakens no contradictory or paralyzing suggestions, and consequently produces immediately its appropriate motor effects. It was our acute consciousness of both the warmth and the cold during the period of struggle, which paralyzed our activity then and kept our idea of rising in the condition of *wish* and not of *will*. The moment these inhibitory ideas ceased, the original idea exerted its effects.

The social sciences have not escaped. CT models have been applied to stock market crashes, union decisions to strike, abrupt shifts of public opinion about anything, panic behavior of crowds, revolutions, alterations of social status by marriage, the fall of Rome. The sudden rise of the stock market last April 14 was an economic catastrophe. Robert Holt, a political scientist at the University of Minnesota, believes that the beginning and end of World War I can be usefully analyzed with catastrophe models. *Behavioral Science* devoted its September 1978 issue to CT.

No mathematician, it is important to realize, disputes the elegance of Thom's models or denies their value as descriptive metaphors. It is one thing, however, to describe nature in a novel way, quite another to apply models that lead to significant explanations and predictions. It is not hard to understand why the absence of such results, especially in CT applications to the "soft" sciences, led to a second catastrophe—a bitter backlash of opposition on the part of both mathematicians and scientists.

In this country the first report on the backlash was Gina Bari Kolata's hard-hitting article in *Science* in 1977 (vol. 196, April 15, 287 ff.). Writing on "Catastrophe Theory: The Emperor Has No Clothes," Kolata found that large numbers of eminent mathematicians held low opinions about applied CT. There are, they pointed out, well-established branches of mathematics, such as the theory of shock waves, quantum mechanics, and bifurcation theory, that handle complex natural discontinuities quite well. Applying CT to biology and human behavior is little more, they said, than describing a familiar structure in a colorful new terminology. The new descriptions are too vague and nonquantitative to lead to worthwhile insights. They tell us, said the critics, nothing we don't already know.

One thinks of Kurt Lewin, the German Gestalt psychologist who became so enamored in the thirties with topological diagrams that he applied them to hundreds of human behavioral events. Like CT, Lewin's "topological psychology" made temporary converts, and there was even a school of topological sociology. I recently reread some of the pros and cons of this debate and was surprised by how much the rhetoric resembles that of today's CT controversy. Even the behavior space of CT has its analogue in Lewin's "life space." His diagrams seemed promising at the time, but it soon became evident that they were little more than sterile restatements of the obvious.

This is exactly the criticism now being leveled at the more complicated topological diagrams of CT. "I do not see that fitting the surface of a cusp to a phenomenon involving discontinuity is an enormous conceptual breakthrough" is how topologist John Guckenheimer, one of the milder critics, has put it. Stephen Smale (like Thom he has won the Fields Medal, the highest honor a mathematician can receive) declared: "In a sense I reject catastrophe theory completely. It is more of a philosophy than mathematics, and even as a philosophy it doesn't apply to the real world." Marc Kac, another top mathematician, called Zeeman's *Scientific American* article "the height of scientific irresponsibility."

Hector Sussmann and Raphael Zahler, both of Rutgers, are CT's most outspoken critics. They do not deny that CT has many useful applications in the physical sciences, but they are dubious about its applications to the biological and social sciences. "The proponents of catastrophe theory," they declared recently, "have overwhelmed the public with a mass of claims, but reality fails to substantiate them. The great achievements of catastrophe theory are no more than delusions. . . . We find that the proponents . . . systematically make fundamental mathematical errors, reach conclusions that are either false, or vague, or meaningless, or trivial, and often misrepresent empirical evidence." A long and detailed attack on applied CT, by Sussmann and Zahler, will appear in the forthcoming (volume 37) issue of *Synthèse*.

Thesis, antithesis. It is too early to know what form the synthesis (if any) will take, but most mathematicians are now on the side of the critics. Laymen with no knowledge of CT but interested in this acrimonious fracas can do no better than read *Catastrophe Theory,* the little book by biologist Woodcock and science writer Davis. It is nontechnical, and although the authors are strong defenders of applied CT they try to give a fair account of the opposition. They agree that CT has so far accomplished little except in physics, but they are optimistic. "Someday," they write, "it may be as natural to speak of a 'cusp situation' or a 'butterfly compromise' as it is today to speak of the 'point of diminishing returns' or a 'quantum jump.'"

In contrast to the popularly written Dutton book, the volume by Poston and Stewart is a ponderous textbook of 491 pages intended mainly for scientists who are firmly grounded in calculus and interested in applying CT to their own fields of research. The book is handsomely illustrated, clearly written, and astonishingly up to date considering the speed with which papers on CT are proliferating. A valuable bibliography lists more than 400 references.

Half the book is on the mathematics of CT, half on applications. The emphasis is on applications in physics, with special attention to areas where CT is moving from its earlier, purely qualitative analysis toward quantitative methods. There are sections on the application of CT to the stability of ships (with some recent results on vertical-sided vessels such as oil drilling platforms), mirages, sonic booms, fluid mechanics, ocean waves, thermodynamics, magnetism, and laser physics. Scant attention is paid to biological applications, although the authors believe that CT will play a major role in this area. The behavioral sciences are discussed in a final chapter. "If *any* mathematical methods can aid the growth of such wisdom," the book's last sentence predicts, "then catastrophe theory will be part of them."[1]

Poston and Stewart make no effort to reply in detail to the criticisms of Sussmann and Zahler. Neither appears in the index, although their forthcoming paper in *Synthèse* is mentioned as having "enjoyed a certain notoriety, but its usefulness is seriously marred by repeated errors." Throughout the volume CT is treated as a powerful new tool that eventually will be useful in all the sciences. The book is dedicated "To Christopher Zeeman, at whose feet we sit, on whose shoulders we stand"—a curious position that seems not very stable.

For anyone interested in applications of CT to biology and the social sciences, the collection of papers by Zeeman is indispensable. The book is a photo-offset of the original articles, plus a bibliography and general index. Thom's book, which started it all, is best read only after one is familiar with CT's basic ideas. Its idiosyncratic mix of technical mathematics, science, and philosophical musings is likely to inspire either admiration, scorn, or frustration over its opacities.

Although Thom is not easy to understand, the center of his vision is clear enough. It is the opposite of a famous statement by Paul Dirac which Thom quotes as follows: "The main object of physical systems is not the provision of pictures, but the formulation of laws governing phenomena, and the application of these laws to the discovery of new phenomena. If a picture exists, so much the better; but whether a picture exists or not is a matter of only secondary importance."

For Thom the picture is of primary importance. "I am certain," he writes, "that the human mind would not be fully satisfied with a universe in which all phenomena were governed by a mathematical process that was coherent but totally abstract. Are we not then in wonderland?" Simply to describe a paradoxical state of affairs, as for example in so many areas of quantum mechanics, and leave it at that is, for Thom, to "sink into resigned incomprehension"—a habit that stifles scientific progress because it leads to indifference. We must, Thom urges, constantly seek geometrical models. "The dilemma posed by all scientific explanation is this: magic or geometry." On a deeper level, even successful geometry is magic in the sense that it miraculously fits the external world. And all successful magic, Thom adds, is geometry.

CT is thus viewed by Thom as a new way of modeling what may be an infinitely complex and basically unknowable reality. By breaking away from quantitative models and by making use of the topology of higher- and even infinite-dimensional spaces, he believes that for the first time in history we have a method that ultimately can model everything. "There is no doubt it is on the philosophical plane that these models have the most immediate interest," Thom writes. "They give the first rigorously monistic model of the living being, and they reduce the paradox of the soul and the body to a single geometric object. Likewise on the plane of the biological dynamic, they combine causality and finality into one pure topological continuum, viewed from different angles."

"Whither away?" as Poston and Stewart head their final chapter. It is conceivable that CT, like information theory and game theory, will become a valuable tool for the behavioral sciences. It is also possible that, like Lewin's topological pictures, it is fated to die away as another premature effort to apply geometry in areas where it is either trivial or wrong. More likely its trajectory will take a duller road that winds somewhere topologically in between.

NOTE

1. Poston and Stewart were aware of the humor in this dedication. Their contorted position, Poston later informed me, was felt to be appropriate in dedicating a book to a knot theorist.

Postscript

Considering the nature of my review, I received unusually cordial letters about it from Tim Poston and Ian Stewart, and from Monte Davis. Davis thought my comparison of CT to Kurt Lewin's topological psychology was not apt, since Lewin was not a mathematician and there was no new mathematical content in his rather haphazard applications of "watered-down and even then out-of-date topology." All three said they thought that most mathematicians were neither for nor against CT but were taking a wait-and-see attitude, perhaps a bit embarrassed by Zeeman's enthusiasms yet at the same time distressed by the angry tone of CT's critics. I had followed Davis in likening the seven elementary catastrophes to the Platonic solids. Davis thought a better analogy was with the four conic-section curves, particularly since the transition of, say, a comet from an elliptical orbit to a hyperbolic trajectory is a catastrophe. Good points all.

Shortly after my review appeared, I watched a television play by Tom Stoppard about political controls over professors in Czechoslovakia. In one scene a windbag philosopher discourses on the applications of CT to ethics. Philosophers and theologians long ago appropriated game theory, and I suppose it is only a matter of time before Protestant theologians, eager to be up to date, will be applying CT to the Bible. Maybe CT can help explain why Jehovah decided so abruptly to drown all the world's men, women, children, and animals except for Noah and his family. The Cusp of Calvary? The Resurrection Swallowtail?

John Paulos, instead of writing a book about CT and jokes, produced a small volume of a more general nature called *Mathematics and Humor* (University of Chicago Press, 1980). Chapter 5 is called "A Catastrophe Theory Model of Jokes and Humor." The jokes are excellent. But aside from reminding us that a joke's punch line triggers that crazy convulsion we call laughter, I found Paulos's applications of CT to his jokes singularly unenlightening and almost as funny as his jokes.

36

"The Dancing Wu Li Masters" and "Lifetide"

Gary Zukav is a young writer who became so entranced by the psyche-delic marvels of relativity and quantum mechanics (QM) that he felt compelled to write a popular book about them called *The Dancing Wu Li Masters* (Morrow, 1979). The result is in most ways admirable. Zukav is such a skillful expositor, with such an amiable style, that it is hard to imagine a layman who would not find his book enjoyable and informative.

Now a word of caution. Most of the friends Zukav thanks for initiat-ing him into the mysteries of modern physics are deeply involved with Eastern philosophy and/or parapsychology. As a result the book through-out is highly colored by points of view held by only a small minority of physicists. A few years ago physicist Fritjof Capra wrote a similar book, *The Tao of Physics,* around the leitmotiv that theoretical physics is mov-ing rapidly toward the insights of Eastern philosophy. The dance of Shiva, in Hindu mythology, was Capra's dramatic metaphor. For Zukav the metaphor shifts to the rhythmic body movements of *T'ai-chi. Wu Li* is a Chinese phrase for physics. Zukav sees today's physicists as no longer trying to explain ultimate reality, the unknowable Tao, but only enjoying a kind of *Wu Li* dance in harmony with the universe's dancing waves and particles.

In both books the dancing is tinged with Eastern idealism and subjec-tivism. On the microlevel a quantum system is radically altered whenever it is observed. An electron, for example, does not have a definite position until it is measured. Not until the moment of measurement does nature

---

Reprinted with permission from *Newsday,* May 27, 1979.

"decide" what position to give it, assigning the position in an entirely random and uncaused way, but in accord with the probabilities given in the electron's wave function. From this it is an easy jump to the view that physics is somehow essentially a study of consciousness, that beautiful buzz-word of the counterculture. It is a "me generation" approach to physics. One of Zukav's chapters actually is entitled "The Role of 'I.'" "Do not be surprised," he writes, "if physics curricula of the 21st Century include classes in meditation."

No physicist denies that the measurement of quantum systems disturbs them, but it does not follow that there isn't something "out there," independent of our minds, to be disturbed. It is a distinction Zukav constantly blurs. The majority of today's physicists, as well as most philosophers of science, are "realists" who will find Zukav's subjective epistemology as vaguely defined as it is irritating.

Zukav is good in explaining the classic QM paradoxes that seem to suggest a subjective view of reality, such as the double-slit experiment, the paradox of "Schrodinger's cat" (which is neither alive nor dead until someone looks at it), and the paradox of "Wigner's friend" in which Eugene Wigner (a famous physicist), who looks at the cat, is himself not "real" until a friend observes *him,* and so on into an infinite regress of observers. It is regrettable that only a small footnote gives the standard view that when an irreversible event occurs (such as a cat's death, a geiger-counter click, or the photograph of an electron track) the event acquires a structure as independent of observation as a tree or a star. The book is fun to read, but in my opinion it is as wrong in drawing metaphysical conclusions from QM as the earlier writings of Sir James Jeans and others were wrong in drawing similar conclusions from relativity.

Lyall Watson's *Lifetide* (Simon and Schuster, 1979) is not fun to read unless you are as gullible as he in believing fifty impossible things before breakfast. According to the jacket, Watson has a doctorate from the London Zoo, and this is his fourth book on the paranormal. The title, *Lifetide,* derives from Freud's remark that occultism is a "black tide." Watson sees nothing black about it. To him it is the *elan vital* of philosopher Henri Bergson, the life force that permeates all nature and is responsible for evolution and (buzz) consciousness. Its tides surge through our unconscious, providing the hidden force behind ESP, precognition, psychokinesis, and all other psychic wonders. It makes the universe a single organism, "ever-moving and alive, spiritual and material at the same time."

Lifetide is the familiar Watson potboiler into which he tosses anything he comes across that he thinks will titillate his readers. The beginning is typical. A little girl in Venice takes a tennis ball, does something to it, and suddenly she is holding a ball turned inside out! Watson is unable to see a hole or crack. The ball bounces. Cutting it open he finds inside the fuzzy exterior. Later she repeats the trick and Watson says he kept the miraculous ball on his mantel for two days.

What happened to it? Did Watson have the sense to take it to a laboratory? If genuine it would be an artifact of stupendous import. If fake it could easily be detected. Alas, the everted ball vanishes from Watson's narrative and he is off on something equally preposterous.

The book reaches crowning heights of claptrap in a section on flying saucers. Taking his cue from Carl Gustav Jung (Jung is the book's hero—his name seems to pepper every page), Watson is convinced that UFOs are psychic projections of the collective unconscious, not nuts-and-bolts craft from outer space. Do not suppose this means they are ethereal. If enough people see them they take on the solidity of real objects. They knock down trees, scorch grass, melt asphalt, fuse tin roofs.

Watson buys everything he hears, and by assuming the awesome powers of the "lifetide" he can explain everything. Take the winged "fairies" in the photographs taken by two little girls that Sir Arthur Conan Doyle wrote a book about many years ago. The notion that the two girls faked their photos is beyond Dr. Watson's meager comprehension. To Watson, the girls projected thought-forms onto the film in the same way that Ted Serios, a now-forgotten Chicago bellhop, once projected thought-pictures onto Polaroid film—until his simple method was exposed in a 1960 issue of *Popular Photography*.

But there is more. The collective unconscious can also fabricate monsters such as sea serpents and abominable snowmen that, like UFOs, also become physically real if enough people see them. Watson quotes with approval an Alabama self-declared physicist: " . . . a family of plesiosaurs is going to wind up living in Loch Ness."

Quick, Watson, the acupuncture needle! I want to find out if I am asleep and dreaming about your crazy book, or whether it actually was published by Simon and Schuster, who say here on the jacket that it is a "brilliant eclectic approach" to new frontiers of science by a "gifted and perceptive author."

Postscript

Of the many reviews of Zukav's book, one of the best was by Jeremy Bernstein (*New Yorker,* October 8, 1979). He summed up both the Capra and Zukav volumes perfectly by saying: "A physicist reading these books might feel like someone on a familiar street who finds that all the old houses have suddenly turned mauve."[1]

Lyall Watson first hit the paranormal jackpot with *Supernature* (Doubleday, 1973). He followed it with *The Romeo Error* (Doubleday,

1974) and *Gifts of Unknown Things* (Simon and Schuster, 1977). *The Romeo Error* defends reincarnation, out-of-body experiences, and other wonders related to survival after death. The 1977 book reveals such things as how a little girl on the volcanic island of Nus Tarian, in Indonesia, had such powerful gifts of psychic healing that she raised a man from the dead. To escape persecution by local Muslims she turned herself into a porpoise. The book's full-page ads in the *New York Times Book Review* had rave quotes about the book from Adam Smith and Colin Wilson.

Watson always puts a Ph.D. after his name on his introductions. I was astonished to discover that he really does have one. Westfield College, of the University of London, actually gave him a Doctor of Philosophy degree in zoology in 1964. May the farce be with him.

NOTE

1. For Bernstein's critical review of Capra's book see "A Cosmic Flow," in *American Scholar*, Winter 1978/79, pp. 6–9.

37

"Broca's Brain"

Popularizers of science are a patchwork breed. On rare occasions a great scientist whose work is a pivot point in history, Charles Darwin for instance, has been a skillful writer of books that laymen could read with pleasure. But most scientific geniuses find popular writing difficult, and though publishers sometimes persuade them to try, the results are seldom notable. Einstein's best book for laymen was a collaboration with Leopold Infeld. Niels Bohr struggled to explain quantum mechanics to ordinary mortals but his style was almost impenetrable. At the other extreme are writers like the legendary Isaac Asimov, who, though trained in science, recognize that their talent lies not in making discoveries but in writing about science with such enthusiasm, and such obedience to that admirable maxim "eschew obfuscation," that their books have done more for public understanding of science than twenty universities.

Somewhere in the middle are those who pursue a distinguished scientific career and also have a flair for colorful writing. One thinks of T. H. Huxley, the great science popularizer of his day, and of a long chain of British astronomers—Robert Ball, Arthur Stanley Eddington, James Jeans, Fred Hoyle, Dennis Sciama, to name a few—who took time off from professional labors to explain astronomy to the public.

In the United States the two best-known astronomers now writing books for a general audience are Robert Jastrow and Carl Sagan. Jastrow,

This review, which appeared in the *New York Review of Books,* June 14, 1979, and the letters quoted in the Postscript are reprinted with permission from the New York Review of Books. © 1979 NYREV, Inc.

however, has limited his attention to astronomy, whereas Sagan is all over the lot. *Dragons of Eden,* for which he received a Pulitzer Prize, is about the evolution of human intelligence. His just-published book, *Broca's Brain* (Random House, 1979), contains twenty-five short essays, dazzling in their range and eloquence.

The title essay which opens the book is a typical example of Sagan's ability to mix science with philosophy. In the "innards" of the Musée de l'Homme, in Paris, hidden from the public, he is startled to come upon a large collection of human brains. It had been started by Paul Broca, a famous French neurologist and anthropologist, and the father of brain surgery. "Broca's area," on the cerebral cortex near the left temple, is a part of the brain that controls speech. Broca's discovery of this area, Sagan reminds us, was one of the first discoveries of functional differences between the brain's left and right hemispheres.

A bottle label catches Sagan's eye: "P. Broca." Yes, Broca's own brain has found its way into the collection. Holding the container in his hands starts a sequence of fantastic thoughts in Sagan's living brain. Is there a sense in which Broca is still *in* there? Will science someday find a way to scan a preserved brain and extract its memories? Would such an "ultimate breach of privacy" be a good thing?

And now Sagan asks himself if there are scientific questions that ought not to be asked. He thinks of a good example. If humanity were to destroy itself in nuclear warfare would it not be better had no questions been asked about atomic energy? All scientific inquiries carry risks, but Sagan finally decides it is best in the long run to ask everything. The chapter ends with Sagan wondering whether this idea itself might "be sitting there still, sluggish with formalin, in Broca's brain."

In the book's second essay the microstructure of a grain of salt starts Sagan meditating on how much of the universe is knowable. It is no coincidence that the next chapter is about Albert Einstein because Einstein thought deeply on just this question, and in an autobiographical sketch wrote about how his attempt to understand a minute portion of the huge universe, glowing out there like a "great eternal riddle," had been the dominant passion of his life. (Sagan's essays are independent of one another but, as he tells us, carefully arranged.)

"In Praise of Science and Technology" is a chapter in which Sagan sees no way to reverse direction on the perilous road along which science is propelling us. "We are the first species to take evolution into our own hands." With this power comes also, as we know only too well, the power to self-destruct. Which fork we take depends in part on public understanding of science. It is here, Sagan feels strongly, that the best agents for such education—television, motion pictures, and the press—have failed us. Not only is their science "often dreary, inaccurate, ponderous, grossly caricatured," at times it is even hostile to science. This indictment carries us into the book's next section, "The Paradoxers."

Paradoxer is a gentle, old-fashioned term. Scientists today, talking among themselves, do not hesitate to speak of cranks and crackpots. In more public discourse they call them pseudoscientists. These are not scientists with eccentric ideas—such ideas are published constantly in "establishment" journals—but eccentric ignoramuses who work in far-out fringe areas where extraordinary claims are loudly trumpeted with an extraordinary absence of evidence. At the moment our country is experiencing two major paradoxical trends.

One is a by-product of the new wave of Protestant fundamentalism. It is generating a flood of crank books and periodicals attacking evolution, and reviving ancient beliefs in witchcraft, poltergeists, and demon possession. The great success of *The Exorcist* and its imitations reflects this shabby trend, and now we have *The Late Great Planet Earth,* narrated by (of all people) Orson Welles. This film, based on Hal Lindsey's preposterous book of the same name (sales: 10 million copies!), is about how our old Earth will soon end with the rise of anti-Christ, the Battle of Armageddon and the Second Coming. America's leading apocalyptic faiths—Seventh-Day Adventism, Jehovah's Witnesses, Mormonism, and assorted one-man denominations who stress the impending Parousia (e.g., Billy Graham)—are flourishing as never before. (Herbert Armstrong's Worldwide Church of God, which grew fabulously rich by ridiculing evolution and preaching a trivial variant of Seventh-Day Adventism, seems to have suffered a setback since the old man excommunicated his Priapic son and heir, Garner Ted.)

Galloping alongside the fundamentalist awakening is the "occult explosion"—the public's obsession with astrology,[1] pseudoscience, and all things paranormal. The two trends overlap because Christians who believe in Satan can accept all of the alleged outrageous phenomena and attribute most of it to fallen angels rather than to God or science. Few scientists care to speak out on either trend, and it is not hard to understand why.

Consider the curious case of Immanuel Velikovsky, whose books are closely linked to the fundamentalist revival. A devout believer in orthodox Judaism, Dr. Velikovsky (he was trained in psychoanalysis) set himself the task of revising the laws of astronomy and physics, and rewriting vast globs of ancient history, to spin an incredible tale about the planet Venus that would "explain" the major miracles of the Old Testament. Macmillan's lavish advertising for Velikovsky's first book, *Worlds in Collision,* made no secret of how the book supported the historicity of Old Testament miracles. There is no question that the book would never have found a major publisher, would never have become a best-seller, had it not had a strong appeal to old-time religionists.

About 1500 B.C., Velikovsky claims, an enormous comet that had earlier been thrown out of Jupiter grazed the earth on one or two occasions, thereby accounting for Joshua's success in making the sun and moon stand still, Moses' success in parting the Red Sea, the origin of manna from heaven, the plague of extraterrestrial flies (the flies were carried by the comet), the flood of Noah, and many other wondrous Old Testament events, not to mention the formation of mountains and oil deposits known to be millions of years old. Eventually the giant comet, by some means equally unknown to astronomers, abruptly changed its eccentric elliptical orbit to an almost circular one, and became Venus. All this just a few thousand years ago!

To astronomers and physicists, without exception, Velikovsky's scenario is so crazy that most of them saw no reason why they should waste time even reading him. It is to Sagan's credit that he perceived the rise of the Velikovsky cult (in drum-beating articles in *Harper's*, *Reader's Digest*, miserable little occult magazines such as *Fate*, pseudoscholarly periodicals devoted to Velikovsky, and so on) as symptomatic of a deplorable trend. Unlike most of his colleagues, even to the dismay of some who felt he was risking his reputation, Sagan took the time not only to read Velikovsky but to spell out, in a language anyone could comprehend, the fundamental howlers of Velikovsky's central scenario.

Sagan's chapter "Venus and Dr. Velikovsky" is a masterpiece of anti-antiscience rhetoric. I can think of no job in English to rival it except perhaps H. G. Wells's little book *Mr. Belloc Objects,* in which Wells ripped to shreds Hilaire Belloc's ill-informed effort to disprove evolution. Velikovsky's admirers will, naturally, nitpick at Sagan's essay, and fume with indignation about the dogmatism of establishment science and the "persecution" of a modern Galileo. Even Belloc replied to Wells with a book called *Mr. Belloc Still Objects,* in which he revealed himself to be more ignorant of science than he had formerly appeared. No one with a modicum of knowledge about astronomy and physics can read Sagan's chapter without grasping the fact that there *is* no Velikovskian challenge to astronomy, never was one, and never will be. The big noise is just the sound of an ignorant public being had.

Although Sagan devotes more space to Velikovsky than to any other paradoxer, his chapter on "Night Walkers and Mystery Mongers" is an amusing survey of the still-growing public interest in the paranormal. It opens with an account of Alexander, the Uri Geller of the days of Marcus Aurelius, a handsome psychic charlatan about whom Lucian wrote a famous exposé. Before he is through, Sagan has touched on Spiritualism, astral projection, reincarnation, precognitive dreams, Erich von Däniken's infantile theories, mind-reading horses, P. T. Barnum's Cardiff Giant (a famous fake of a fake), the Bermuda Triangle, astrology, extraterrestrial UFOs, and parapsychology. Leading parapsychologists will object strenuously to being lumped together with most of these things; yet they themselves contribute to bizarre books and periodicals in which all these notions are stirred together. To the general public, alas, the voice of Sagan is much less impressive than the voices of Burt Lancaster and Raymond Burr, who each narrated one of the two worst television documentaries ever made about the paranormal. The big networks, responding to public hunger for the occult, thus furnish the feedback that accelerates the trend.

Leading book publishers, with rare exceptions, are providing the same irresponsible feedback. Sagan was wryly amused when the editor in chief of a top publishing house, prompted by a discussion of H. R. Haldeman's account of Watergate, declared: "We believe a publisher has an obligation to check out the accuracy of certain controversial nonfiction works. Our procedure is to send the book out for an objective reading by an independent authority in the field."

"This," adds Sagan, "is by an editor whose firm has in fact published some of the most egregious pseudoscience of recent decades." Sagan does not list them, but it would be easy to name fifty books of worthless science that in recent years have earned fortunes for authors and publishers, not one of which was sent out for evaluation by an expert. Lippincott's book on cloning by David Rorvik ("a fraud and a jackass" was how Rorvik was described by the distinguished geneticist Beatrice Mintz) is only one of the latest examples of this sleazy, money-grubbing genre.

One of Sagan's paradoxer essays is about a recent flap (occasioned by Robert Temple's book, *The Sirius Mystery*) over the Dogon, an African tribe whose legends include some amazingly accurate astronomical information they could not have acquired without telescopes. Temple's thesis is that Dogon legends support the idea of contacts by extraterrestrial astronauts. Sagan argues the far more plausible view that the myths derive from recent contacts with visiting Europeans. Another paradoxer chapter analyzes the ingenious numerology of Norman Bloom, leader of the Children of God cult. Bloom has invented several new proofs of God based on such astronomical coincidences as that the sun and moon, seen from earth, have disks of almost identical size.

The paradoxer section closes with a nostalgic tribute to early science fiction and high praise of some contemporary SF authors. Like Ray Bradbury, Sagan was hooked as a boy on the Mars novels of Edgar Rice Burroughs, only to become disenchanted when he later realized how little science Burroughs understood. He ticks off some SF themes that were favorites of pioneer writers but are no longer usable: the twilight zone around Mercury, the steamy jungles of Venus, the canals on Mars.

The essays of Section 3 report on recent solar-system discoveries. Sagan still hopes that primitive organisms lurk somewhere on Mars. He argues that life forms may float in the atmospheres of Jupiter and Saturn, and perhaps flourish on Titan, Saturn's largest moon. Titan is larger than Mercury and almost as big as Mars, and its atmosphere is more like ours than that of any other body in the solar system.

Section 4, "The Future," discusses the increasing speeds of travel on earth and in space, the life of Robert Goddard (pioneer of rocket propulsion), the rapidity with which astronomy is becoming an experimental science, ways of searching for extraterrestrial intelligence, and the growing power of machine intelligence. Three chapters on "Ultimate Questions" close the volume.

"Sunday Sermon," the most philosophical of the three, begins with a marvelous anecdote that Bertrand Russell liked to recall. When Russell entered prison for his pacifist sentiments during the First World War, a gate warder had to ask some questions. Russell called himself an agnostic. The warder wanted to know how to spell it, then sighed and said, "Well, there are many religions but I suppose they all worship the same God." This remark, said Russell, kept him cheerful for a week.

Sagan, too, is an agnostic. Like Huxley, who not only called himself an agnostic but even invented the word, Sagan does not deny the possibility of a creator God, but, as did Einstein, he prefers Spinoza's pantheistic deity, the totality of being. As we learn more and more about the universe, writes Sagan, we find less and less for the traditional God to do. When a quasar explodes, a million planets, many perhaps with intelligent life, are obliterated. On such a scale human events seem inconsequential. To all this a traditional theist might reply that science is revealing more and more for God to do, that small size does not something inconsequential make, and that a quasar explosion is no harder to reconcile with a personal God than an earthquake that snuffs out thousands of human lives.

But no matter. Sagan closes with a statement so fair-minded that I cannot imagine anyone, from dogmatic atheist to conservative Catholic, disagreeing. "My deeply held belief is that if a god of anything like the traditional sort exists, our curiosity and intelligence is provided by such a God. We would be unappreciative of that gift . . . if we suppressed our passion to explore the universe and ourselves. On the other hand, if such a traditional god does not exist, our curiosity and our intelligence are the essential tools for managing our survival. In either case, the enterprise of knowledge is consistent with both science and religion, and is essential for the welfare of the human species."

"Gott and the Turtles" takes its title from J. Richard Gott and the endless stack of turtles that uphold the earth in some Asian myths. In 1974 Gott and his associates published strong evidence that the universe lacks sufficient matter to provide enough gravity to halt its expansion. But as Sagan puts it, "The issue is still teetering." The missing matter may yet be found. and in the next few decades we "shall see if Gott knows."

The book's last chapter, "The Amniotic Universe," leaps from Gott to Grof—Stanlislav Grof, a Czechoslovakian psychiatrist now in the United States, and the author of two recent books on the unconscious mind. Along with some of Freud's early disciples Grof believes that we retain dim unconscious memories of our birth. Sagan speculates on the roles that such memories, if indeed they exist, might play in the evolution of religious mythologies, perhaps even in cosmologies. Is it possible, he wonders, that the originators of the now-discredited steady-state theory (in which the universe placidly maintains its overall structure throughout eternity) were born by Caesarian section? If so, they would have escaped those traumatic birth stages that could have predisposed them to the big bang!

This is Sagan in his most whimsical mood. But no one could be more seriously exhilarated by the frontiers of tomorrow's science. Near the end of his book's introduction Sagan predicts that in the next few decades astronomers may even learn the answer to that awesome question: How did our cosmos get started?

If Sagan means no more than deciding between a one-big-bang and an oscillating model that endlessly repeats bangs and crunches, he may be right, but if he means solving the ultimate riddle of the universe's origin I must respectfully demur. Neither model touches the metaphysical problem of genesis. On *this* question one cannot even imagine an advance in cosmology that could put science in a better position to answer the riddle than could Plato or Aristotle.

"In all of the four-billion-year history of life on our planet," Sagan concludes his introduction, "in all of the four-million-year history of the human family, there is only one generation privileged to live through that unique transitional moment: that generation is ourselves."

On the continuum of scientific progress one can indeed mark off unique periods, but will our generation be unique in throwing light on "ultimate questions"? Or unique in the awesomeness of scientific questions it answers? There is an old joke about Adam and Eve as they walked hand in hand out of the (amniotic?) Garden of Eden. "We are living, my dear," said Adam, "in a time of great transition."

NOTE

1. The January 1979 issue of the *Author's Guild Bulletin* reports that the highest price ever paid for paperback reprint rights to a book of nonfiction was the $2.25 million just given by Fawcett to Harper & Row for *Linda Goodman's Love Signs*. Bantam had offered $1 million, and three other paperback houses had bid beyond $1.7 million. Houghton Mifflin is currently advertising Jeane Dixon's *Horoscopes for Dogs* as "the book of the year for everyone who loves a dog!"

Postscript

It is impossible to write anything disparaging about Velikovsky without being deluged by dozens of apoplectic letters from Velikovsky admirers. C. Leroy Ellenberger, a chemical engineer for whom defending Velikovsky has become a mania, sent the most vitriolic blast. It was filled with personal insults, and much too long for the *NYR* to print. They did print (October 25, 1979) a short letter from Lynn E. Rose, a professor of philosophy at the State University of New York at Buffalo, and a longer letter from Daniel L. Kline, of the University of Cincinnati Medical Center.

> For over a quarter of a century, Martin Gardner has been taking cheap shots, grossly ill-informed and maliciously irresponsible, at the work of Dr. Immanuel Velikovsky.
>
> One of Gardner's favorite tactics is to describe Velikovsky as a fundamentalist. Velikovsky's own words on this are emphatic: "I am not a fundamentalist at all, and I oppose fundamentalism." Nothing could be clearer, if Gardner would but listen.
>
> Gardner's latest diatribe repeats some of his old nonsense about fundamentalism, and shows that he still doesn't know even the main points of the theory that he is talking about. To give just one example, Gardner says that, according to Velikovsky's theory, Venus, at that time a giant comet, caused such events as "the flood of Noah."
>
> Actually, Velikovsky regards the Deluge as one of the effects of a nova-like explosion of *Saturn*. Venus didn't even exist yet!
>
> Isn't it about time Gardner got his facts straight?
>
> Lynn E. Rose

My answer to Rose follows:

Professor Rose, an intrepid contributor to Velikovskian publications, is right on one count, wrong on another. I have never called Velikovsky a fundamentalist. How could I, since the term labels a Protestant movement? My review accurately characterized Velikovsky as a "devout believer in orthodox Judaism," and correctly pointed out that the fantastic sales of his first book (72 printings by 1974) were "closely linked" to today's astonishing revival of fundamentalism.

In my list of Old Testament miracles I should not have attached Noah's name to the floods produced by close Earth-Venus encounters. I hope Rose will forgive this terrible blunder. As Rose knows, the Deluge was an earlier, greater flood that Velikovsky believes occurred about nine thousand years ago when Earth and Moon passed through a cosmic cloud of water. For several centuries both Earth and Moon were completely covered with water.

I find it amusing that after wrongly blasting me for calling Velikovsky a fundamentalist, Rose reminds us that Velikovsky accepts literally the Genesis account of the entire Earth being under water in historic times. I suppose

It is too much to expect the Buffalo philosopher, when he teaches his next course on Velikovskianism, to let his students know some of the overwhelming geological evidence against such hoary balderdash.

Martin Gardner

This is what Kline had to say:

Martin Gardner continues the Punch and Judy game of taking turns with Carl Sagan in bashing Immanuel Velikovsky over the head. Since Velikovsky's cosmic theory that Venus was ejected from Jupiter and caused perturbations on earth which were recorded in folk history is shot through with childish errors of chemistry and physics, one wonders why they continue their game instead of permitting the theory to succumb to its own errors. They justify their attacks on the basis that their real target is the anti-intellectualism that they feel is behind the widespread appeal of Velikovsky's book, *Worlds in Collision,* which is still attracting attention twenty-five years after publication. However, I find it difficult to accept this as a complete explanation of their savage personal attacks on Velikovsky ("charlatan," "fraud") when he is obviously a sincere, if mistaken, individual.

It has struck me, after a careful reading of the criticisms of Velikovsky's book, that the critics have taken a Fundamentalist position which may explain their excessive emotionalism. According to them, the book can be of no value since the author indulges in ridiculous confusions of hydrocarbons with carbohydrates and shows a complete lack of understanding of the laws governing the heat of vaporization of solids. The same conclusion, of course, can be applied to the Bible. Should we discard this ancient text because it is full of unscientific statements? Is it possible that Velikovsky may have hit upon an insight into the origin of some of our folklore even though his explanation is scientifically unsound?

Fortunately, some sober and less emotionally involved scientists are beginning to take a new look at Velikovsky's cosmic theory and are asking, as objective scientists, whether such a stimulating idea could be modified so that the basic laws of nature would not be violated. In the November 9, 1979, issue of *Nature,* one of the most prestigious scientific journals in the world, Dr. Michael Rowan-Robinson, a mathematician, dares to suggest that such revisions are possible. As an example, he writes that if a massive long period comet were first noticed toward Jupiter and then appeared as a bright evening or morning star, this would eliminate almost all of the refutations offered against Velikovsky. What would remain is his evidence from ancient texts that this event produced catastrophic effects on the earth, the most interesting part of *Worlds in Collision.* One might even picture a number of such comet-like bodies originating from Jupiter itself.

It is not easy for me, a scientist, to defend Velikovsky when his cause has been taken up by people who believe in UFOs, that plants can communicate, and similar nonsense. However, scientists must consider possible explanations even when they arise from unlikely primordial ooze. Personal attacks have no place in scientific debate.

Daniel L. Kline

My reply:

I have just reread my review of Carl Sagan's new book and I can find no place where I call Velikovsky a charlatan or a fraud, two nouns that Dr. Kline puts inside quote marks as if I or Sagan or both of us had used them. Nor have I been able to find a use of either epithet by Sagan.

I fully agree—and I think I can also speak for Sagan—that Velikovsky is "obviously a sincere, if mistaken, individual." One could say the same thing about Trofim Lysenko or Francis Gall, the father of phrenology. All the great pseudoscientists of the past, who won a popular following, were sincere and mistaken. Indeed, sincerity is the main attribute that distinguishes a pseudoscientist from a mountebank.

As for Dr. Kline's second point, I readily admit that Velikovsky is not always wrong. Considering the scope of science and history covered in his books, it would be astonishing if he did not occasionally get something right. Gall was right in believing that parts of the brain have different functions, but that doesn't turn phrenology into a science. Astrologers are experts in predicting where planets will be, but that doesn't make astrology a science.

On Dr. Kline's final sentence, it all depends on what "personal attack" means. It is one thing to attack a pseudoscientist's character or lifestyle, quite another to attack his scientific ignorance. I have never read a line by Sagan or myself that I would consider a "savage personal attack" on Velikovsky. Would Dr. Kline consider Thomas Huxley's debate with Bishop Wilberforce a savage personal attack, or Clarence Darrow's rhetoric at the Scopes trial a savage personal attack on William Jennings Bryan? Vigorous attacks, yes, but they were directed toward a man's ignorance, not his integrity.

Martin Gardner

Professor Rose and I exchanged letters again in the *NYR* of March 6, 1980:

Martin Gardner's reply serves only to muddy the waters and to add several new items to the long list of erroneous claims that he has made about Dr. Immanuel Velikovsky.

The word "fundamentalist" was mine, and was correctly used to refer generally to anyone who holds that Scripture is literally true. Gardner's practice of restricting the word "fundamentalist" to Protestants (even though more general senses are also in common use) does not alter the fact that Gardner *has* repeatedly accused Velikovsky of accepting Scripture as literally true. Nor does it explain his inattention to Velikovsky's protestations that he is *not* a fundamentalist, and that he does *not* accept Scripture as literally true.

The continuing and growing interest in Velikovsky's theories is due to the unprecedented record of confirmation that those theories have enjoyed. Such interest is not "closely linked" to any "revival of fundamentalism."

Velikovsky has repeatedly stressed: "I am not a fundamentalist at all, and I oppose fundamentalism." Does Gardner think that fundamentalists are so stupid as to take an avowed opponent for an ally? Perhaps he does, but he is wrong.

I find it amusing that Gardner could say that "Rose reminds us that Velikovsky accepts literally the Genesis account of the entire Earth being under water in historic times." Actually, my letter stressed that Velikovsky is not a fundamentalist and thus does not accept Genesis or any other ancient text literally!

I challenge Gardner to provide any evidence that Velikovsky has ever said that for "several centuries" Earth was "completely covered with water." This "balderdash," as Gardner calls it, seem to be one of Gardner's own more recent fantasies.

Isn't it about time Gardner got his facts straight?

Lynn E. Rose

My reply:

Rather than squabble over the word "fundamentalist," which I have never heard applied to anyone except Protestants (and in recent months to Moslems), let me confine my reply to Rose's final challenge. Writing on "Are the Moon's Scars Only Three Thousand Years Old?" in *Velikovsky Reconsidered* (Warner Books), the late Velikovsky said:

> In my understanding, less than ten thousand years ago, together with the Earth, the Moon went through a cosmic cloud of water (the Deluge) and subsequently was covered for several centuries by water, which dissociated under the ultraviolet rays of the sun with hydrogen escaping into space.

On page 89 of the same book Velikovsky writes: "Some nine thousand years ago water was showered on Earth and Moon alike (deluge). But on the Moon all of it dissociated . . . water covered it only for a very limited time (following the deluge) counted in hundreds of years."

I took this to mean that both Earth and moon were equally drenched by the water cloud which, by the way, had its origin in an explosion of Saturn! Since the water of Noah's flood would have had much more difficulty evaporating from Earth than from the moon, it is hard to see how it could have remained longer on the moon than on Earth. However, if Velikovsky meant that only the moon was completely covered for several centuries, and that Earth's present oceans are the product of the Deluge, then I stand both corrected and amazed; amazed because it would mean that Velikovsky adheres even closer to the biblical account of the Deluge than I had supposed.

Velikovsky unambiguously states in the same book (page 257) that all the features we now see on the moon were carved "less than three thousand years ago." This hypothesis, made necessary by Velikovsky's water-cloud-from-Saturn fantasy, is contradicted by so much evidence that only Professor Rose and his fellow Velikovskians can regard it as serious science.

Martin Gardner

If Professor Rose sent a third letter, I did not see it; so I cannot say just how long he thinks Velikovsky thought that the Great Deluge kept the earth under water. As for Velikovsky's accepting every passage of the Old Testament literally, I have always assumed he did not. Modern fundamentalists are not that stupid either. There are biblical passages that, taken literally, say the earth is flat, but not even Oral Roberts believes in a flat earth. Many fundamentalists today think that the "days" of Genesis refer to long geological epochs and do not for a moment insist that Eve was literally made from Adam's rib or that Lot's wife turned into salt.

Velikovsky, let it be said plainly, did not take the Old Testament stories literally. But he was a devout man, who kept a kosher house and believed that the Old Testament miracle tales reflected actual historical events. He did not take "literally" the assertion that the sun and moon once stood still. He took this to mean that the earth stood still. I find it astonishing that some Velikovsky supporters are unable to perceive how Velikovsky's religious convictions strongly shaped his astronomy. There is, of course, no doubt that Velikovsky's publishers have in their advertising exploited to the full the fact that Velikovsky's theories support fundamentalist doctrines. It was this support, not the scientific merit of the theories, that was primarily responsible for the fantastic sales of Velikovsky's crank books.

Velikovsky died in November 1979 at the age of 84, but you can be sure Doubleday will do everything it can to keep his books alive and the great "controversy" burning. Indeed, in 1980 Doubleday was promoting all of the master's books with shameless ads. I call them shameless because they said that Velikovsky's views are rapidly gaining strong support from recent scientific discoveries, which simply is not true. For every trivial prediction by Velikovsky that has been confirmed, there are hundreds that have been falsified, and hundreds more that betray his feeble comprehension of modern astronomy, physics, chemistry, geology, and archaeology. James Oberg has listed a few of Velikovsky's biggest whoppers in his article on "Ideas in Collision" in the *Skeptical Inquirer* (Fall 1980). In the same issue you will find Henry H. Bauer's balanced overview of the Velikovsky affair and Kendrick Frazier's listing of the outright lies in the 1980 Doubleday advertisements.

According to these ads Velikovsky left the manuscripts of several new books. I have no doubt that Doubleday or some other house will soon be hustling them, to the great delight of Ellenberger, Rose, and all the other Velikovsky buffs whose knowledge of modern science can be put inside a thimble.

38

Two Books on Talking Apes

Remember the great tumult during the sixties about talking dolphins? Because a dolphin brain is larger than ours, could it be that porpoises are potentially as bright as we are, maybe more so? John C. Lilly seriously tried teaching English to these clever little whales and for a time actually believed he had taught dolphins to mimic human speech. Like the black races of Africa, Lilly once said, porpoises are on the brink of becoming Westernized, a revolution with unpredictable consequences. "If dolphins come to understand our cold war," he warned, "we don't know how they will proceed to operate."

After Lilly became convinced that several of his Florida porpoises had committed suicide, he abandoned his watery research to wander off into the jungles of parapsychology and Eastern mysticism. He reported fantastic encounters with extraterrestrial intelligence. He told reporters that dolphins were using ESP to "infiltrate" human minds. Eventually it became clear to almost everybody, as it had been all along to "establishment" biologists, that Lilly's research was hopelessly flawed and that the whale mind, though wondrous and unique, is not much more so, if at all, than the mind of a pig or an elephant. The lovable dolphins swam away from the press and television, leaving in their wakes a batch of careless books and TV documentaries, and surely the worst movie (*The Day of the Dolphin*) ever directed by Mike Nichols.

This review, which appeared in the *New York Review of Books,* March 20, 1980, and the letters quoted in the Postscript are reprinted with permission of the New York Review of Books. © 1980 NYREV, Inc.

As the dolphin flap faded, a new media enthusiasm began to gather momentum. At the University of Nevada, Allen and Beatrice Gardner succeeded in teaching ASL (American Sign Language) to an infant female chimpanzee named Washoe. For the first time in history, it was loudly proclaimed, a lower primate had mastered a language in which it could talk to humans.

"Talk" and "language" are, of course, fuzzy terms with wide spectrums of meaning. A bluejay "talks" to other birds when it warns them of a cat. A cat "talks" when it asks to be fed by rubbing against your calf. Dogs communicate by barking, growling, whimpering, wagging their tails, and leaving symbolic messages on fire hydrants. Even so, the world was astounded by Washoe's ability to understand hundreds of sign gestures, especially by her ability to combine signs in ways that suggested a rudimentary grasp of grammar.

The best-known instance of Washoe inventing a phrase was when her teacher, Roger Fouts, had taken her out in a rowboat and a swan glided by. Fouts signed "What's that?" Washoe, knowing the signs for water and bird, responded with "water bird." There were many other two-word combinations mastered by Washoe: *Washoe sorry, Roger tickle, you drink,* and so on.

Other researchers soon were teaching other visual languages to young chimps. In California David Premack symbolized words with plastic tiles of different shapes and colors. His star pupil, Sarah, became almost as famous as Washoe. Like Washoe, Sarah seemed to create significant phrases. David's wife, Ann, wrote a book called *Why Chimps Can Read.*

In Georgia, Duane Rumbaugh tried a new tack. He had a computer built with a console of keys bearing patterns that represented words. A chimpanzee named Lana was taught to speak in this computer language of "Yerkish," named for the Yerkes Primate Center in Atlanta where Rumbaugh did his work. Lana, too, apparently combined signs in meaningful sequences. She called a cucumber a *green banana.* She called an orange an *orange apple.*

The achievements of Washoe, Sarah, and Lana have now been surpassed, so it is claimed, by the fabulous linguistic feats of Koko, a female gorilla trained since 1972 by a psychologist, Francine ("Penny") Patterson, in Stanford. It is not hard to understand why Penny—young, pretty, with long blond hair—has received such enormous publicity. What could be more dramatic than color photographs of Beauty and the Beast, heads together, raptly chattering to one another? Patterson wrote a cover story (the cover photo of Koko was snapped by Koko) for *National Geographic* (October 1978) titled "Conversations with a Gorilla." The pair graced the cover of the *New York Times Magazine* (June 12, 1977). In *Koko. A Talking Gorilla,* a stirring film documentary that opened last

December in Manhattan, Koko does a fine job of acting like a gorilla, but otherwise the film is mostly flimflam.

There is another reason for Penny's growing fame. Her claims for ape intelligence far exceed those of any other trainer. For one thing, Koko loves to make up rhymes: *Squash wash, do blue, bear hair,* and so on. (She has learned English vocalizations by hearing Penny repeat them, and by using a typewriter speech synthesizer designed by the Stanford mathematician Patrick Suppes.) Once Koko made up the poem: *Flower pink, fruit stink—fruit pink stink.* Here is a sampling of Koko's skill in inventing clever metaphors: *Elephant baby* (for a Pinocchio doll), *eye hat* (mask), *finger bracelet* (ring), *white tiger* (zebra), *fake mouth* (nose).

A reporter asked Koko who she liked best, Penny or her assistant. According to Penny, Koko looked back and forth, then diplomatically signed, "Bad question." On another occasion Penny asked, "What are you afraid of?" Koko: "Afraid alligator." Koko had never seen a live alligator. Penny thinks this shows how researchers can learn new facts about apes now that they can ask them questions.

According to Eugene Linden (who wrote a popular book about talking apes), in a wildly laudatory article ("Talk to the Animals," *Omni*, January 1980), when Koko was asked where you go when you die she signed, "Comfortable hole bye." Once when Penny became exasperated by the number of toys Koko had broken Penny muttered, "Why can't you be normal like any other kid?" Koko, says Linden, signed "Gorilla."

From the beginning large numbers of experts on animal behavior have been deeply skeptical of these extraordinary claims, but their animadversions appeared only in technical journals. Now the secret is out. Two books have been published, one popular, one technical, that give a strong case for the view that apes do not comprehend sign sequences in any way essentially different from a dog's understanding of such commands as "Sit up and shake hands" or "Go get the newspaper."

Nowhere on the jacket of *Nim* (Knopf, 1979) or in the book's advertising does the publisher so much as hint that the book severely criticizes practically all earlier work with talking apes. Even the author, Herbert Terrace, a psychologist at Columbia University, plays down his doubts at the start of his book, though there is a reason. When he began training Nim Chimpsky, a baby male chimp named in honor of Noam Chomsky, he had high hopes of confirming the earlier findings. His book is an informal narrative, with marvelous photographs, about four years that he and his many assistants spent in teaching ASL to Nim. Not until Chapter 13, after Nim has returned to his birthplace in Oklahoma, does Terrace see the light.

Terrace's complete disenchantment did not descend until he began to study his own extensive videotapes. Here are some of the things he learned:

Nim rarely initiated signing. Ninety percent of his signing was in response to gestures by teachers.

Half of Nim's signs imitated part or all of what a teacher had just signed. In many cases his teachers were astonished to see how often they had unconsciously started a sign that Nim had noticed.

If Nim wanted something he first grabbed, signing only when the grab failed. He never initiated signs except when expecting such rewards as food, hugs, and tickling.

Most of Nim's phrases were random combinations of signs, usually involving *me, hug,* and *Nim*—signs that fitted with almost all other signs, and which he had learned were likely to elicit favorable reactions.

Unlike children when they start to talk, Nim constantly interrupted teachers. He never learned the two-way nature of conversation. Researchers have attributed such interruptions to an ape's eagerness to talk.

Nim's mistakes were more often the confusing of signs similar in form rather than similar in meaning.

When Nim began to extend sentences beyond two or three words he simply added a string of nonsense words, usually repeating earlier signs. For example: "Give orange me give eat orange me eat orange give me eat orange give me you." This in contrast to the longer utterances of children which expand the sense of shorter ones.

Nim never signed to another chimpanzee who knew ASL unless a teacher was present to coax him.

Nim Chimpsky finally convinced Terrace that Noam Chomsky, the most distinguished of skeptical linguistics experts, was right. Although apes have a remarkable memory that enables them to master hundreds of visual signs, Terrace believes there is no evidence yet that they understand any kind of syntax. Of course this may be true also of very young children, but children quickly go on to form sentences that require a firm grasp of the rules of form. When an ape learns to put a few signs together there is no reason, says Terrace, to suppose it is doing anything essentially different from a pigeon that has been taught to obtain food by pecking four differently colored buttons in a specific order regardless of how the buttons are arranged.

When Terrace examined the videotapes of other researchers he found the same disturbing features. In many cases of film released for public viewing and for fund raising, episodes had been edited so that initial promptings were not seen. A Nova documentary called *The First Signs of Washoe* consistently followed this practice. Uncut versions of the same episodes showed that every one of Washoe's multi-sign statements came after similar signs by teachers.

"Can an Ape Create a Sentence?" is the title of Terrace's report in *Science* (November 23, 1979). His reluctant answer is no. "Apes can learn many isolated symbols (as can dogs, horses, and other nonhuman species), but they show no unequivocal evidence of mastering the conversational, semantic, or syntactic organization of language." Of the earlier researchers only Rumbaugh so far seems impressed by Terrace's analysis. His own work, he told the *New York Times* (October 21, 1979) has been pushing him toward similar views.

Speaking of Apes (Plenum, 1980) edited by linguist-semiotician Thomas A. Sebeok and anthropologist Donna Jean Umiker-Sebeok, is a much needed anthology of important articles on both sides of the intensifying controversy over the capacities of apes for language. It is impossible to discuss such a wide variety of papers so I will concentrate mainly on the long introductory article, "Questioning Apes," by the Sebeoks. Both are at Indiana University's Research Center for Language and Semiotic Studies, of which Thomas Sebeok is chairman. Their introduction is the most powerful indictment in print of the early work on talking apes.

Psychologists have a term, "experimenter effect," that covers all the insidious ways a researcher's strong convictions can unwittingly distort data. The Sebeoks first remind us of obvious ways that scientists in any field can be unconsciously motivated to get positive results. The stronger the results the faster their career advances and the more likely will their work attract funding. Assistants are strongly motivated to please an employer who pays their salary, and success often advances their own careers. If the work is controversial there is a tendency for research teams to form a cluster of insiders deeply suspicious of outsiders. They become, as the Sebeoks put it, a "dedicated group of enthusiastic workers, one that constitutes a tightly knit social community with a solid core of shared beliefs and goals in opposition to outside visitors. . . . In fact, it is difficult to imagine a skeptic being taken in as a member of such a 'team.'"

Within this frame the Sebeoks see a variety of curious ways in which talking-ape results are easily twisted in the direction of belief. Consider, for example, the "Clever Hans effect." The term comes from a classic 1907 study by Oskar Pfungst, a German psychologist, of a famous performing horse of the day who could answer difficult questions, including arithmetical problems, by pawing the ground. In most cases of such performing animals (there have also been "learned" dogs, pigs, and even geese) a trainer tells the animal when to stop by secret cueing, such as a slight sniff, but in the case of Hans, Pfungst was able to prove by ingenious tests that the horse had learned to respond to subliminal cueing on the part of spectators.

Talking-ape researchers have tried to exclude the Clever Hans effect, but the Sebeoks show convincingly that the effect is omnipresent. There is no evidence, they maintain, that successful teachers had any training in controlling unconscious facial movements, breathing rhythms, bodily tensions and relaxations, and so on. Some reactions, such as eye-pupil size, are probably uncontrollable. Pfungst reported his inability to avoid cueing Hans no matter how hard he tried.

Talking apes seldom perform well for strangers. Believers explain this by an ape's emotional attachment to certain teachers, but it is as readily explained by assuming that over the years apes develop a special sensitivity to unconscious reactions peculiar to a loved human and which they naturally fail to perceive if someone new tries to talk to them. Could it be, the Sebeoks ask, that the best trainers are those most expressive in unconscious cueing? A study of unedited films shows that ape teachers are, in the authors' words, "anything . . . but stone faced." Even uncropped still photos reveal obvious cueing. The Sebeoks cite some horrendous instances in the photos illustrating Patterson's *National Geographic* article, and in Mrs. Premack's book.

Concerning the famous Washoe-swan incident, both Terrace and the Sebeoks point out what should have been obvious at once. Washoe may simply have signed "water," then noticed the bird and signed "bird." It is unlikely that Fouts could have concealed his elation. Washoe, observing this social reward, would henceforth associate the double sign with a swan.

There is no solid evidence that an ape has ever invented a composite sign by understanding its parts. In the course of several years an ape will put together signs in thousands of random ways. It would be surprising if it did not frequently hit on happy combinations that would elicit an immediate Clever Hans response. No teacher has bothered to record all the nonsense combinations produced by an ape, but every lucky hit is sure to be reinforced by cues of approval and to go into a teacher's records, reports, books, and lectures.

Even when an ape has memorized a sign it often makes errors in reproducing it. When this happens, the Sebeoks point out, ape teachers have a battery of excuses. Instead of a mistake it becomes a joke or a lie or an insult. Patterson is especially prone toward this kind of subjective evaluation. She asks Koko to sign drink. Koko touches her ear. Koko is joking. She asks Koko to put a toy under a bag. Koko raises it to the ceiling. Koko is teasing. She asks Koko what rhymes with sweet. Koko makes the sign for red, a gesture similar to the one for sweet. Koko is making a gestural pun. She asks Koko to smile. Koko frowns. Koko is displaying a "grasp of opposites." Penny points to a photograph of Koko and asks, "Who gorilla?" Koko signs "Bird." Koko is being "bratty."

The *National Geographic* article reproduces a crayon picture by Koko that is captioned "Representational Art." Its black squiggles, says Penny, are spiders. An orange scrawl is Koko's drinking glass. A similar tendency to overhumanize ape behavior, though less blatant, infects all the earlier work. It is little different from the firm belief of sentimental pet owners that a beloved cat, or even a parrot, understands almost everything you say to it.[1]

It is possible, of course, that apes do have a feeble talent for creating meaningful composite signs, but by the principle of Occam's razor, the Sebeoks insist, should we not accept simpler explanations first? So far there is no reason to suppose that Koko's remarkable utterances are anything more than responses to unwitting cueing on Penny's part, or to Penny sifting out from thousands of nonsense combinations those that make sense to *her,* not to Koko. An objective evaluation of a phrase like "bad question" cannot be made without a videotape of the scene to make sure the details are correctly recalled, and without knowledge of how many of the ape's unlearned and spontaneous two-word combinations are nonsense. Otherwise we have nothing more than a collection of anecdotes.

Some researchers, especially Premack, have tried "double blind" tests to rule out Clever Hans effects, and whenever these controls were tight the ape's ability dropped almost to chance. Much is made of the slight deviations from chance, but the Sebeoks list numerous ways in which bias could have slipped into these efforts to exclude it. We are not told what controls were placed on photographers. Reports often fail to note the presence of others who happened to be around but were deemed too irrelevant to mention. One-way windows eliminate visual cues but not sound cues. Details are sparse about randomizing procedures and the rules followed in scoring.

There are religious beliefs—in the West notably those of the Catholic Church and conservative Protestantism—that make it necessary to assume that human beings have an immortal soul denied to the beasts that perish. Mortimer J. Adler wrote a book a few years ago called *The Difference of Man and the Difference It Makes,* in which he enlarges on Thomist arguments that the ability to understand syntax is one of the main ways a human mind differs from the mind of a beast. At the time Adler wrote his book the dolphin language flap was in full swing, and Adler made much of the fact that if we ever succeed in conversing with a whale his thesis will be undermined. If the book has a new edition you can be sure Adler will underplay porpoises and concentrate his dialectical fire on apes.

It is good to understand that this sort of metaphysical objection to talking apes, reinforced by Revelation, is not behind the views of Chomsky, Terrace, the Sebeoks, or any other major critic. What they are saying is much simpler. Contemporary humans and apes are terminal branches on the tree of evolution. Transitional types that flourished over the millennia during which human beings acquired the ability to talk are no longer available for study. Chomsky believes that evolution gave to humans, as it did not give to any living lower primates, a capacity for language that is deeply interlocked with the inherited structure of their brains.

Little is gained by quibbling over the meaning of "language." As Chomsky says in his contribution to the Sebeok anthology, this is a conceptual not a scientific question. If you define flying, he writes, as rising into the air without the aid of special equipment and landing some distance away, then human broadjumpers can fly about thirty feet. Chickens do slightly better—about three hundred feet.

Suppose, Chomsky continues, we label the four colors pecked by pigeons with four words: *Please-give-me-food.* "Do we want to say that pigeons have the capacity for language, in a rudimentary way? This is much like the question whether humans can fly almost as well as chickens though not as well as Canada geese. The question is not clear or interesting enough to deserve an answer."

The central empirical question can be simply put. Do apes have the ability to link visual signs in ways that justify saying they are using syntax? Yes, say most of the researchers and many outsiders. Jane H. Hill, an American anthropologist, closes her contribution to *Speaking of Apes* by writing: "It is unlikely that any of us will in our lifetime see again a scientific breakthrough as profound in its implications as the moment when Washoe . . . raised her hand and signed 'come-gimme' to a comprehending human."

No, say some researchers and a growing number of outsiders. If no, Chomsky concludes, then a study of ape signing can be expected to cast as little light on human language or conversely as a study of human jumping can cast light on the mechanism of bird flight or conversely. One can teach two pigeons to bat a ball, writes Ms. Hill, quoting a familiar aphorism, but is it ping-pong? She thinks it unjust to apply this skepticism to talking-ape research. Chomsky holds the opposite opinion.

No one can rule out the hope that as talking-ape research continues, under better controls, it may turn out that apes do have a dim awareness of syntax. If so, then the researchers will have made a point even though it may not be a big one. At the moment, however, the situation seems little different from that which confronted biologists a century ago. Here is how Darwin summed it up in a section on language in *The Descent of Man:*

> As the voice was used more and more, the vocal organs would have been strengthened and perfected through the principle of the inherited effects of use; and this would have reacted on the power of speech. But the relation between the continued use of language and the development of the brain has no doubt been far more important. The mental powers in some early progenitor of man must have been more highly developed than in any existing ape, before even the most imperfect form of speech could have come into use. . . .

NOTE

1. For two pathetic examples of this "pathetic fallacy," see *The Language Barrier: Beasts and Men* (Holt, 1968) by Thomas Mann's daughter, Elizabeth Mann Borgese, in

which she tells of her experiences with a dog, an elephant, and a chimpanzee; and *Look Who's Talking* (T. Y. Crowell, 1978), by Emily Hahn.

Postscript

Penny Patterson's response to my review appeared in the *NYR,* October 9, 1980:

> As the target of many of its taunts, I am responding to Martin Gardner's review of *Nim* and *Speaking of Apes.* Gardner suggests that all research into two-way communication with animals is faddish and uses the two books reviewed to support his point. As an active researcher in the field, I would like to suggest that debunking can go both ways. Citing examples out of context and stripping them of supporting experimental documentation, the Sebeoks and Gardner are guilty of one of the oldest forms of journalistic deception. We are not, as the Sebeoks charge, basing our reports on "nothing more than a collection of anecdotes." We are striving to outline the boundaries, to define the differences as well as the similarities, between human and nonhuman language use.
>
> To say that the gorilla's use of sign language is virtually identical to that of the human child is wrong; but to say that the gorilla's use of sign language is uncreative, repetitious, and forced or cued is equally wrong.
>
> The first book reviewed is *Nim* by Herbert Terrace, which states that "there is no evidence yet that apes understand any kind of syntax." Terrace is evidently unaware of my publication of experimental evidence (controlled for cueing) of the comprehension of novel sign and spoken English sequences by a gorilla (AAAS Selected Symposium 16, 1978).
>
> Gardner remarks that Terrace's disenchantment with the chimpanzee Nim's abilities arose when he studied his own extensive videotapes. These "extensive videotapes" are 3½ hours of Nim's signing sampled under artificial and high-pressure conditions which very likely contributed to the high levels of interruption and imitation he observed. Terrace's extended conclusions on the language abilities of gorillas are based on fifty (50) seconds of footage produced for popular media consumption.
>
> The second book Gardner reviews is *Speaking of Apes,* edited by Donna Jean and Thomas Sebeok. Gardner seems to accept indiscriminately everything they say. I'd like to review several critical points.
>
> The Sebeoks' comments are opinions and conjecture, not factual statements. They have not examined my data and have neither research experience with apes nor expertise in American Sign Language.
>
> Nonverbal cues are omnipresent in human communication as well as in ape-human communication. Speculating that perhaps "the best trainers are those most expressive in unconscious cueing," the Sebeoks comment that "the apes' teachers are anything but stone faced." Facial movements and expressions are an integral part of sign language—they frequently function

grammatically. A "stone-faced" teacher would not be a good model for sign language acquisition for either child or ape. Contrary to the Sebeoks' assertion, it is easy to control for cues such as eye-pupil size and direction of gaze by wearing mirrored sunglasses. We have employed this and a variety of other controls. The result is that the gorillas' spontaneous and appropriate signing continues unabated. Live observation and review of our videotapes indicate that in test situations requiring forced choices between objects or other materials, the gorillas are looking down at the materials, rather than searching our faces for cues, almost without exception. If we drop deliberate miscues by positioning, touching, leaning, or looking toward the wrong choice, the gorillas respond to the questions, not to the cues. We have restructured certain situations and our possible cues so as to deliberately mislead the gorillas. Instead of asking the usual "Where is your ear," and so on, the experimenter asks, "Is this your ear?" pointing to her nose, and looking at the gorilla's nose. In a recent test, the gorilla Michael responded in each case by correcting the questioner and not by following the cues.

The Sebeoks again reveal their ignorance of sign language structure in their arguments designed to dismiss sign modulations as mistakes. Errors in reproducing signs are not random for either human or gorilla users of sign language. Rather, a circumscribed set of parameters is systematically varied and many possible variations never occur. Neither Koko nor Michael routinely makes articulation errors in using signs such as *drink*. Errors are of a conceptual nature—*eat* or *sip* may be emitted where *drink* is appropriate, but the *drink* sign does not drift randomly about the signing space as the Sebeoks contend. Errors of articulation are made between signs whose location, configuration, and motion are similar and are recorded as such. By moving the *drink* sign to her ear instead of to her mouth, Koko altered its articulation in a way that did not conform to any of the standard error patterns. Koko had never before this incident nor ever since placed the thumb of her fisted hand to her ear. The Sebeoks assume that because Koko had refused to use this sign on this particular day with this particular assistant, even when the teacher repeatedly demonstrated ("cued!") the sign, that she had not mastered it; however *drink* was a sign Koko had then used reliably on a daily basis for several years. When Koko finally complied, a grin accompanied the distorted sign.

Koko's humor, like that of the young child, is based on discrepant statements about overlearned relations. If the distortion is taken out of context, and the above constraints on articulation errors are disregarded, then the "drink in the ear" sign is perhaps best interpreted as an error. But given the contexts of the situation, of Koko's behavior prior to and during the incident, the sign's acquisition history and pattern of errors, the nature of infantile humor, and of the gorilla temperament (all of which the Sebeoks fail to take into account), it would be a mistake to categorize such a response as an error.

There are numerous (almost fifty) misstatements, non sequiturs, misleading elliptical quotations, and kindred erroneous remarks in the Sebeoks' chapter. (Gardner's review is based solely on this first chapter, not on any material by animal researchers.) One in particular evidently impressed a reporter from *Time*. He asked me half a dozen times if I had scored Koko's inappropriate responses as errors in a blind test of her vocabulary. The Sebeoks state that the types of responses Koko gave to avoid the double-blind test were not included in the four categories of errors I listed, so that "we may assume they are in fact not represented in the sixty percent score" she achieved. I could not believe my eyes reading this—those responses are indeed in the list of errors which is on the very same page as a quotation they include from that report!

Gardner asserts that "no teacher has bothered to record all the nonsense combinations produced by an ape. . . ." I routinely record everything Koko signs in three formats: written notes, audiotaped samples, and video-taped samples. Most of Koko's signed communications are appropriate to the situations in which they occur. (Gardner calls these "lucky hits.")

According to Gardner, one of my exceptional "claims" is that Koko has a capacity to rhyme the sounds of English words using signs. The full context of Koko's "Flower pink, fruit stink, fruit stink pink" rhyme was a dinnertime discussion of broccoli with two teachers, one of whom responded, "You're rhyming, neat!" to which Koko replied "Love meat sweet." Following this incident, her ability to rhyme was tested. Koko successfully performed a task requiring her to produce signs whose English translations rhyme with the English translation of other signs. For example, Koko responded "do" to the word "blue" spoken by the experimenter and "wash" to "squash." Note that these word pairs are *not* examples of Koko's rhymes, as Gardner mistakenly assumes, but responses to test questions. She also demonstrated an ability to select from an array of objects those with rhyming English names or with names rhyming with an English word spoken by the experimenter.

Another of my "exceptional claims," according to Gardner, is that Koko generates utterances which are innovative in a way paralleling metaphor. After documenting numerous instances of such novel descriptive phrases, we assessed the gorillas' ability to appreciate metaphor using a test devised by Harvard psychologist Howard Gardner. Administered blind, the test involved the assignment by the gorillas of polar adjectives (such as loud-quiet and hard-soft) to pairs of colors. The level of performance of both gorillas (90 percent metaphoric matches) was on a par with that of seven-year-old human children (82 percent).

Gardner's grand finale is a quote from a book by Darwin published in 1871 presented as evidence that the Terrace and Sebeok pronouncements on apes fit neatly with evolutionary theory. Very few gorillas had been successfully maintained in captivity and no attempts had been made to assess their mental faculties at that point in time. Darwin's statement was based on the mistaken notion that language is synonymous with vocal speech.

One cannot trace the evolution of language from an armchair in Indiana. By studying the cognitive capacities of man's closest relative we may come a step closer to discovering what our ancestors' language abilities were like five million years ago when humans and gorillas set out on separate evolutionary paths.

In summary:
Blind and double-blind tests have been administered, and the gorillas' level of performance is significantly above chance.

The gorillas sign spontaneously and appropriately to themselves and to each other.

The gorillas sign to strangers.

The gorillas sign and respond to questions appropriately even when we try to mislead them with nonverbal cues.

The gorillas frequently initiate signed communication; the majority of their utterances are meaningful.

Francine Patterson

President, The Gorilla Foundation
Woodside, California

Patterson's letter provoked replies from both Terrace and the Sebeoks, which ran in the *NYR,* December 4, 1980:

In her reply to Martin Gardner's review of my book, *Nim,* Francine Patterson questions my negative conclusions regarding an ape's linguistic competence. She does so without coming to grips with the facts which led me to reverse my original interpretation of Nim's multi-sign sequences as sentences.

Nim's signing, and that of the other signing apes as well, appears to be motivated more by a desire to obtain some object, or to engage in some activity, than a desire to exchange information for its own sake. First the ape tries to obtain what it wants directly—without signing. When reminded by its teacher that it must sign, the ape often signs until the teacher complies with its request. The critical question is whether the ape is generating sentences or simply running on with its hands until it gets what it wants.

Careful scrutiny of the ape's utterances favor the latter interpretation. Consider, for example, a typical exchange in which the teacher signed *you play cat?,* and Nim responded by signing *me Nim cat play.* Two of the teacher's signs are combined with two general purpose signs, signs I refer to as "wild cards" because of their universal relevance. These and other features of an ape's discourse with its teachers were discovered by painstaking frame-by-frame analyses of videotapes. Unlike the sentences of a child, Nim's combinations amounted to unstructured mixtures of signs. Some are imitative of the teacher's prior utterance; others are selected unsystematically from a small group of "wild card" signs.

Patterson rejects this interpretation of an ape's sequences of signs on the grounds that three and a half hours of videotape of Nim signing with his teachers provided too small a sample and that the data we did collect were artifacts of the pressure Nim's teachers exerted in getting him to sign. Both of these arguments are at odds with statements and data that appear in Patterson's dissertation (Stanford University, 1979), a document that impresses me as the most thorough of Patterson's publications about Koko's signing.

Psycholinguists are in general agreement that a child's production, as opposed to his comprehension, of language provides the most telling evidence of grammatical competence. Consider Patterson's assessment of Koko's production of signs. "The majority of Koko's utterances were not spontaneous, but solicited by questions from her teachers and companions. My interactions with Koko were often characterized by frequent questions such as 'What's this?'" (p. 153).

Patterson's dissertation contains five one-hour transcripts of videotapes of Koko signing with her teachers (the only such transcripts that have been published to date). I found no evidence that Koko's use of sign language differed from Nim's. Just as Nim was prone to produce long utterances such as *give orange me give eat orange me eat orange give me eat orange give me you,* Koko produced utterances of structurally unrelated signs such as *mess red thirsty mouth thirsty* (p. 339) and *please milk please me like apple bottle* (p. 345).

Until Patterson publishes data to support her view that my videotapes of Nim signing with his teachers were obtained under "artificial and high-pressure conditions which very likely contributed to the high levels of interruption and imitation," I see no way to evaluate that claim. I also suggest that Patterson perform a discourse analysis of Koko's signing as shown in a documentary film in which she participated, *Koko, a Talking Gorilla* (New Yorker Films). The scenes of that film, in which both Koko and her teacher are visible, left me (and many other viewers) with the clear impression that the teacher initiated most of the signing and that Koko's signing was highly imitative of the teacher's utterances.

That Koko could perform at better than chance levels on a comprehension test of novel utterances in sign language and in spoken English is not evidence of grammatical competence. As I have argued elsewhere (*Journal of the Experimental Analysis of Behavior,* vol. 31, pp. 161–175), the type of problem that was administered in this test can be solved by applying nongrammatical strategies.

Much of Patterson's letter is devoted to defending what Martin Gardner describes as her "exceptional claims" of Koko's signing. While not directly germane to Patterson's comments concerning my conclusions, they do raise questions about her criteria for characterizing Koko's signing as language. Totally absent from Patterson's description of Koko's ability to rhyme, to use metaphors and such abstract concepts as *because* and *imagine* is the kind of training needed to establish such linguistic skills. No mention is made of just how a gorilla, who can't produce human phonemes, learns to identify English words that rhyme with one another. Claims that a gorilla is as competent to produce metaphors as a seven-year-old human child cannot be evaluated without knowing what kind of training was used to get the gorilla to appreciate a metaphorical use of language. Without such information, one is left with the impression that Patterson is simply projecting onto her gorillas' hand movements what a human child might do in similar circumstances. These and other aspects of the superficial assessments Patterson makes of the "linguistic" achievements of her gorillas can be overcome only by a rigorous description of their training and testing histories.

H. S. Terrace

Francine Patterson's letter, inherently chaotic enough, complicates matters in that it seems to be addressed, pell-mell, to Herbert Terrace and the two undersigned, whose books are being reviewed, and to Martin Gardner, who reviewed them. Her letter contains scattered quotations, as in her opening paragraph, ascribed to "the Sebeoks," but some of these are not, in fact, from our book; they are the words of Mr. Gardner. Her complaint also features a number of bizarre denials of dramatic assertions which, to our knowledge, no one has ever made—certainly not the two of us. An example of the latter is Patterson's indignant rejection: "To say that the gorilla's use of sign language is virtually identical to that of the human child is wrong. . . ." The contrary has never been asserted by any scholar, not even the most enthusiastic scribbler.

Patterson's repeated whimper, "They have not examined my data," is counterfactual. Indeed, we have checked over every scrap of information—such as it is—that she has disclosed through normal scientific outlets and have also surveyed all available vulgarizations of her data, presuming that it was she who authorized their public circulation. Her dissertation, long delayed, became accessible to us only after *Speaking of Apes* was typeset; it will, accordingly, be critically dealt with in our forthcoming article, "Clever Hans and Smart Simians, The Self-Fulfilling Prophecy and Kindred Methodological Pitfalls," now in press and due to appear, early in 1981, in a leading anthropological journal. To anticipate, however, we must point out here that there are basic and very disturbing discrepancies between her data as reported in her thesis and as published in her scattered articles.

Patterson argues in her letter that "Nonverbal cues are omnipresent in human communication as well, as in ape-human communication. . . . Contrary to the Sebeoks' assertion, it is easy to control for cues such as eye-pupil size . . . by wearing sunglasses." In fact, Sebeok says virtually the same thing, on p. 420 of *Speaking of Apes*. Patterson has not, however, previously troubled to report this form of control and still neglects scores of other sources of leakage, many of which we enumerated in our study. In Patterson's reports, the precise training methods or test situations she employed are not usually described, and we are asked to accept her assertions about her apes' performances on faith. When experimental conditions are specified, controls are often so feeble as to defy belief. Her facile statement about the importance of nonverbal cues belies her continual treatment of the Clever Hans effect as a minor methodological irritation instead of recognizing it—in the face of the vast amount of scientific evidence attesting to its pervasive influence—as a global application of the self-fulfilling prophecy, even in situations where experimenters are not as emotionally committed to their experimental subjects as Patterson is, by all accounts, to her gorillas.

Patterson censures us for our "ignorance of sign language structure," but the shoe is on the other foot. She has never produced a shred of evidence that her apes' gestures are, in fact, *signs,* in the technical acceptation of this basic semiotic unit, as tellingly put forth by Petitto and Seidenberg (*Brain & Language* 8:162–83, 1979). Terrace's findings, which have now been supplemented by a discourse analysis of the data presented in Patterson's dissertation, fully confirm our long-held suspicion that the roughly duplicative gorilla gestures that she persists in calling *signs* are scarcely more than "signifiers" without any "signification" in the human sense.

Patterson's repeated overinterpretations of her subjects' behavior as jokes, apologies, puns, and now English rhymes (!) are clear examples of the Pathetic Fallacy, with which we have dealt at length in our study. In the case of the sign for drink, which she focuses on in her letter, we would still like to know what other locations were used by Koko with the hand configuration in question. Assuming that, as Patterson reports, the trainer had been trying for some time to persuade Koko to make this sign, we may guess that the ape in fact moved her hands in various directions during the session, and with various accompanying facial expressions. How were all of these other "signs" and expressions interpreted? The possibilities, given the lack of the kind of information any thoughtful person would demand, are limitless. Patterson would profit immensely from mastering the principles of biology—notably the writings of Jacob von Uexküll and his *Umweltlehre*—and the best of contemporary linguistic theory—e.g., Noam Chomsky's *Rules and Representations,* chapter 6—not to mention a number of classic circus training manuals.

Patterson complains that there are "numerous" erroneous remarks in our introductory chapter. In her letter, she offers but one example, which happens to be false. We cannot repeat here our detailed critique of the cautionary method of the so-called "double-blind" test, a magic device in which Patterson seems to place touching faith, but which, as we and a number of others have demonstrated, is all too often embarrassingly inadequate. To take a single example from Patterson's work, descriptions and illustrations in her published articles show that the box she used for double-blind testing was sufficiently small so that Koko could have moved it around at will. Patterson's experimental design for such tests by no means rules out certain guessing strategies on the part of ape and the "blind" experimenter, given the small universe of stimuli which was available for use and the familiarity of both ape and human with these materials, as well as with one another's facial expressions, body movements, and the like. It should be said, finally, that these double-blind tests have been used only rarely by Patterson. We have found published reports of only one series, administered to Koko in September, 1975. Koko, Patterson has admitted in print, was extremely reluctant to perform under these conditions.

Patterson claims that "one cannot trace the evolution of language from an armchair in Indiana." Our view is that, to the contrary, a *Gedankenexperiment* can never be separated from its technical realizations in any field of science, since an understanding of nature can only be obtained by the informed and careful application of both. Patterson's lack of methodological sophistication is precisely traceable to her innocence of fundamental theoretical advances in fields adjacent to her own. Her fellow psychologists who are knowledgeable about such issues will have to judge the extent to which she is a victim of self-deception and why, when some of the more prominent "pongists" have publicly renounced this line of investigation, she persists, against the weight of versant opinion and the laws of probability, in pursuing the will-o'-the-wisp that apes are capable of language-like performances. Those who are ignorant of the millennial history of the Clever Hans phenomenon are doomed to replicate it endlessly with one animal form or another, whether embodied in birds, horses, pigs, porpoises, the great apes, or, most recently, the wondrous tortoises of Milwaukee.

Jean Umiker-Sebeok
Thomas A. Sebeok

Postscript

In case you think I made up my quotations from John Lilly, you'll find them all (and others even more preposterous) in Lilly's several books about dolphins and in interviews published in *Psychology Today* (December 1971) and in the *Village Voice* (April 19, 1976). The notion that dolphins communicate with humans by ESP is now commonplace among believers, and there are even organizations seeking funds for research on whale ESP. This is hardly surprising. J. B. Rhine firmly believed that Lady Wonder, a performing horse, read his mind, and many top parapsychologists are convinced that performing animals like Clever Hans reacted less to visual or auditory cues than to ESP.

When Hans's owner died, the horse was acquired by Karl Krall, at Elberfeld, who trained several other horses to perform like Hans. His big book about this, *Denkende Tiere* ("Talking Animals") was published in Leipzig in 1912. Maurice Maeterlinck wrote an incredibly naive chapter about Krall's horses in his book *The Unknown Guest*. Discussing these horses in *Fate* (April 1980), D. Scott Rogo stupidly assumes that because Krall trained a blind horse, and put blindfolds on other horses, this ruled out all cueing! Rogo says nothing about the exhaustive demolition of Krall's claims by Stefan von Máday in his *Gibt es denkende Tiere?* (Leipzig, 1914).

Before and after Hans there have been scores of stage animals that solved mathematical problems, spelled out answers, and so on—not only horses (as I said earlier), but also dogs and pigs; even a "learned goose" that performed in London in the late eighteenth century and two "Curious Birds" that did staggering mental tricks in England in the early nineteenth century. Methods for cueing such stage animals have been described in several books on animal training. The cueing is usually by a sound too weak for human ears to hear. It can be produced in many subtle ways, such as making a click with the nails of the thumb and middle finger, or a slight nose sniff. Parapsychologists, notoriously uninformed about such things, tend to look for some obvious hand or body movement by the trainer. It seldom occurs to investigators of animal ESP that when the owner of a "psychic" animal leaves the room, the signaling can be taken over by a friend posing as an astonished spectator. For two recent books about horse ESP see *Talking with Horses* (1975) and *Thinking with Horses* (1977), both by Henry N. Blake and published in London by Souvenir Press.

When I wrote my review of the talking ape books I had not read Pfungst's book on Clever Hans. I have since obtained a copy (Holt, Rinehart and Winston reprinted it in 1965), and I must say I am impressed by the thoroughness of Pfungst's testing. He makes a convincing case for the hypothesis that Hans's owner and trainer, a retired mathematics teacher named Wilhelm von Osten, did not consciously cue Hans, but that the horse responded to involuntary movements of the person questioning him. The cue was primarily a slight upward motion of the head, extremely difficult to suppress even when told to do so, that occurs when a performing horse has hoof-tapped a correct number of times. The motion was magnified by the large-brimmed hat that von Osten habitually wore.

Pfungst discloses little familiarity with methods of deception used by trainers of "thinking" animals, and I finished the book with the impression that von Osten was less innocent than Pfungst believed, or perhaps wanted to say in print. In any case there seems little doubt that Hans had been trained to respond to visual cues unconsciously given by most people, rather than to a secret signal. This would explain why Pfungst's work has never been repeated. Hans may have been unique among performing animals in having been trained in this way, and replication will have to wait for someone willing to spend years training an animal in a similar manner. What trainer of a performing animal will bother to do this when it is so much easier to teach the animal to respond to a deliberate but subtle cue? That von Osten did it the hard way supports Pfungst's opinion that the training was unwitting on his part, though when the tests demonstrated so clearly the nature of the cueing it is hard to believe von Osten would not have accepted the findings. At any rate, he angrily terminated the experiments, maintaining until he died that Hans was capable of "inner speech" and the ability to comprehend syntax and do mathematical calculations.

The current controversy between talking-animal researchers and their skeptical opponents is growing in bitterness. *Time* concluded its coverage of the controversy (March 10, 1980) with the following statement by Chomsky: "It's about as likely that an ape will prove to have language ability as that there is an island somewhere with a species of flightless birds waiting for human beings to teach them to fly." See also articles in *Science News* (May 10, 1980), *Science-80* (July/August 1980), and *Science* (vol. 207, March 21, 1980; vol. 208, June 20, 1980.)

A "Conference on the Clever Hans Phenomenon: Communication with Horses, Whales, Apes and People" was held at the Roosevelt Hotel in Manhattan, May 6–7, 1980. Allen and Beatrice Gardner were scheduled to speak but did not attend, assuming that because Sebeok had organized the conference the cards would be stacked against them. Among the leading researchers, only the Rumbaughs showed up. Angry words were exchanged at the sessions, and each side accused the other of lying. One woman ape-trainer, skilled in sign language, stood up to protest the theory of unconscious cueing. As she spoke, her hands were constantly and unconsciously signing what she said. I am told that the proceedings of the conference will be published in 1981 by the New York Academy of Sciences, which sponsored the meeting, as one of their annals.

For some recent Penny-publicity, see Garry Hanauer's article about her research in the November 1980 issue of *Penthouse.* There are excellent photos of Koko in the nude.

Index